Instant Notes in GENETICS

The INSTANT NOTES series

Series editor
B.D. Hames
School of Biochemistry and Molecular Biology, University of Leeds, Leeds, UK

Biochemistry
Animal Biology
Molecular Biology
Ecology
Genetics

Forthcoming titles
Microbiology
Chemistry for Biologists
Immunology
Neuroscience
Psychology

Instant Notes in GENETICS

P.C. Winter
Department of Haematology,
Royal Victoria Hospital, Belfast,
UK

and

G.I. Hickey & H.L. Fletcher
School of Biology & Biochemistry,
The Queen's University of Belfast,
Belfast, UK

First published 1998

Reprinted 2000, 2002

A CIP record for this book is available from the British Library.

ISBN 1 85996 166 5

BIOS Scientific Publishers Ltd
9 Newtec Place, Magdalen Road, Oxford OX4 1RE, UK.
Tel. +44 (0) 1865 726286. Fax +44 (0) 1865 246823
World Wide Web home page: http/www.bios.co.uk/

Published in the United States of America, its dependent territories and Canada by Springer-Verlag New York Inc., 175 Fifth Avenue, New York, NY 10010-7858, in association with BIOS Scientific Publishers Ltd.

Published in Hong Kong, Taiwan, Singapore, Thailand, Cambodia, Korea, The Philippines, Indonesia, The People's Republic of China, Brunei, Laos, Malaysia, Macau and Vietnam by Springer-Verlag Singapore Pte Ltd, 1 Tannery Road, Singapore 347719, in association with BIOS Scientific Publishers Ltd.

Production Editor: Andrea Bosher.
Typeset and illustrated by The Florence Group, Stoodleigh, UK.
Printed by Biddles Ltd, www.biddles.co.uk

CONTENTS

ABBREVIATIONS

3D	three-dimensional
5BU	5-bromouracil
ADA	adenine deaminase
AIDS	acquired immune deficiency syndrome
ATP	adenosine triphosphate
BAC	bacterial artificial chromosome
bp	base pairs
BrdU	bromdeoxyuridine
bZIP	basic leucine zipper
CAP	catabolite activator protein
cAMP	cyclic adenosine monophosphate
cdk	cyclin dependent kinase
CFTR	cystic fibrosis transmembrane conductance regulator
cM	centiMorgan
DIG	digoxigonin
DNA	deoxyribonucleic acid
dATP	2′-deoxyadenosine 5′-triphosphate
dCTP	2′-deoxycytosine 5′-triphosphate
ddNTP	dideoxynucleotide triphosphate
dGTP	2′-deoxyguanosine 5′-triphosphate
dNTP	deoxynucleotide triphosphate
dTTP	2′-deoxythymidine 5′-triphosphate
ds	double-stranded
ES	embryonic stem
EST	expressed sequence tags
ETS	external transcribed spacers
F	fertility
FISH	fluorescent *in situ* hybridization
GDP	guanosine diphosphate
GTP	guanosine triphosphate
HFr	high frequency recombination
HIV	human immunodeficiency virus
HLH	helix-loop-helix
hnRNA	heterogeneous nuclear RNA
HUGO	human genome organization
ICR	internal control region
ITS	internal transcribed spacers
kb	kilo base
kbp	kilo base pairs
LINE	long interpersed nuclear elements
LTR	long terminal repeat
MCS	multiple cloning site
mDNA	mitrochondrial DNA
mRNA	messenger RNA
NOR	nuclear organizer region
ORF	open reading frame
PAC	P1 artificial chromosome
PCR	polymerase chain reaction
PKU	phenylketonuria
pms	postmeiotic segregation
QTL	quantitative trait loci
R	resistance
RF	replicative form
RFLP	restriction fragment length polymorphism
RNA	ribonucleic acid
ROS	reactive oxygen species
rRNA	ribosomal RNA
SAR	scaffold attachment regions
SCE	sister chromatid exchanges
SINE	short interspersed nuclear elements
snRNP	small nuclear ribonucleoproteins
ss	single-stranded
SSB	single-strand binding
STS	sequence tagged sites
TF	transcription factor
tRNA	transfer RNA
TIC	transcription initiation complex
VNTR	variable number tandem repeats
YAC	yeast artificial chromosome

PREFACE

This text aims to provide a comprehensive set of basic notes in genetics. Genetics is like riding a bike, easy when you know how, but impossible until you try it. Genetics is considered by some students to be the most difficult aspect of biology. This is often because you have to think about it. Insects have six legs. A simple observable fact of no further consequence, unless they start wearing shoes. The rules of segregation of genetic material, first discovered by Mendel, apply to most, perhaps all, of the 50 000 plus human genes. The rules of genetics are similar for all of the several million species on Earth and have consequences at all levels of life. Rules are conceptual but the facts of chromosome structure and genetic transmission which cause the rules are almost invisible. One understood, however, the behavior of all the genes is accessible. Thus, genetics is not suitable for rote learning, but like a new language, understanding the basics of genetics opens a book which gives an insight and coherence to all life.

Understanding requires a mental image. DNA is a long thin thing with a series of coded instructions on it. These are the plans for the molecular machines which make and run cells, and in turn read the DNA code and complete the cycle of interdependence. Groups of cells make organs and organisms, and the appearance (phenotype) of these reveals the instructions on their DNA (their genotype) to the outside world. These levels encompass all genetics from the structure of DNA and its mutation, through the workings and interactions of the proteins and RNAs it encodes, to evolutionary changes in the mixture of individual genotypes in populations.

The science of genetics has exploded since the development of techniques to manipulate and sequence DNA and RNA have enabled researchers to look at, and deliberately create, individual base changes in DNA. It is now possible to manufacture a DNA sequence to code for any interesting sequence of amino acids and so produce designer proteins. The gene can be given a selective promoter, which can be switched on or off as the researcher requires. Mutations which cause disease can be identified and the normal function of the gene investigated, with the intention of designing a drug to replace it. It may even be possible to replace the gene itself with the correct DNA sequence. An inhaler spray carrying the sequence of the gene whose absence causes cystic fibrosis is already in use. These exciting developments have led to genetics textbooks doubling in size. These things are interesting, but can make it difficult for the beginner to find the rules of the game amongst the advanced applications. Full appreciation of how genetic profiles can incriminate rapists needs an understanding of both the molecular genetic techniques used to produce the profile and the population genetics which tells how unlikely it is that the suspect is innocent.

We have attempted to cover the basics, without exhaustive detail or repetitive examples, and have used our experience of student's mistakes to draw attention to some common misconceptions and sources of confusion. The Keynotes at the start of each topic follow the series' successful format by presenting the absolute basics, but students should know why these brief statements are true, not just learn them. The breadth of the text was determined by attempting to include all the essential requirements for a foundation genetics course, accommodating the fact that different courses will have different requirements and bias, some molecular, some medical, some biological, some ecological. Omitted topics were generally considered to be appropriate to a more advanced level. Depth was determined by the accessibility of the exciting advances. If they were not readily understandable, they were left out. We hope that the combination will provide broad support to most genetics lecturers and accesible information for genetics students.

The sequence of presentation is also arbitrary. Many texts run historically, starting with Mendel. We choose to run logically, starting with DNA and the genetic code, Understanding these gives the explanation for Mendels laws. The first section covers the chemistry of DNA, the genetic code and protein synthesis. The second covers the organisation of DNA into chromosomes at the cellular level. Thirdly the transmission of DNA between generations is discussed, with an exploration of the interaction between alleles, genes and their products, and how these determine the phenotype of the organism. The emphasis is on diploids, so is especially relevant to humans. Section D, population genetics and evolution, takes genetics to the next level, from examining the causes for the high level of cystic fibrosis in Europeans to the changes in allele frequencies which are evolution, and eventually cause speciation. Section E describes some of the techniques used in the molecular genetics revolution, and section F introduces some of the recent applications of genetics, and implications for the future. We hope this will encourage students to appreciate that genetics has an important position in society, and that the effort of understanding it is well worth while.

P.C. Winter, G.I. Hickey and H.L. Fletcher

A1 DNA STRUCTURE

Key Notes

Nucleotides

DNA is a polymer containing chains of nucleotide monomers. Each nucleotide contains a sugar, a base and a phosphate group. The sugar is 2′-deoxyribose which has five carbons named 1′ (prime) 2′ etc. There are four types of base: adenine and guanine have two carbon–nitrogen rings and are purines; thymine and cytosine have a single ring and are pyrimidines. The bases are attached to the 1′ carbon of the deoxyribose. A sugar plus a base is termed a nucleoside. A nucleotide has one, two or three phosphate groups attached to the 5′ carbon of the sugar. Nucleotides occur as individual molecules or polymerized as DNA or RNA.

DNA polynucleotides

Nucleotide triphosphates of the four bases are joined to form DNA polynucleotide chains. Two phosphates are lost during polymerization and the nucleotides are joined by the remaining phosphate. A phosphodiester bond forms between the 5′ phosphate of one nucleotide and the 3′ hydroxyl of the next nucleotide. The polynucleotide has a free 5′ phosphate at one end (5′ end) and a free 3′ OH (3′ end) at the other end. The sequence of bases encodes the genetic information. It can be read 5′→3′ or 3′→5′. Polynucleotides are extremely long. It is possible to have 4^n different sequences.

The double helix

DNA molecules are composed of two polynucleotide strands wrapped around each other to form a double helix. The sugar–phosphate part of the molecule forms a backbone. The bases face inwards and are stacked on top of each other. The two polynucleotide chains run in opposite directions. The double helix is right-handed and executes a turn every 10 bases. The helix has a major groove which mediates interactions with proteins. Variant DNA structures have been identified including Z DNA which has a left-handed helix.

Complementary base pairing

Hydrogen bonds between bases on the two DNA strands stabilize the double helix. The available space between the strands restricts the bases that can interact such that a purine always interacts with a pyrimidine. Thus, A interacts only with T and G only with C. This is called complementary base pairing. The restriction on base pairing means that the sequence of bases on the two strands are related to each other, such that the sequence of one determines and predicts the sequence of the other. This allows genetic information to be preserved during replication of the DNA and expression of the genes. Disruption of the hydrogen bonds between the bases by heat or chemicals or by the action of enzymes causes the strands of the double helix to separate.

RNA structure

In RNA thymine is replaced by uracil and 2-deoxyribose by ribose. RNA normally exists as a single polynucleotide strand however, short stretches of base pairing may occur between complementary sequences.

Related topics

Gene transcription (A4)
DNA replication (A9)
DNA mutation (B5)

Nucleotides

The ability of DNA to carry the genetic information required by a cell to reproduce itself is closely related to the structure of DNA molecules. DNA is a polymer and consists of a long chain of monomers called **nucleotides**. The DNA molecule is said to be a polynucleotide. Each nucleotide has three parts: a sugar, a nitrogen containing ring-structure called a **base**, and a phosphate group. The sugar present in DNA is a five carbon pentose called 2′-deoxyribose in which the –OH group on carbon 2 of ribose is replaced by hydrogen (*Fig. 1*). The carbon atoms in the sugar are numbered 1–5. The numbers are given a dash (′) referred to as **prime** to distinguish them from the numbers of the atoms in the base. The numbering is important because it indicates where other components of the nucleotide are attached to the sugar.

Nucleotides contain one of four bases: **adenine**, **guanine**, **thymine** or **cytosine** (*Fig. 2*). These are complex molecules containing carbon and nitrogen ring structures. Adenine and guanine contain two carbon–nitrogen rings and are known as **purines**. Cytosine and thymine contain a single ring and are called **pyrimidines**. The bases are attached to the sugar by a bond between the 1′ carbon of the sugar and a nitrogen at position 9 of the purines or position 1 of the pyrimidines. A sugar plus a base is called a **nucleoside** (*Fig. 3a*).

Nucleotides contain phosphate groups (PO_4) attached to the 5′ carbon of the sugar (*Fig. 3b*). A nucleoside is called a nucleotide when a phosphate group is attached, the attachment can consist of one, two or three phosphate groups joined together. The phosphate groups are called α, β and γ, with α directly attached to the sugar. Nucleotides may exist in cells as individual molecules (nucleotide triphosphates play an important role in cells as the carriers of energy used to power enzymatic reactions) or polymerized as nucleic acids (DNA or RNA).

DNA polynucleotides

Nucleotide triphosphates are joined together to give polynucleotides. There are four used to synthesize DNA polynucleotides, 2′-deoxyadenosine 5′-triphosphate (dATP or A), 2′-deoxythymidine 5′-triphosphate (dTTP or T), 2′-deoxycytosine 5′-triphosphate (dCTP or C) and 2-deoxyguanosine 5′-triphosphate (dGTP or G). The β and γ phosphates are lost during polymerization and the nucleotide units are joined together by the remaining phosphate. The 5′ phosphate of one nucleotide forms a bond with the 3′ carbon of the next nucleotide eliminating the –OH group on the 3′ carbon during the reaction. The bond is called a **3′–5′ phosphodiester bond** (C–O–P) (*Fig. 4*). The polynucleotide chain has a free

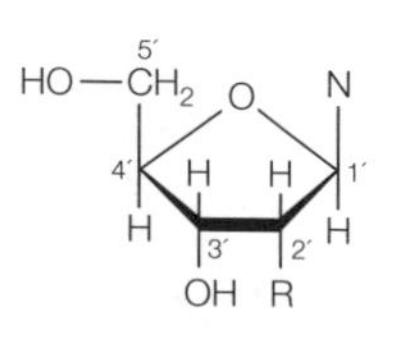

2′–Deoxyribose

Fig. 1. Structure of 2′-deoxyribose.

Adenine (A)

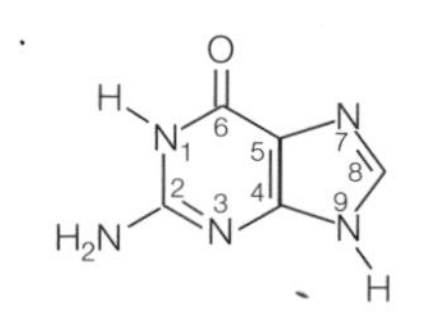

Guanine (G)

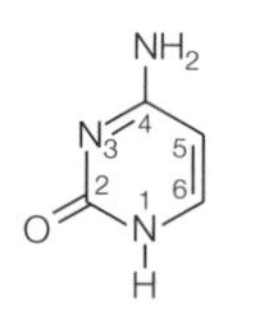

Cytosine (C)

Thymine (T)

Fig. 2. Bases in DNA.

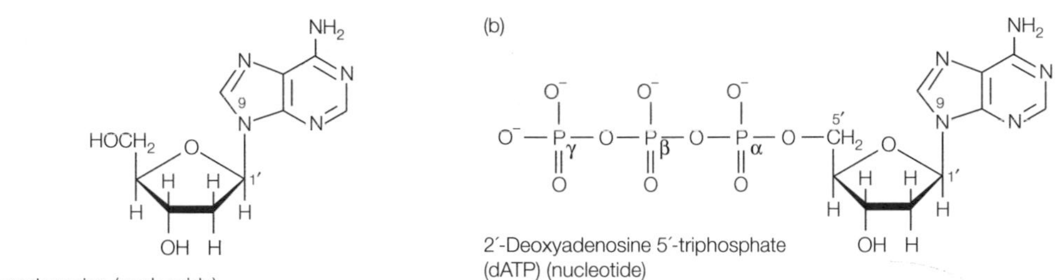

Fig. 3. Structure of (a) nucleosides, (b) nucleotides.

5′ triphosphate at one end known as the 5′ end and a free 3′ hydroxyl group at the other end called the 3′ end. This distinction gives the DNA polynucleotide polarity so that a DNA molecule can be described as running 5′→3′ or 3′→5′.

It is the sequence of the bases in the DNA polynucleotide that encodes the genetic information. This sequence is always written in the 5′→3′ direction (polymerase enzymes copy DNA molecules in this direction). Polynucleotides can be extremely long with no apparent limit to the number of nucleotides and no restrictions on the sequence of the nucleotides. The maximum number of possible base sequences for a polynucleotide is 4^n , where n is the number of nucleotides. This is an enormous number. For example, a polynucleotide containing just six bases could be arranged as $4^6 = 4096$ different sequences.

The double helix

DNA molecules have a very distinct and characteristic three-dimensional structure known as the double helix (*Fig. 5*). The structure of DNA was discovered

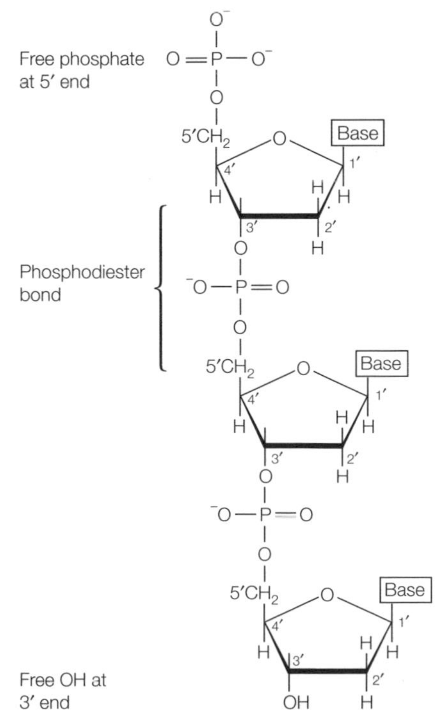

Fig. 4. Phosphodiester bonds join nucleotides in a DNA polynucleotide.

in 1953 by Watson and Crick working in Cambridge using X-ray diffraction pictures taken by Franklin and Wilkins. DNA exists as two polynucleotide chains wrapped around each other to form the double helix. The sugar–phosphate part of the molecule forms a spine or backbone which is on the outside of the helix. The bases, which are flat molecules, face inwards towards the center of the helix and are stacked on top of each other like a pile of plates.

X-ray diffraction pictures of the double helix show repeated patterns of bands that reflect the regularity of the structure of the DNA. The double helix executes a turn every 10 base pairs. The pitch of the helix is 34Å so the spacing between bases is 3.4Å. The diameter of the helix is 20Å. The double helix is said to be 3 **antiparallel**. One of the strands runs in the 5′→3′ direction and the other 3′→5′ direction. Only antiparallel polynucleotides form a stable helix. The double helix is not absolutely regular and when viewed from the outside a **major groove** and a **minor groove** can be seen. These are important for interaction with proteins, for replica-

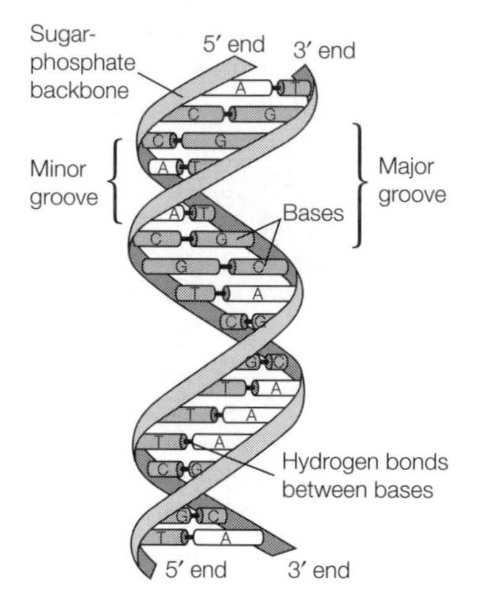

Fig. 5. The double helix.

tion of the DNA and for expression of the genetic information. The double helix is right-handed. This means that if the double helix were a spiral staircase and you were climbing up, the sugar–phosphate backbone would be on your right.

A number of variant forms of DNA occur when crystals of the molecule are formed under different conditions. The form present in cells is called the **B form**. Another form called the A form has a slightly more compact structure. Other forms that exist are C, D, E and Z, which is striking because it exists as a left-handed helix. Regions in chromosomes containing nonstandard structures such as Z-DNA have recently been identified.

Complementary base pairing

The bases of the two polynucleotide chains interact with each other. The space between the polynucleotides is such that a two-ring purine interacts with a single-ring pyrimidine. Thus, thymine always interacts with adenine and guanine with cytosine. Hydrogen bonds form between the bases and help to stabilize the interaction. Two bonds form between A and T and three between G and C. Thus, G–C bonds are stronger than A–T bonds. The way in which the bases form pairs between the two DNA strands is known as **complemen-**

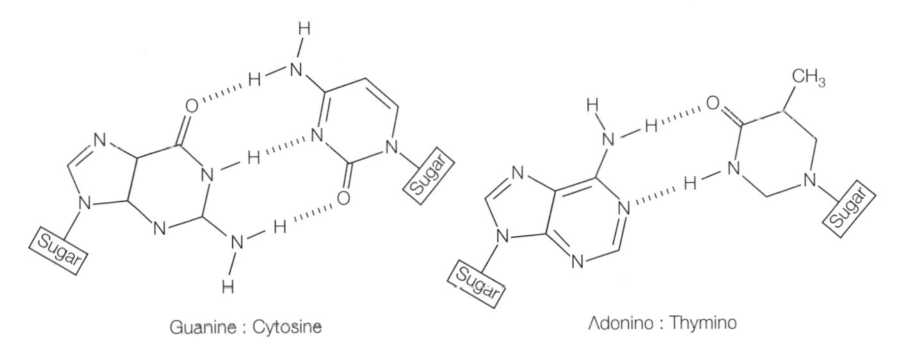

Fig. 6. *Complementary base pairing. Hydrogen bonds are shown as dashed lines.*

tary base pairing and is of fundamental importance (*Fig. 6*). Combinations other than G–C and A–T do not work because they are too large or too small to fit inside the helix or they do not align correctly to allow hydrogen bond formation. Because G must always bond to C and A to T the sequences of the two strands are related to each other and are said to be complementary with the sequence of one strand predicting and determining the sequence of the other. This means that one strand can be used to replicate the other. This is a vital mechanism for retaining genetic information and passing it on to other cells following cell division. Complementary base pairing is also essential for the expression of genetic information and is central to the way DNA sequences are transcribed into mRNA and translated into protein.

The double helix is stabilized by hydrogen bonds between the base pairs. These can be disrupted by heat and some chemicals. This results in separation of the double helix into two strands and the molecule is said to be denatured. In cells enzymes can separate the strands of the double helix for the purposes of copying the DNA and for expression of the genetic information.

RNA structure

The structure of RNA is similar to that of DNA but a number of important differences exist. In RNA ribose replaces 2′-deoxyribose and the base thymine is replaced by another base, uracil, which can also base pair with adenine (*Fig. 7*). In addition, RNA molecules normally exist as a single polynucleotide strand and do not form a double helix. However, it is possible for base pairing to occur between complementary parts of the same RNA strand resulting in short double-stranded regions.

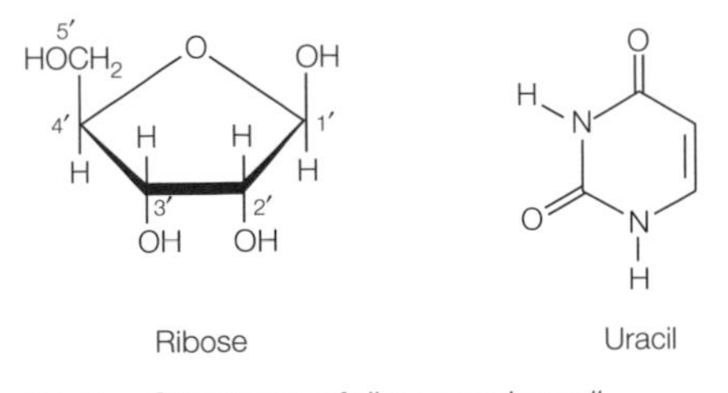

Fig. 7. *Structures of ribose and uracil.*

A2 Genes

Key Notes

Structure of genes

A gene is a unit of information and corresponds to a discrete segment of DNA that encodes the amino acid sequence of a polypeptide. Human cells contain 50–100 000 genes arranged on 23 chromosomes. The genes are dispersed and are separated by noncoding intergenic DNA. Information is encoded on the template strand which directs the synthesis of an RNA molecule. Both DNA strands can act as the template strand. DNA molecules have an enormous capacity to store genetic information.

Gene families

Some genes are arranged as clusters known as operons and multigene families. Operons occur in bacteria and contain coregulated genes with a related function. Multigene families occur in higher organisms and contain genes that are identical or similar that are not regulated coordinately. Simple multigene families contain identical genes whose product is required in large amounts. Complex multigene families contain genes that are very similar and encode proteins with a related function.

Gene expression

The biological information encoded in genes is made available by gene expression. In this process, an RNA copy of a gene is synthesized which then directs the synthesis of a protein. The central dogma states that information is always transferred from DNA to RNA to protein. The functioning of cells is dependent on the coordinated activity of many proteins. Gene expression ensures that proteins are synthesized in the correct place at the correct time.

Gene promoters

Gene expression is highly regulated. Not all of the genes present in a cell are active and different types of cell express different genes. The expression of a gene is regulated by a segment of DNA upstream of the coding sequence called the promoter, this binds RNA polymerase and associated transcription factor proteins and initiates synthesis of an RNA molecule.

Introns and exons

The coding sequence of a gene is split into a series of segments called exons which are separated by noncoding sequences called introns which usually account for most of the gene sequence. The number and sizes of the introns vary between genes. Introns are removed from RNA transcripts by a process called splicing prior to protein synthesis. Introns are not usually present in bacteria.

Pseudogenes

Copies of some genes exist which contain sequence errors acquired during evolution that prevent them from producing proteins. These are called pseudogenes and they represent evolutionary relics of original genes. Examples include the globin pseudogenes.

Related topics

Regulation of gene expression in prokaryotes (A10)
Regulation of gene expression in eukaryotes (A11)
The human genome (B4)

Structure of genes

The biological information needed by an organism to reproduce itself is contained in its DNA. The information is encoded in the base sequence of the DNA and is organized as a large number of genes, each of which contains the instructions for the synthesis of a polypeptide. In physical terms, a gene is a discrete segment of DNA with a base sequence that encodes the amino acid sequence of a polypeptide. Genes vary greatly in size from less than 100 base pairs to several million base pairs. In higher organisms the genes are present on a series of extremely long DNA molecules called **chromosomes**. In humans there are an estimated 50–100 000 genes arranged on 23 chromosomes. The genes are very dispersed and are separated from each other by sequences that do not appear to contain useful information; this is called **intergenic DNA**. The intergenic DNA is very long, such that in humans gene sequences account for less than about 30% of the total DNA. Only one of the two strands of the DNA double helix carries the biological information: this is called the **template strand** and it is used to produce an RNA molecule of complementary sequence which directs the synthesis of a polypeptide. The other strand is called the **nontemplate strand**. Both strands of the double helix have the potential to act as the template strand: individual genes may be encoded on different strands. Other terms are used to describe the strands of the double helix as alternatives to template and nontemplate. These include **sense/antisense** and **coding/noncoding**: the terms antisense and noncoding are equivalent to the template strand.

The capacity of DNA molecules to store information is enormous. For a DNA molecule n bases long, the number of different combinations of the four bases is 4^n. Even for very short DNA molecules the number of different sequences possible is very large. In practice, there are limitations to the sequences that can contain useful information. However the capacity to encode information remains vast.

Gene families

Most genes are spread out randomly along the chromosomes, however some are organized into groups or clusters. Two types of cluster occur: these are **operons** and **multigene families**.

Operons are gene clusters found in bacteria. They contain genes that are regulated in a coordinated way and encode proteins with closely related functions. An example is the *lac* operon in *E. coli* which contains three genes encoding enzymes required by the bacterium to break down lactose. When lactose is available as an energy source, the enzymes encoded by the *lac* operon are required together. The clustering of the genes within the operon allows them to be switched on or off at the same time allowing the organism to use its resources efficiently (*Fig. 1*).

In higher organisms, operons are absent and clustered genes exist as multigene families. Unlike operons, the genes in a multigene family are identical or are very similar and are not regulated coordinately. The clustering of genes in multigene families probably reflects a requirement for multiple copies of that

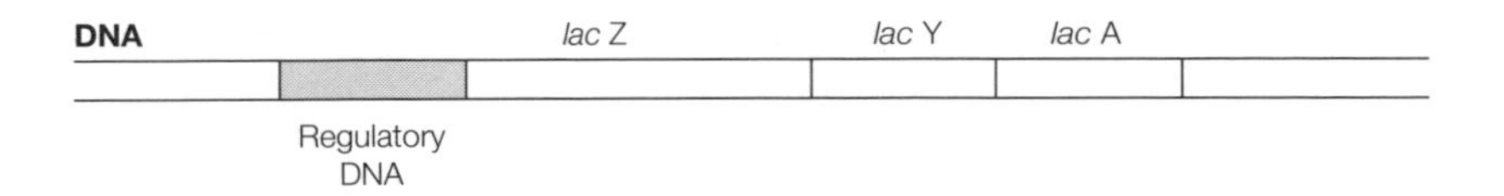

Fig. 1. The lac *operon. Three genes* (lac Z, Y and A) *are arranged and regulated together.*

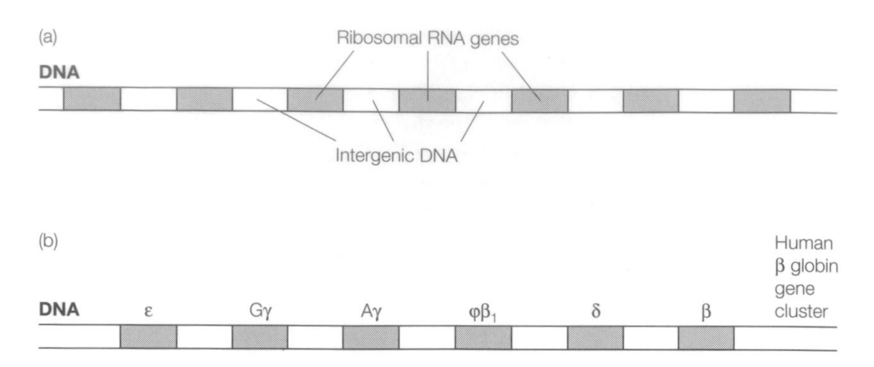

Fig. 2. (a) Simple multigene family; (b) complex multigene family.

gene which was fulfilled by duplication during evolution. Some multigene families exist as separate clusters on different chromosomes; this probably arose by rearrangements of the DNA during evolution which resulted in the breaking up of clusters. Multigene families may be simple or complex. In **simple multigene families** the genes are identical. An example is the gene for the 5S ribosomal RNA. In humans, there are about 2000 clustered copies of this gene reflecting the high demand of cells for the gene product (*Fig. 2a*). **Complex multigene families** contain genes that are very similar but not identical. An example is the globin gene family that encodes a series of polypeptides (α β γ ε ζ globins) that differ from each other by just a few amino acids. Globin polypeptides form complexes with each other and with a cofactor molecule called heme to give the adult and embryonic forms of the oxygen carrying blood protein, hemoglobin (*Fig. 2b*).

Gene expression

The biological information in a DNA molecule is contained in its base sequence. Gene expression is the process by which this information is made available to the cell. The use of the information is described by the **central dogma**, originally proposed by Crick, which states that information is transferred from DNA to RNA to protein (*Fig. 3*). During gene expression, DNA molecules copy their information by directing the synthesis of an RNA molecule of complementary sequence. This process is known as **transcription**. The RNA then directs the synthesis of a polypeptide whose amino acid sequence is determined by the base sequence of the RNA. This process is known as **translation**. The amino acid sequence of the protein determines its three-dimensional structure which in turn dictates its function. The central dogma states that the transfer of information can only occur in one direction – from DNA to RNA to protein – and cannot occur in reverse. An exception to this rule is found in retroviruses which have an enzyme called **reverse transcriptase** which can copy RNA into DNA. The functioning of cells, and in turn of living organisms, is dependent on the

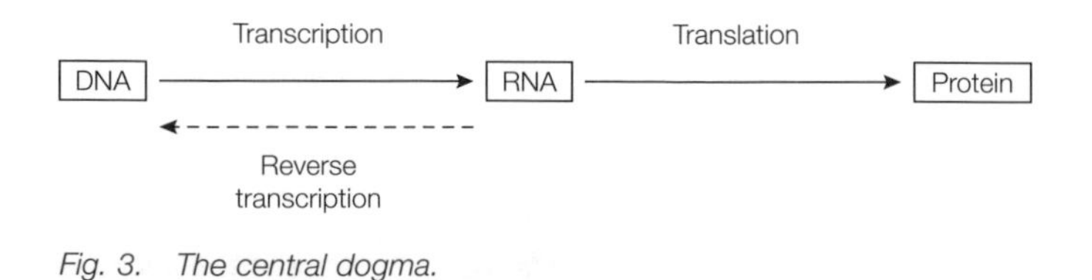

Fig. 3. The central dogma.

coordinated activity of many different proteins. The biological information contained within the genes acts as a set of instructions for synthesizing proteins at the correct time and in the correct place.

Gene promoters

The expression of the biological information present in genes is highly regulated. Not all the genes present in a cell's DNA are expressed and different genes are active in different cell types. The overall complement of genes that are active determines the characteristics of a cell and its function within the organism. Thus, for example, many of the genes that are active in muscle cells are different from those that are active in blood cells. Expression of genes is regulated by a segment of DNA sequence present upstream of the coding sequence known as the **promoter**. Conserved DNA sequences in the promoter are recognized and bound by the RNA polymerase and other associated proteins called **transcription factors** that bring about the synthesis of an RNA transcript of the gene. The expression of a gene in a cell is determined by the promoter sequence and its ability to bind RNA polymerase and transcription factors.

Introns and exons

One of the more surprising features of genes is that in higher organisms the coding information is usually split into a series of segments of DNA sequence called **exons**. These are separated by sequences that do not contain useful information called **introns** (*Fig. 4*). The number of introns varies greatly, from zero to more than 50 in some genes. The length of the exons and introns also varies but the introns are usually much longer and account for the majority of the sequence of the gene. Before the biological information in a gene can be used to synthesize a protein, the introns must be removed from RNA molecules by a process called **splicing** which leaves the exons and the coding information continuous. Introns are a feature of higher organisms only and are not usually found in bacteria.

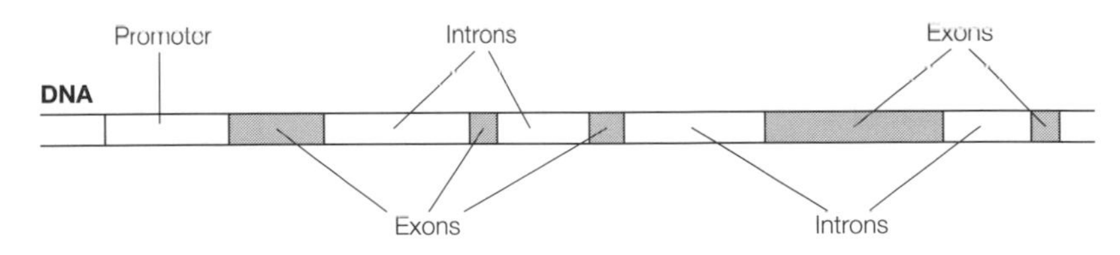

Fig. 4. Structure of a gene.

Pseudogenes

Some genes exist which resemble other genes but examination of their base sequence shows errors that make it impossible for them to contain useful biological information. These are called pseudogenes and they represent genes that have acquired errors or mutations in their DNA sequence during evolution causing their biological information to be scrambled so that they are no longer able to direct the synthesis of a protein. As such, pseudogenes are evolutionary relics. During evolution, the initial base changes causing loss of biological information are followed by more rapid changes so that the sequence of the pseudogene eventually deviates substantially from the original gene. Examples include several globin pseudogenes that are present in the globin gene clusters.

A3 THE GENETIC CODE

Key Notes

Gene expression

Genetic information is encoded in the base sequence of DNA molecules as a series of genes. Gene expression is the term used to describe how cells decode the information to synthesize proteins required for cellular function. The expression of a gene involves the synthesis of a complementary RNA molecule whose sequence specifies the amino acid sequence of a protein. The DNA sequence of the gene is colinear with the amino acid sequence of the polypeptide.

Genetic code

Amino acids are encoded by 64 base triplets called codons which encode the 20 amino acids. Most amino acids have more than one codon. This is known as the degeneracy of the genetic code and it helps to minimize the effect of mutations. Codons that specify the same amino acid are called synonyms and differ at their third base, known as the 'wobble' position. AUG is the initiation codon and encodes methionine. There are three stop codons: UAG, UGA and UAA.

Reading frames

Three possible sets of codons can be read from any sequence depending on which base is chosen as the start of a codon. Each set of codons is known as a reading frame. The initiation codon determines the reading frame of the protein coding sequence. Other reading frames tend to contain stop codons and are not used for protein synthesis. An open reading frame is a sequence of codons bounded by start and stop codons.

Universality of the code

The genetic code applies universally with all organisms using the same codons for each amino acid. However, some exceptions to the standard codon usage occur in mitochondrial genomes and in some unicellular organisms.

Related topics Transfer RNA (A5) Translation (A8)

Gene expression

The information required by an organism to reproduce itself is carried by its DNA, encoded in the base sequence and organized as a series of genes. Gene expression is the term used to describe the process by which cells decode and make use of this information to synthesize the proteins that are responsible for cellular function. During gene expression, information is copied from DNA to RNA by the synthesis of an RNA molecule whose base sequence is complementary to that of the DNA template. The RNA then directs the synthesis of a protein whose amino acid sequence is specified by the base sequence of the RNA. For every gene the DNA sequence is colinear with the amino acid sequence of the polypeptide it encodes such that the 5′→3′ base sequence of the coding strand specifies the amino acid sequence of the encoded polypeptide from the amino to the carboxy terminus.

Genetic code

The genetic code describes how base sequences are converted into amino acid sequences during protein synthesis. The DNA sequence of a gene is divided into a series of units of three bases. Each set of three bases is called a **codon** and specifies a particular amino acid. The four bases in DNA and RNA can combine as a total of $4^3 = 64$ codons which specify the 20 amino acids found in proteins (*Table 1*). Because the number of codons is greater, all of the amino acids, with the exceptions of methionine and tryptophan, are encoded by more than one codon. This feature is referred to as the **degeneracy** or the **redundancy** of the genetic code. Codons which specify the same amino acid are called **synonyms** and tend to be similar. For example, ACU, ACC, ACA and ACG all specify the amino acid threonine. Variations between synonyms tend to occur at the third position of the codon, which is known as the **wobble position**. The degeneracy of the genetic code minimizes the effects of mutations so that alterations to the base sequence are less likely to change the amino acid encoded and possible deleterious effects on protein function are avoided. Of the 64 possible codons, 61 encode amino acids. The remaining three, UAG, UGA, and UAA, do not encode amino acids but instead act as signals for protein synthesis to stop and as such are known as **termination codons** or **stop codons**. The codon for methionine, AUG, is the signal for protein synthesis to start and is known as the **initiation codon**. Thus all polypeptides start with methionine although this is sometimes removed later.

Table 1. The genetic code

First position (5′ end)	Second position				Third position (3′ end)
	U	C	A	G	
U	Phe UUU	Ser UCU	Tyr UAU	Cys UGU	U
	Phe UUC	Ser UCC	Tyr UAC	Cys UGC	C
	Leu UUA	Ser UCA	Stop UAA	Stop UGA	A
	Leu UUG	Ser UCG	Stop UAG	Trp UGG	G
C	Leu CUU	Pro CCU	His CAU	Arg CGU	U
	Leu CUC	Pro CCC	His CAC	Arg CGC	C
	Leu CUA	Pro CCA	Gln CAA	Arg CGA	A
	Leu CUG	Pro CCG	Gln CAG	Arg CGG	G
A	Ile AUU	Thr ACU	Asn AAU	Ser AGU	U
	Ile AUC	Thr ACC	Asn AAC	Ser AGC	C
	Ile AUA	Thr ACA	Lys AAA	Arg AGA	A
	Met AUG	Thr ACG	Lys AAG	Arg AGG	G
G	Val GUU	Ala GCU	Asp GAU	Gly GGU	U
	Val GUC	Ala GCC	Asp GAC	Gly GGC	C
	Val GUA	Ala GCA	Glu GAA	Gly GGA	A
	Val GUG	Ala GCG	Glu GAG	Gly GGG	G

Reading frames

In addition to identifying the start of protein synthesis, the initiation codon determines the reading frame of the RNA sequence. Depending on which base is chosen as the start of a codon, three possible sets of codons may be read from any base sequence. In practice, during protein synthesis, normally only one reading frame contains useful information; the other two reading frames usually contain several stop codons which prevent them from being used to

Reading frame 1. 5′ – AUG ACU AAG AGA UCC GG –3′
Met Thr Lys Arg Ser

Reading frame 2. 5′ – A UGA CUA AGA GAU CCG G –3
Stop Leu Arg Asp Pro

Reading frame 3. 5′ – AU GAC UAA GAG AUC CGG –3′
Asp **Stop** Glu Ile Arg

Fig. 1. Every DNA sequence can be read as three separate reading frames depending on which base is chosen as the start of the codon.

direct protein synthesis (*Fig. 1*). A set of codons that runs continuously and is bounded at the start by an initiation codon and at the end by a termination codon is known as an **open reading frame** (ORF). This characteristic is used to identify protein coding DNA sequences in genome sequencing projects.

Universality of the code

Initially the genetic code was believed to apply universally, that is all organisms would recognize individual codons as the same amino acids. However, it has now been shown that some variation in the code exists, although this is rare. For example, mitochondria have a small DNA genome containing about 20 genes in which deviations from the genetic code occur. Changes are mostly associated with start and stop codons. For example, UGA, which is normally a termination codon, codes for tryptophan whereas AGA and AGG which normally encode arginine are termination codons, and AUA, normally isoleucine, specifies methionine. It is thought that these changes tend to be viable because the mitochondrion is a closed system. A few examples of nonstandard codon usage have now been found outside mitochondrial genomes in unicellular organisms. For example UAA and UAG which are normally stop codons, encode glutamic acid in some protozoa.

A4 GENE TRANSCRIPTION

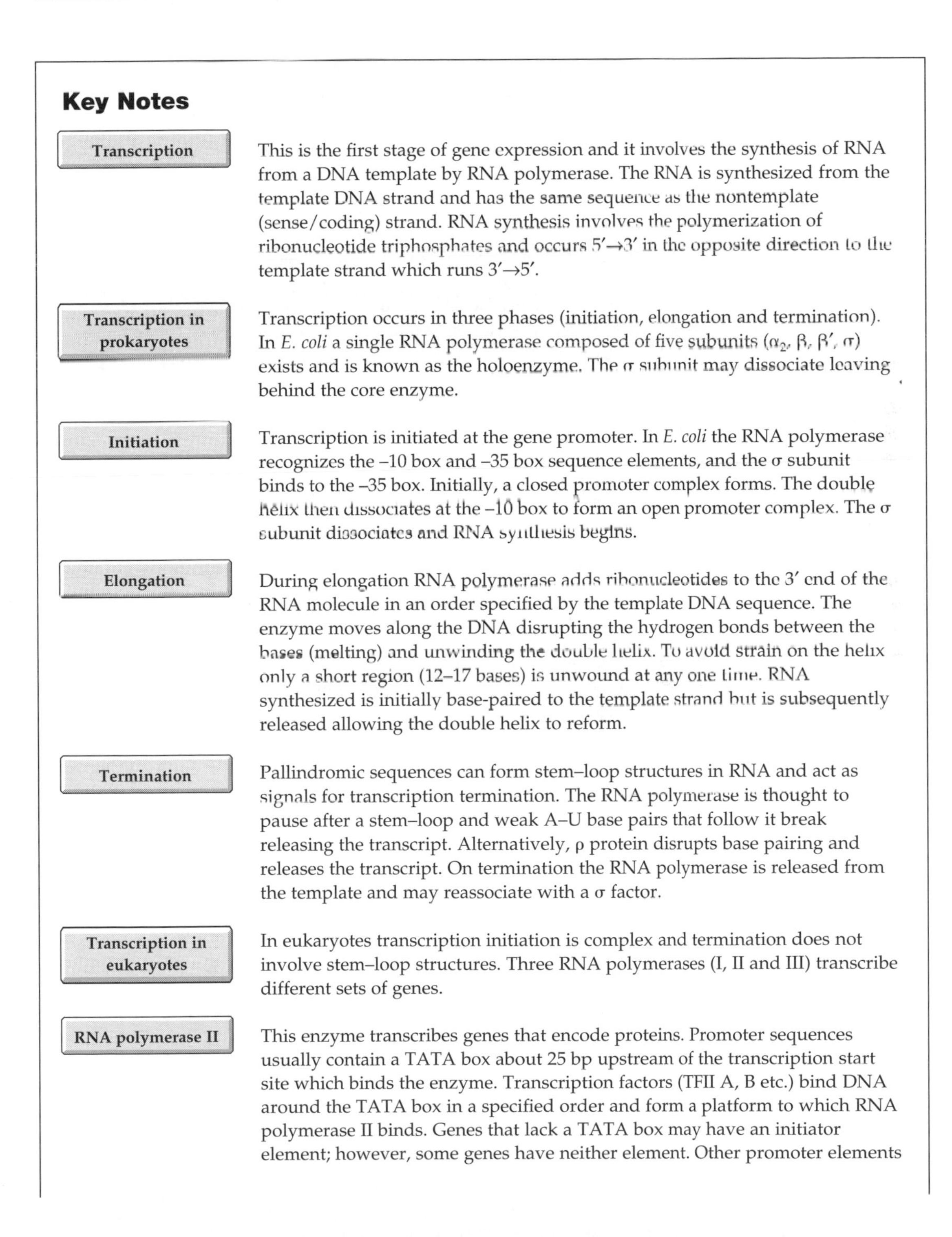

Key Notes

Transcription

This is the first stage of gene expression and it involves the synthesis of RNA from a DNA template by RNA polymerase. The RNA is synthesized from the template DNA strand and has the same sequence as the nontemplate (sense/coding) strand. RNA synthesis involves the polymerization of ribonucleotide triphosphates and occurs 5′→3′ in the opposite direction to the template strand which runs 3′→5′.

Transcription in prokaryotes

Transcription occurs in three phases (initiation, elongation and termination). In *E. coli* a single RNA polymerase composed of five subunits (α_2, β, β', σ) exists and is known as the holoenzyme. The σ subunit may dissociate leaving behind the core enzyme.

Initiation

Transcription is initiated at the gene promoter. In *E. coli* the RNA polymerase recognizes the –10 box and –35 box sequence elements, and the σ subunit binds to the –35 box. Initially, a closed promoter complex forms. The double helix then dissociates at the –10 box to form an open promoter complex. The σ subunit dissociates and RNA synthesis begins.

Elongation

During elongation RNA polymerase adds ribonucleotides to the 3′ end of the RNA molecule in an order specified by the template DNA sequence. The enzyme moves along the DNA disrupting the hydrogen bonds between the bases (melting) and unwinding the double helix. To avoid strain on the helix only a short region (12–17 bases) is unwound at any one time. RNA synthesized is initially base-paired to the template strand but is subsequently released allowing the double helix to reform.

Termination

Pallindromic sequences can form stem–loop structures in RNA and act as signals for transcription termination. The RNA polymerase is thought to pause after a stem–loop and weak A–U base pairs that follow it break releasing the transcript. Alternatively, ρ protein disrupts base pairing and releases the transcript. On termination the RNA polymerase is released from the template and may reassociate with a σ factor.

Transcription in eukaryotes

In eukaryotes transcription initiation is complex and termination does not involve stem–loop structures. Three RNA polymerases (I, II and III) transcribe different sets of genes.

RNA polymerase II

This enzyme transcribes genes that encode proteins. Promoter sequences usually contain a TATA box about 25 bp upstream of the transcription start site which binds the enzyme. Transcription factors (TFII A, B etc.) bind DNA around the TATA box in a specified order and form a platform to which RNA polymerase II binds. Genes that lack a TATA box may have an initiator element; however, some genes have neither element. Other promoter elements such as the CAT box act as binding sites for other transcription factors that influence the rate of transcription initiation. Distant elements called enhancers and silencers also influence the transcription rate. The signals that mediate transcription termination are uncertain.

RNA polymerase I

This enzyme transcribes 18S, 28S and 5.8S ribosomal (r) RNAs. The promoter contains two elements essential for transcription: a core element that overlaps the transcription start site and an upstream control sequence at around position –100. An 18 bp termination signal is present approximately 600 bp after the end of the gene.

RNA polymerase III

This enzyme transcribes short genes encoding transfer (t) RNAs and the 5.8S rRNA. The promoter sequences occur within the coding sequence and are called internal control regions (ICRs). The tRNA gene has two important sequence elements, the A box and the B box; transcription of the 5S rRNA gene requires a sequence called the C box. A second sequence called the A box is important. Transcription terminates at a signal sequence containing a run of A residues.

Related topics

Transfer RNA (A5)
Ribosomal RNA (A6)
Messenger RNA (A7)
Regulation of gene expression in prokaryotes (A10)
Regulation of gene expression in eukaryotes (A11)

Transcription

In this process an RNA copy of the DNA sequence of a gene is produced as the first stage of gene expression. The RNA is synthesized by enzymes called **RNA polymerases** using DNA as a template. The two strands of the double helix are called the template and the nontemplate strands. RNA is produced using the **template strand** and the RNA molecule synthesized is a copy of the **nontemplate strand** (*Fig. 1*). Gene sequences usually refer to the nontemplate strand. Other names used to describe the nontemplate strand are the **sense (+) strand** or the **coding strand**. The RNA molecule synthesized is called a **transcript** and may subsequently undergo translation to produce a protein or may be used as ribosomal or transfer RNA.

During transcription RNA is synthesized by the polymerization of ribonucleotide triphosphate subunits (ATP, UTP, GTP, CTP). The 3′-OH of one ribonucleotide reacts with the 5′ phosphate of another to form a phosphodiester bond. The order in which the ribonucleotides are added to the growing RNA chain is determined by the order of the bases in the template DNA. New ribonucleotides are added to the growing chain at the free 3′ end. The transcript is synthesized in the 5′→3′ direction but because the chains must be antiparallel for base pairing the template strand runs in the opposite, 3′→5′, direction.

Transcription in prokaryotes

In prokaryotic organisms transcription occurs in three phases known as **initiation, elongation** and **termination**. RNA is synthesized by a single RNA polymerase enzyme which contains multiple polypeptide subunits. In *E. coli* the RNA polymerase has five subunits: two α, one β, one β′ and one σ subunit ($\alpha_2\beta\beta'\sigma$).

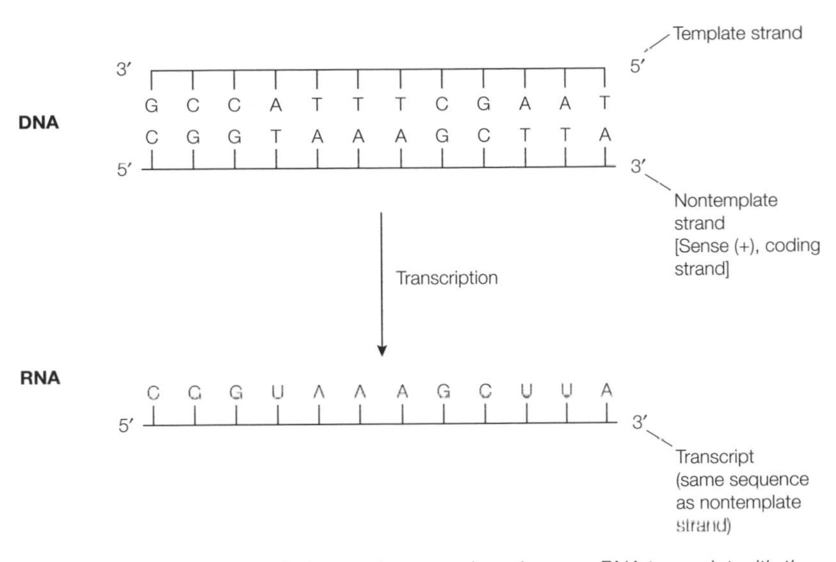

Fig. 1. Transcription of the DNA template strand produces an RNA transcript with the same sequence as the DNA nontemplate strand.

This form is called the **holoenzyme**. The σ subunit may dissociate from the other subunits to leave a form known as the **core enzyme**. These two forms of the RNA polymerase have different roles in transcription.

Initiation

Transcription cannot start randomly but must begin specifically at the start of a gene. Signals for the initiation of transcription occur in the **promoter** sequence which lies directly upstream of the transcribed sequence of the gene. The promoter contains specific DNA sequences that act as points of attachment for the RNA polymerase. In *E. coli*, two sequence elements recognized by the RNA polymerase known as the **–10 sequence** and the **–35 sequence** are present. The exact sequences can vary between promoters but all conform to an overall pattern known as the **consensus sequence**. The σ subunit of the RNA polymerase is responsible for recognizing and binding the promoter, probably at the –35 box. In the absence of the σ subunit the enzyme can still bind to DNA but binding is more random. When the enzyme binds to the promoter it initially forms a **closed promoter complex** in which the promoter DNA remains as a double helix. The enzyme covers about 60 base pairs of the promoter including the –10 and –35 boxes. To allow transcription to begin, the double helix partially dissociates at the –10 box, which is rich in weak A–T bonds, to give an **open promoter complex**. The σ subunit then dissociates from the open promoter complex leaving the core enzyme. At the same time the first two ribonucleotides bind to the DNA, the first phosphodiester bond is formed and transcription is initiated (*Fig. 2*).

Elongation

During elongation the RNA polymerase moves along the DNA molecule melting and unwinding the double helix as it progresses. The enzyme adds ribonucleotides to the 3′ end of the growing RNA molecule with the order of addition determined by the order of the bases on the template strand. In most cases, a **leader sequence** of variable length is transcribed before the coding sequence of the gene is reached. Similarly, at the end of the coding sequence a noncoding

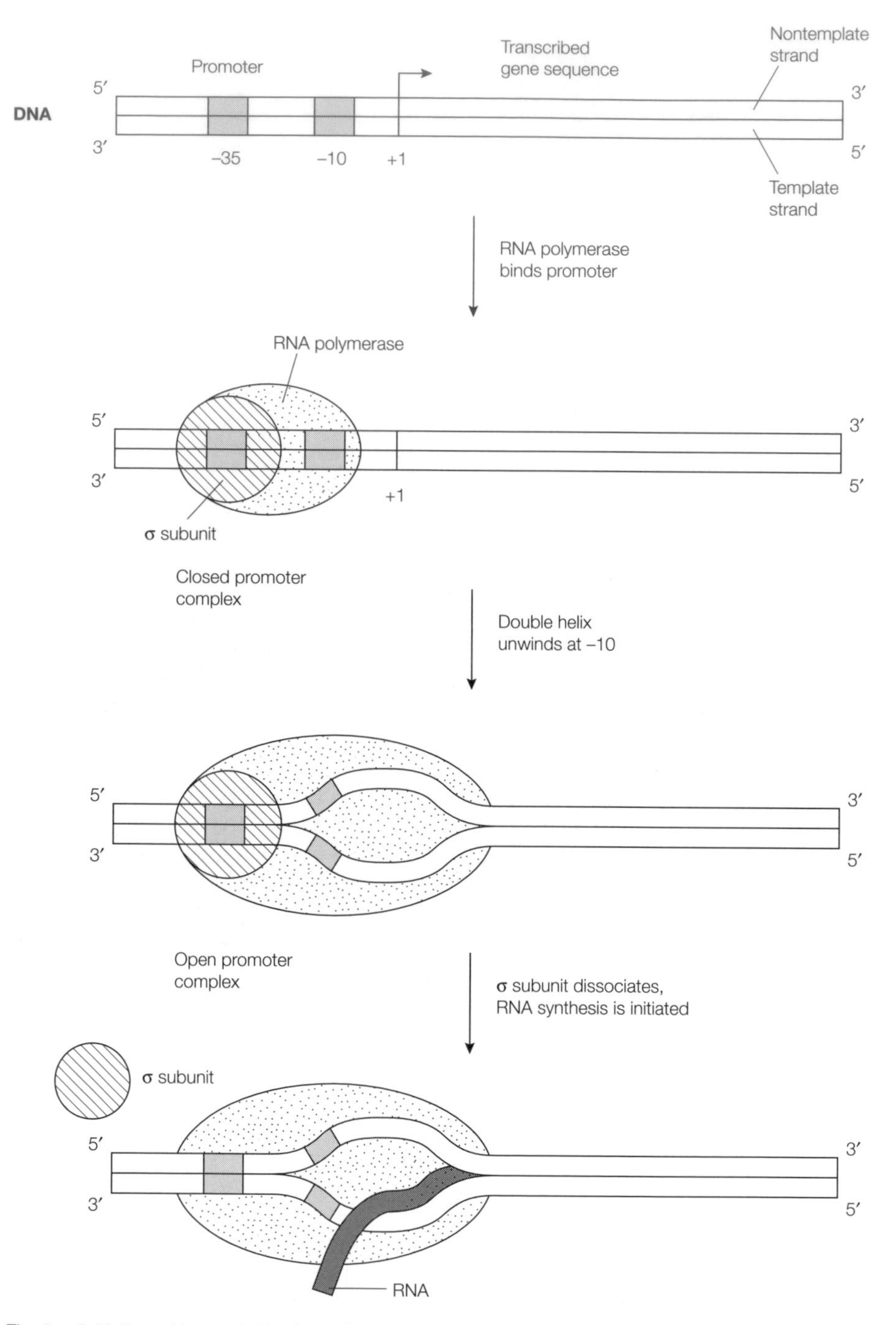

Fig. 2. Initiation of transcription in prokaryotes.

trailer sequence is transcribed before transcription ends. During transcription, only a small portion of the double helix is unwound at any one time. The unwound area contains the newly synthesized RNA base-paired with the template DNA strand and extends over 12–17 bases. The unwound area needs to remain small because unwinding in one region necessitates overwinding in adjacent regions and this imposes strain on the DNA molecule. To overcome this problem, the RNA is released from the template DNA as it is synthesized allowing the DNA double helix to reform (*Fig. 3*).

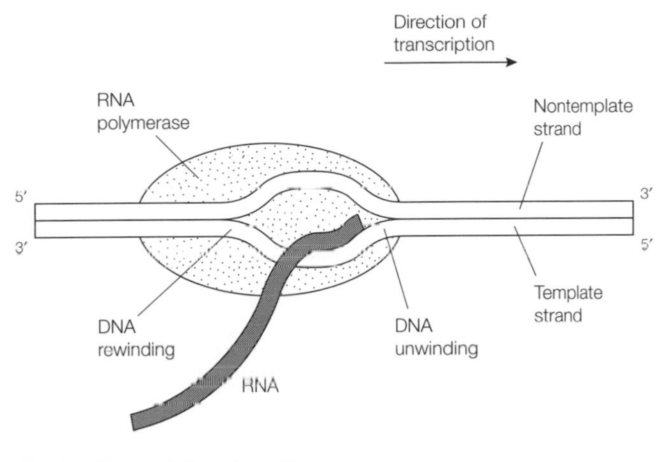

Fig. 3. Transcription elongation.

Termination

The termination of transcription occurs nonrandomly and takes place at specific points after the end of the coding sequence. In *E. coli*, termination occurs at sequences known as **pallindromes**. These are symmetrical about their middle such that the first half of the sequence is followed by its exact complement in the second half. In single-stranded RNA molecules this feature allows the first half of the sequence to base pair with the second half to form what is known as a **stem–loop structure** (*Fig. 4*). These appear to act as signals for termination. In some cases the stem–loop sequence is followed by a run of 5–10 As in the DNA which form weak A–U base pairs with the newly synthesized RNA. It is thought that the RNA polymerase pauses just after the stem–loop and that the weak A–U base pairs break causing the transcript to detach from the template. In other cases the run of As is absent and a different mechanism occurs based on binding of a protein called **Rho (ρ)** which disrupts base-pairing between the template and the transcript when the polymerase pauses after the stem–loop. The termination of transcription involves the release of the transcript and the core enzyme which may then reassociate with the σ subunit and go on to another round of transcription.

Transcription in eukaryotes

Transcription occurs in eukaryotes in a similar way to prokaryotes. However, initiation is more complex, termination does not involve stem–loop structures and transcription is carried out by three enzymes (RNA polymerases I, II, III) each of which transcribes a specific set of genes and functions in a slightly different way.

RNA polymerase II

This enzyme transcribes genes that encode proteins. Binding of RNA polymerase II to its promoter involves several different DNA sequence elements and a number of proteins called **transcription factors**. The promoter usually (but not always) contains a DNA sequence element called the **TATA box** which acts as the attachment site for the RNA polymerase II. This has the consensus sequence

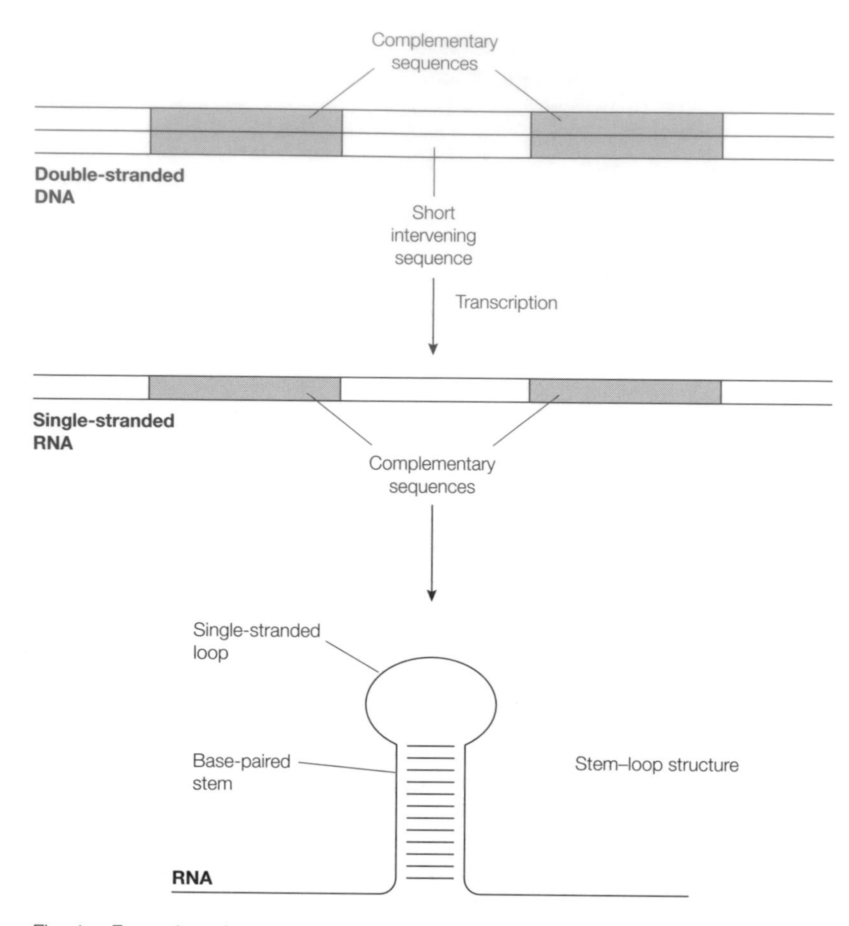

Fig. 4. Formation of a stem–loop structures in RNA.

5′ TATA(A/T)A(A/T) 3′ and is present about 25 bp upstream from the transcription start site. Its function is to locate the RNA polymerase at the start of the gene in the correct position to begin transcription. Attachment of the RNA polymerase at the TATA box is achieved with the help of a series of transcription factors specific to RNA polymerase II, referred to as TFIIA, TFIIB etc. These bind to the DNA around the TATA box and form a platform to which the RNA polymerase II is bound. The transcription factors bind in a specific order. TFIID binds first followed by TFIIA and TFIIB. The RNA polymerase II then binds followed by TFIIF, E, H and J to produce a functional complex capable of initiating transcription (*Fig. 5*).

Genes that lack a TATA box may contain an alternative initiator element around the transcription start site. Other gene promoters lack either a TATA box or an initiator element. These genes are usually transcribed at low levels and transcription initiation can occur at several different points. These genes often contain a GC-rich sequence 100–200 bp upstream of the transcription start site.

A number of other promoter elements with characteristic consensus sequences influence transcription. These sequences act as binding sites for other transcription factors that regulate transcription by stimulating or repressing

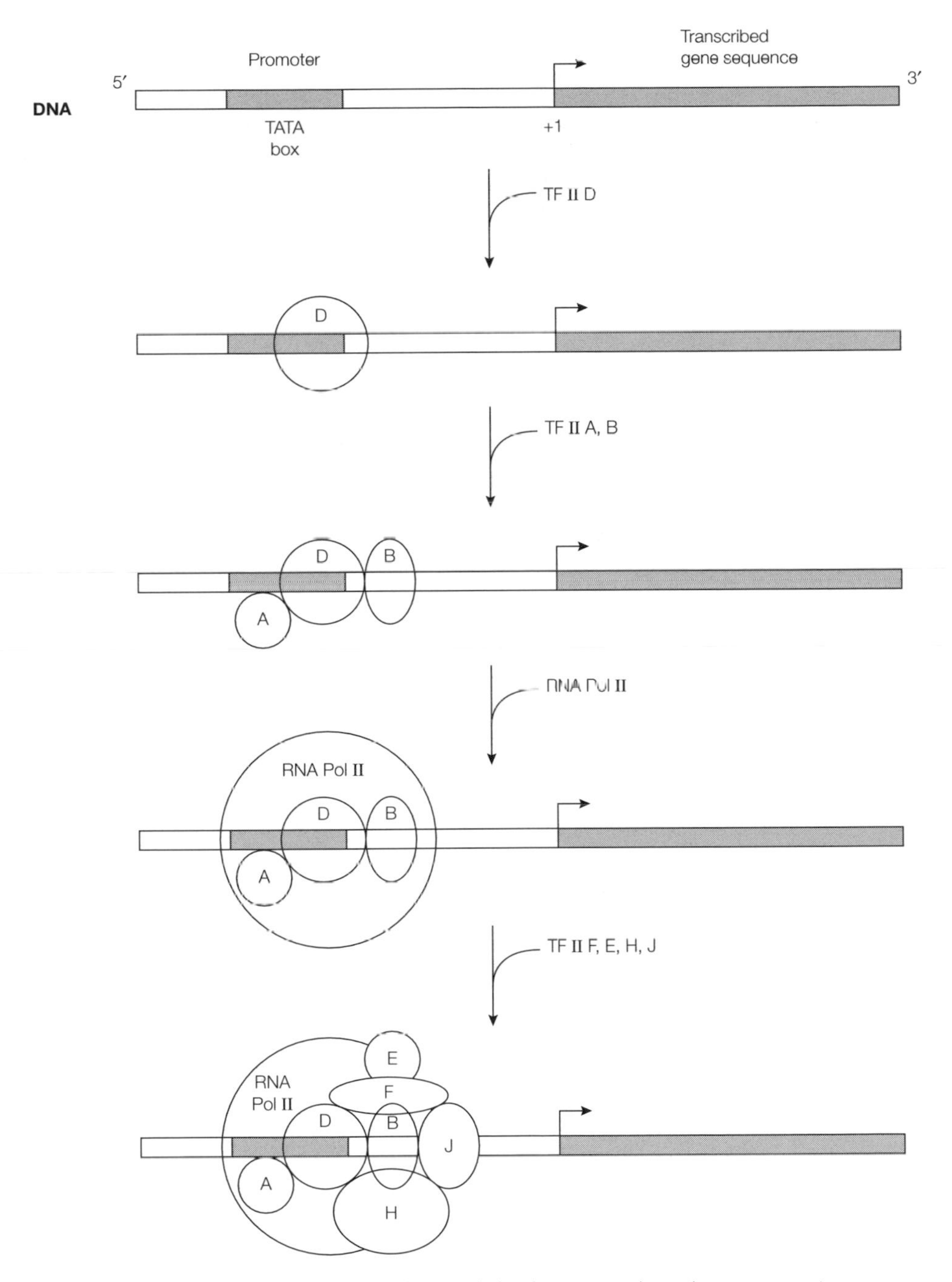

Fig. 5. Binding of RNA polymerase II and transcription factors to eukaryotic gene promoters.

initiation. Examples include the CCAAAT (pronounced CAT) box found upstream of the TATA box in some genes. Many genes also contain sequences called **enhancers** that greatly stimulate transcription. These are located outside the promoter often far away from the gene they influence and work independently of their orientation. Similar sequences known as **silencers** also occur which inhibit transcription.

Termination of transcription by RNA polymerase II occurs at a point after the end of the protein coding sequence by a mechanism which is uncertain. Termination signals are difficult to identify because the 3′ end of the message is removed shortly after it is transcribed. It may be that definite termination signals do not exist but that the dissociation of a transcription factor, at some point, destabilizes the transcription complex causing the RNA polymerase to fall off the template at a later time.

Initiation of transcription by RNA polymerase I and III is similar to RNA polymerase II and involves specific promoter sequences that act as binding sites for the RNA polymerase and associated transcription factors (TIF-I for RNA polymerase I and TFIII A–C for RNA polymerase III).

RNA polymerase I

This enzyme transcribes genes encoding three of the four ribosomal RNAs (18S, 28S and 5.8S). The promoter recognized by RNA polymerase I has two important sequence elements required for efficient transcription: a core element overlapping the transcription start site and an upstream control sequence located approximately 100 bp upstream of the transcription start site. These sequences bind RNA polymerase I and its associated transcription factors. The core sequence is essential for transcription to occur and the upstream control sequence is involved in stimulating the rate of transcription initiation. A termination signal consisting of an 18 bp consensus sequence is present about 600 bp after the end of the gene.

RNA polymerase III

This enzyme transcribes a set of short genes that encode transfer RNAs and the 5S ribosomal RNA. The promoter sequences recognized by RNA polymerase III are unusual in that they occur after the transcription start site within the transcribed sequence of the gene. These are known as internal control regions (ICRs) and they occur within about 100 bases of the transcription start site. The 5S rRNA gene contains a sequence known as the C box which acts as a binding site for transcription factors and RNA polymerase III. A second sequence upstream known as the A box is also important. For the transfer RNA genes the ICR is present as two highly conserved sequences known as the A box and the B box which together act as a binding site for transcription factors and the RNA polymerase III. Termination of transcription by RNA polymerase III occurs at a DNA sequence recognized by the enzyme which contains a run of A residues and which occurs soon after the end of the gene.

A5 Transfer RNA

Key Notes

Role in translation

Transfer RNAs (tRNAs) are small molecules that bring amino acids together for protein synthesis in an order specified by a messenger RNA (mRNA) sequence. Cells contain a number of different tRNAs each of which binds a specific amino acid. Each tRNA also binds a specific codon in the mRNA allowing it to place its amino acid in the correct position.

Structure

Base-pairing of tRNA molecules produces a cloverleaf structure composed of stem-loops called arms. These include: the acceptor arm which is the point of attachment of the amino acid; the anticodon arm which recognizes codons in the mRNA sequence; the DHU arm which contains dihyrouracil; the optional arm; and the TφC arm which contains pseudouracil. Transfer RNAs have conserved nucleotides that correspond to base paired regions. The tertiary structure predicts similar base pairing to the 2D cloverleaf representation and shows the acceptor and anticodon arms at opposite ends of the molecule.

Synthesis and processing

Transfer RNA genes occur as multiple copies each of which encodes several tRNAs. They are transcribed by RNA polymerase III as pre-tRNAs which are processed by ribonucleases to release mature tRNAs. RNAseP contains ribozyme activity. The sequence CCA is added to eukaryotic tRNAs after transcription.

Modification of nucleotides

The nucleotides in tRNA molecules are modified following transcription. Modifications include: methylation, base rearrangements, double bond saturation, deamination, sulfation and addition of larger groups. The function of the modifications is uncertain.

Related topics

Gene transcription (A4)
Ribosomal RNA (A6)
Translation (A8)

Role in translation

Cells contain three types of RNA – transfer, ribosomal and messenger – which are produced by transcription from DNA. Transfer and ribosomal RNAs form part of the machinery of protein synthesis and messenger RNAs act as the template for the synthesis of proteins during translation.

Transfer RNAs (tRNAs) are small molecules that act as adapters during protein synthesis; they link the nucleotide sequence of the messenger RNA (mRNA) to the amino acid sequence of the polypeptide. Cells contain a number of tRNAs each of which can bind a specific amino acid. Each tRNA recognizes a codon in the mRNA allowing it to place its amino acid in the correct position in the growing polypeptide chain as determined by the sequence of the mRNA.

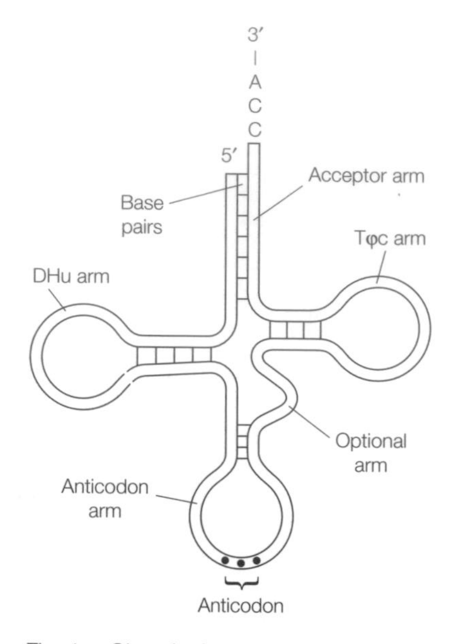

Fig. 1. Cloverleaf structure of transfer RNA.

Structure

Transfer RNA molecules contain between 74 and 95 nucleotides. Base-pairing takes place between complementary parts of the nucleotide sequence resulting in a structure known as the **cloverleaf** which is characteristic of tRNA molecules (*Fig. 1*). The cloverleaf is composed of a series of **stem–loop** structures known as arms. These include:

- The **acceptor arm** which is formed by base-pairing between nucleotides at the 5′ and 3′ ends of the tRNA. The sequence CCA, which occurs at the 3′ terminus, is not base-paired and is the point of attachment for amino acids.
- The **D** or **DHU arm** is a stem–loop structure containing dihydrouracil, an unusual pyrimidine nucleotide.
- The **anticodon arm** is responsible for recognizing and binding codons in the mRNA.
- The **extra, optional** or **variable arm** occurs only in some tRNAs. It may be small containing only 2–3 nucleotides (class I tRNAs) or larger containing 13–21 nucleotides with up to five base pairs in a stem (class II tRNAs).
- The **TφC arm** contains the sequence TφC, where φ is a modified nucleotide called pseudouracil.

Comparison of different tRNAs shows that parts of the nucleotide sequence are conserved. At certain positions the nucleotide present is invariant and is the same in all tRNAs. At other positions, the nucleotide is always a purine or a pyrimidine and is said to be semi-invariant and at other points the sequence is not conserved and the nucleotide present varies.

The cloverleaf structure is a two-dimensional description of the tRNA molecule. A more accurate representation is obtained from its three-dimensional (3D) structure which has been determined by X-ray diffraction (*Fig. 2*). The base-paired nucleotides described in the cloverleaf structure are still present but some nucleotides which appear far apart in the cloverleaf are base-paired in the 3D structure. Many of the nucleotides that are base-paired are invariant

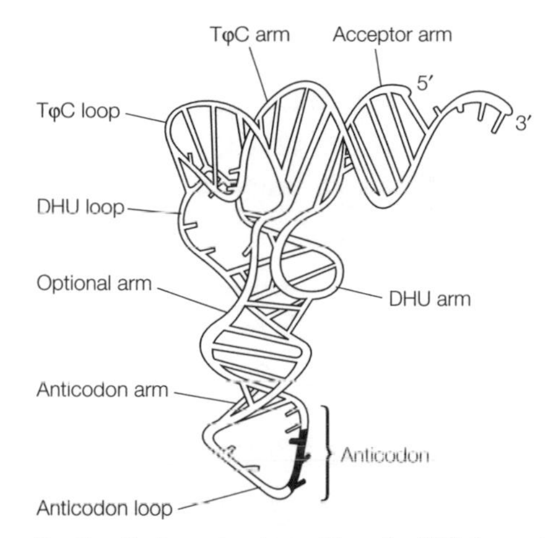

Fig. 2. Tertiary structure of transfer RNA (reproduced from Genetics a Molecular Approach, second edition, T.A. Brown, Chapman and Hall, with permission).

or semi-invariant. In the 3D structure the acceptor arm and the anticodon loop are at opposite ends of the molecule consistent with their role in translation.

Synthesis and processing

Transfer RNAs are synthesized by transcription of tRNA genes by the RNA polymerase III enzyme (see Topic A4). The genes exist as multiple copies, especially in eukaryotic cells, reflecting the large requirement of cells for tRNA. The tRNAs are produced as precursor RNA molecules called **pre-tRNAs** which are processed to give mature tRNAs (*Fig.* 3). Several tRNA genes may be transcribed together as a single pre-tRNA which is then processed by ribonucleases that cleave at the 5′ and 3′ ends of each tRNA sequence. In prokaryotes, processing is carried out in an ordered series of steps by the ribonucleases, RNaseD and P. RNaseP, which is found in prokaryotes and eukaryotes, is unusual in that it has an RNA component with catalytic activity known as a **ribozyme**. Eukaryotic tRNAs differ from their prokaryotic counterparts in that many of them are transcribed containing a short intron which is removed during processing. The sequence CCA is present at the 3′ terminus of all tRNAs and is the point of attachment for amino acids. In eukaryotes, CCA is not present in the DNA of the tRNA gene but is added later by a tRNA nucleotidyl transferase. In prokaryotes, the CCA is present in the coding sequence but is sometimes removed by RNaseD and then replaced by a prokaryotic nucleotidyl transferase.

Modification of nucleotides

Transfer RNAs contain unusual nucleotides produced after transcription by chemical modification. The most common modifications are:

- **Methylation** of the ribose sugar of the nucleotide. For example, guanosine is methylated to 7-methylguanosine.
- **Base rearrangements.** These involve interchanging of the positions of atoms in a purine or pyrimidine ring. An example is the conversion of uridine to pseudouridine.
- **Double-bond saturation.** An example is the conversion of uridine to dihyrouridine.
- **Deamination.** This involves removal of amino groups from bases. For example, guanosine is deaminated to produce inosine.
- **Sulfur substitution.** For example, the oxygen atom of uridine is replaced by sulfur to give 4-thiouridine.

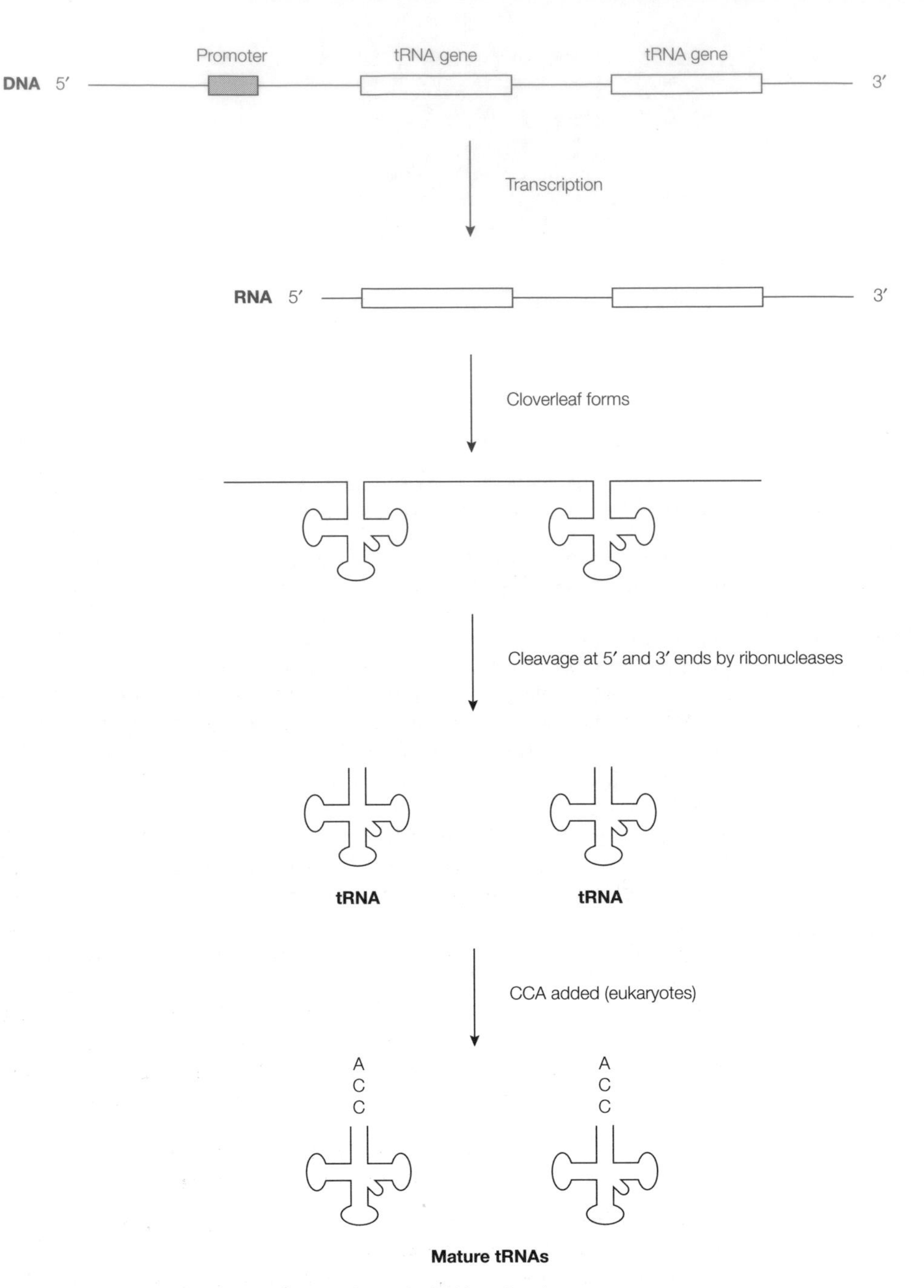

Fig. 3. Transcription and processing of transfer RNA molecules.

- **Addition of larger groups.** An example is the conversion of gaunosine to queosine.

Over 50 types of modification have been described, each carried out by a different tRNA modifying enzyme. The reasons for most of the modifications are unknown but in some cases roles have been assigned for modified nucleotides within the anticodon loop.

A6 RIBOSOMAL RNA

Key Notes

Ribosomes

Ribosomes are macromolecular structures composed of ribosomal RNA (rRNA) and protein. They occur in large numbers in the cytoplasm where they translate messenger RNA into protein. Ribosomes have large and small subunits with characteristic sizes, described in terms of sedimentation (S) values. Prokaryotic ribosomes are 70S with 50S and 30S subunits. They contain three rRNAs (23S, 16S and 5S). Eukaryotic ribosomes are 80S and have 60S and 40S subunits. They contain four rRNAs (28S, 18S, 5.8S and 5S).

Transcription and processing of rRNA genes in prokaryotes

The three rRNAs in *E. coli* are transcribed from a single gene. Seven copies of the gene occur in the genome. A single transcript is produced which is processed to give mature rRNAs. Processing involves folding of the RNA, attachment of ribosomal proteins, methylation of bases and cleavage by ribonucleases.

Transcription and processing of rRNA genes in eukaryotes

Eukaryotic rRNAs are transcribed from a single gene present in multiple copies arranged as a series of clusters. The genes are transcribed in the nucleolus by RNA polymerase I. A single pre-rRNA is synthesized which is processed to give mature 28S, 18S and 5.8S rRNAs. Processing involves folding of the rRNA, attachment of ribosomal proteins, methylation of ribose sugars by small nuclear ribonucleoproteins (snRNPs) and removal of spacer sequences between the rRNAs by ribonucleases. The 5S rRNA is transcribed separately from unlinked genes by RNA polymerase III.

Related topics

Transfer RNA (A5)
Messenger RNA (A7)
Translation (A8)

Ribosomes

Ribosomes are macromolecular structures composed of ribosomal RNA (rRNA) bound to protein. They occur in the cell cytoplasm where they bind to messenger RNA and translate it to produce proteins. Large numbers of ribosomes are required to fulfill the cell's requirement for protein. A typical bacterium contains 20 000 ribosomes which account for 80% of its RNA and 10% of its protein. Because ribosomes are very large, estimates of their molecular weight are difficult to obtain. The size of a ribosome is measured by its **S value** (Svedberg units) which is related to the rate at which it passes through a dense solution such as sucrose when centrifuged at high speed. The S value is determined by the size, shape and the macromolecular structure of the ribosome.

Each ribosome is composed of two parts called the **large and small subunits** (*Fig. 1*). In prokaryotes such as *E. coli*, the ribosome is 70S and is made up of 50 and 30S subunits (S values are not additive). The 50S subunit contains two rRNAs (23S and 5S) complexed with 31 polypeptides. The 30S subunit contains a single rRNA (16S) and 21 polypeptides. In eukaryotes, the ribosome is 80S

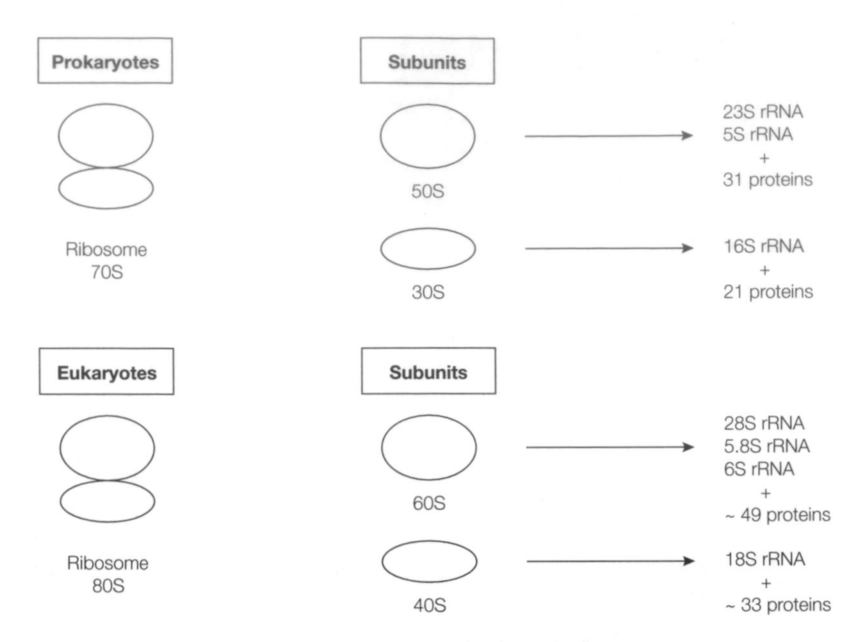

Fig. 1. Composition of typical prokaryotic and eukaryotic ribosomes.

and is composed of 60S and 40S subunits. The 60S subunit contains three rRNAs (28S, 5.8S and 5S) and about 49 polypeptides; the 40S subunit contains one rRNA (18S) and about 33 proteins.

In ribosomes the rRNA molecules adopt a characteristic three-dimensional structure which is stabilized by complementary base-pairing both within and between RNA molecules. The RNA molecules are believed to from a framework to which the proteins, which provide most of the functional activity of the ribosome, are attached. Some of the RNA molecules also have enzymatic activity. These are known as **ribozymes** and they may contribute to the functioning of the ribosome.

Transcription and processing of rRNA genes in prokaryotes

Cells contain large numbers of ribosomes which must be replicated when the cell divides. As a result, cells have a huge requirement for rRNA which is produced by transcription of rRNA genes by RNA polymerase. To ensure that correct numbers of each of the different rRNAs are produced, they are transcribed together from a single gene present in the genome as multiple copies. In prokaryotes such as *E. coli*, seven rRNA genes occur scattered throughout the genome. Each gene contains one copy each of the 16S, 23S and 5S rRNA sequences arranged consecutively. In addition, between one and four transfer RNA (tRNA) sequences are present in each gene. The gene is transcribed to produce a single RNA molecule called **pre-rRNA** (30S) which is processed to produce individual rRNAs and tRNAs (*Fig. 2*). Processing involves a series of defined steps. Following transcription, the RNA molecule folds and complementary parts of the sequence base-pair to give a series of stem–loop structures. The ribosomal proteins then bind to the folded RNA. At this stage some of the bases in the RNA are modified by the addition of methyl groups. Finally, the RNA is cleaved at specific points by the ribonuclease, **RNAse III**, to release the 5S, 23S and 16S rRNAs. Further trimming at the 5′ and 3′ ends by other ribonucleases called M5, M16 and M23 then yields mature rRNAs.

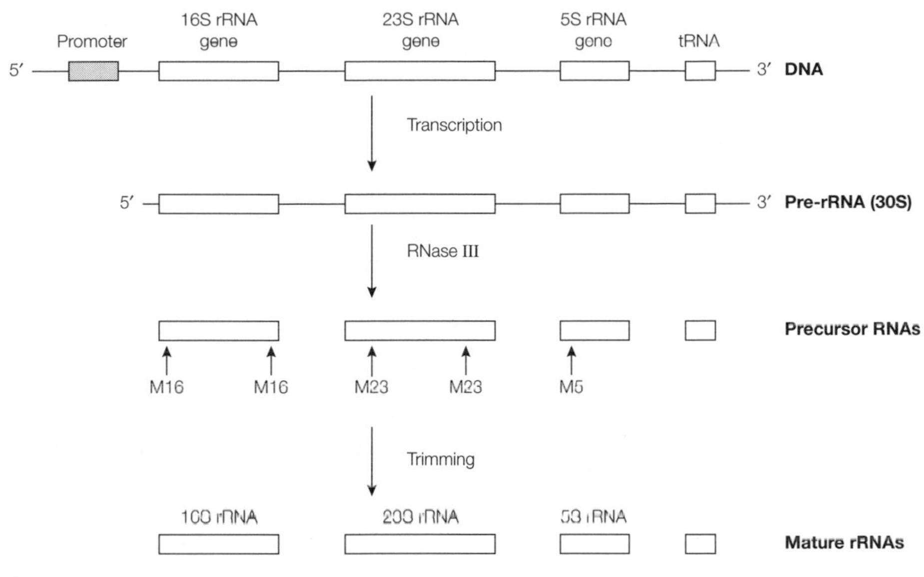

Fig. 2. Transcription and processing of rRNA genes in E. coli.

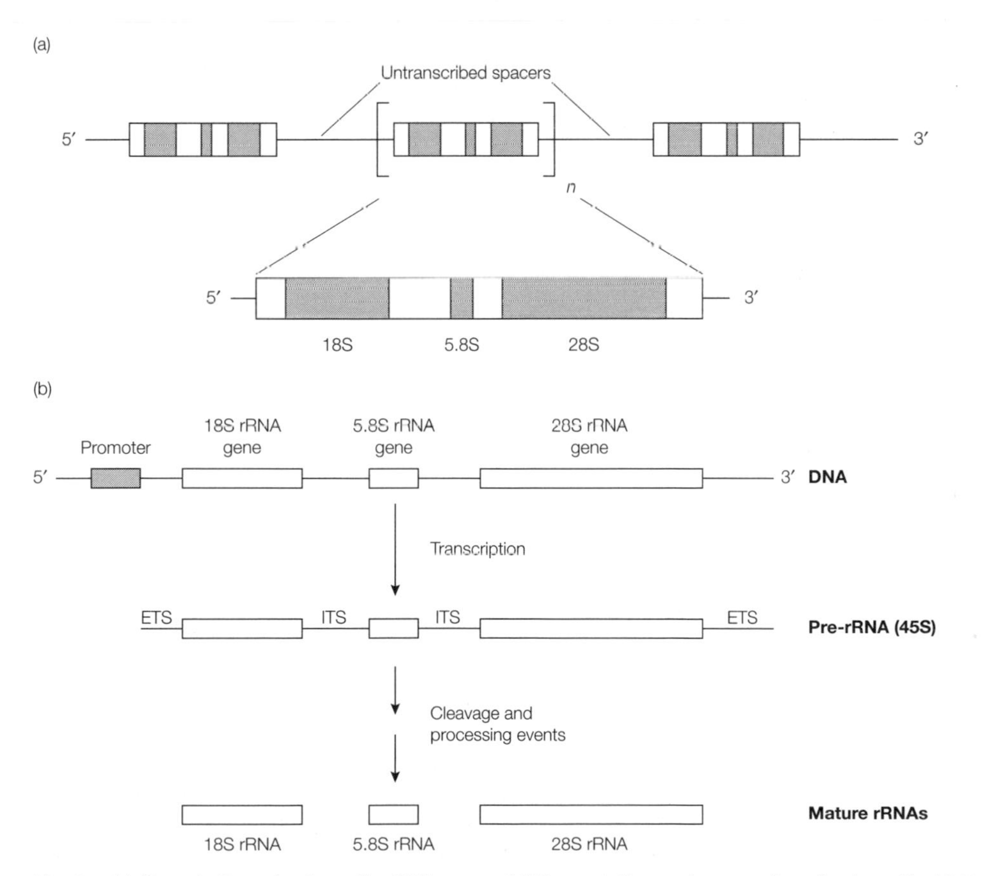

Fig. 3. (a) Organization of eukaryotic rRNA genes. (b) Transcription and processing of eukaryotic rRNA genes.

Transcription and processing of rRNA genes in eukaryotes

In eukaryotes, the sequences of the 28S, 18S and 5.8S rRNAs are present in a single gene which exists as multiple copies separated from each other by short nontranscribed regions (*Fig. 3a*). In humans, there are about 200 genes arranged as a series of five clusters of about 40 genes on separate chromosomes. The genes are transcribed by **RNA polymerase I** in the cell nucleus in a region known as the **nucleolus**. In humans, a single pre-rRNA (45S) is synthesized which is processed to give individual 28S, 18S and 5.8S rRNAs (*Fig. 3b*). The 5S rRNA is transcribed separately by the RNA polymerase III enzyme from unlinked genes as a short 121 base transcript which does not undergo processing. The eukaryotic pre-rRNA is processed in a similar way to its prokaryotic counterpart. Following transcription, the pre-rRNA folds and ribosomal proteins bind to it. Some of the bases are then modified by methylation of the ribose sugar. This reaction is catalyzed by molecules composed of RNA and protein called **small nuclear ribonucleoproteins** (snRNPs, pronounced *snurps*). Mature 28S, 18S and 5.8S rRNAs are then produced by a series of steps in which the pre-rRNA is cleaved by ribonucleases. Initial cleavage of the 45S pre-rRNA occurs in regions known as the **external transcribed spacers (ETSs)**. This is followed by cleavage in regions known as the **internal transcribed spacers (ITSs)** to produce 20S and 32S precursor rRNAs. Further cleavage produces mature 28S, 18S and 5.8S rRNAs. In the final processing step the 5.8S rRNA base pairs with the 28S rRNA.

A7 MESSENGER RNA

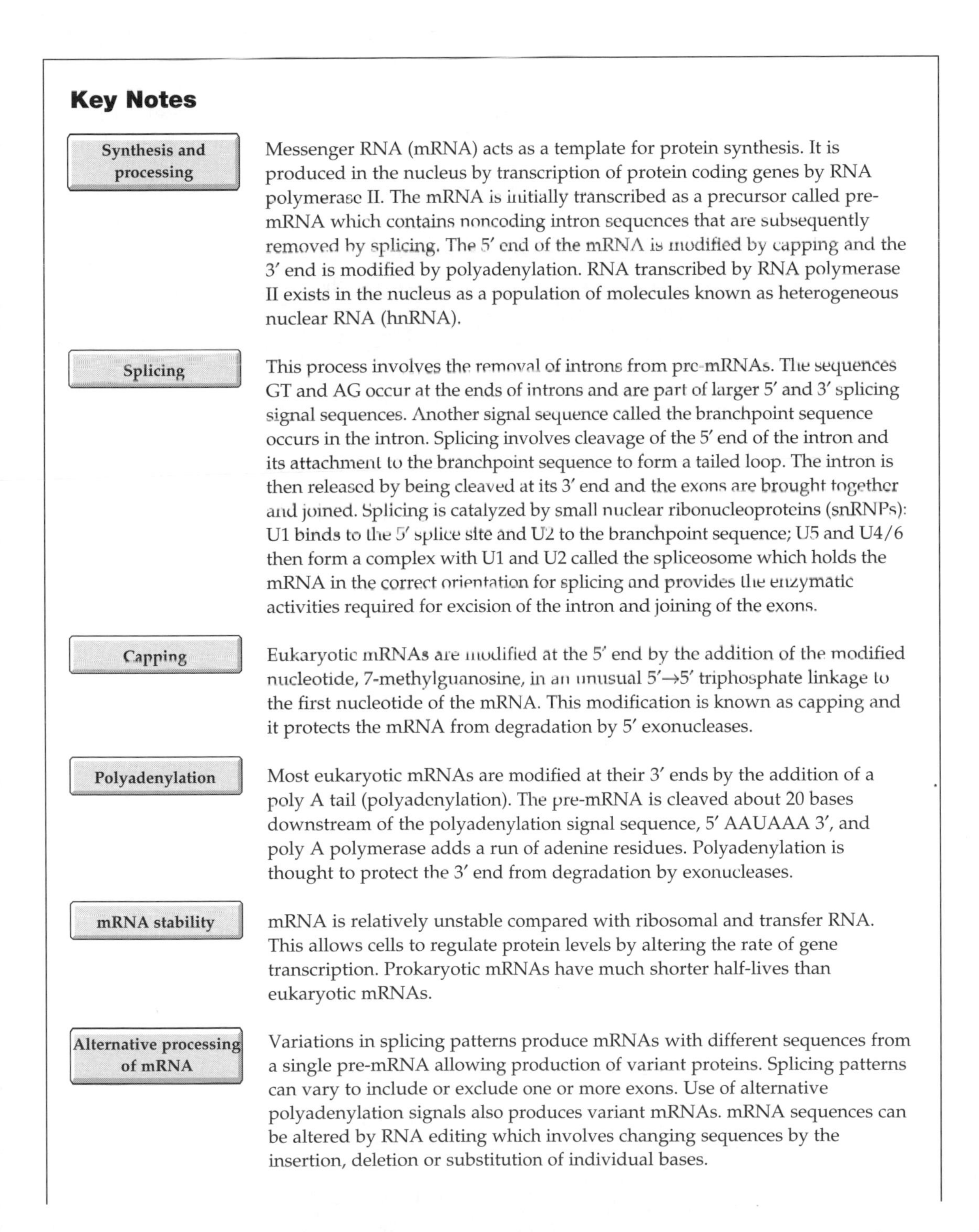

Key Notes

Synthesis and processing

Messenger RNA (mRNA) acts as a template for protein synthesis. It is produced in the nucleus by transcription of protein coding genes by RNA polymerase II. The mRNA is initially transcribed as a precursor called pre-mRNA which contains noncoding intron sequences that are subsequently removed by splicing. The 5′ end of the mRNA is modified by capping and the 3′ end is modified by polyadenylation. RNA transcribed by RNA polymerase II exists in the nucleus as a population of molecules known as heterogeneous nuclear RNA (hnRNA).

Splicing

This process involves the removal of introns from pre-mRNAs. The sequences GT and AG occur at the ends of introns and are part of larger 5′ and 3′ splicing signal sequences. Another signal sequence called the branchpoint sequence occurs in the intron. Splicing involves cleavage of the 5′ end of the intron and its attachment to the branchpoint sequence to form a tailed loop. The intron is then released by being cleaved at its 3′ end and the exons are brought together and joined. Splicing is catalyzed by small nuclear ribonucleoproteins (snRNPs): U1 binds to the 5′ splice site and U2 to the branchpoint sequence; U5 and U4/6 then form a complex with U1 and U2 called the spliceosome which holds the mRNA in the correct orientation for splicing and provides the enzymatic activities required for excision of the intron and joining of the exons.

Capping

Eukaryotic mRNAs are modified at the 5′ end by the addition of the modified nucleotide, 7-methylguanosine, in an unusual 5′→5′ triphosphate linkage to the first nucleotide of the mRNA. This modification is known as capping and it protects the mRNA from degradation by 5′ exonucleases.

Polyadenylation

Most eukaryotic mRNAs are modified at their 3′ ends by the addition of a poly A tail (polyadenylation). The pre-mRNA is cleaved about 20 bases downstream of the polyadenylation signal sequence, 5′ AAUAAA 3′, and poly A polymerase adds a run of adenine residues. Polyadenylation is thought to protect the 3′ end from degradation by exonucleases.

mRNA stability

mRNA is relatively unstable compared with ribosomal and transfer RNA. This allows cells to regulate protein levels by altering the rate of gene transcription. Prokaryotic mRNAs have much shorter half-lives than eukaryotic mRNAs.

Alternative processing of mRNA

Variations in splicing patterns produce mRNAs with different sequences from a single pre-mRNA allowing production of variant proteins. Splicing patterns can vary to include or exclude one or more exons. Use of alternative polyadenylation signals also produces variant mRNAs. mRNA sequences can be altered by RNA editing which involves changing sequences by the insertion, deletion or substitution of individual bases.

Related topics Gene transcription (A4) Ribosomal RNA (A6)
Transfer RNA (A5) Translation (A8)

Synthesis and processing

In eukaryotes, messenger RNA (mRNA) produced by transcription of protein coding genes by the **RNA polymerase II** enzyme acts as a template for protein synthesis during translation. The coding information in eukaryotic genes is discontinuous and is arranged as a series of exons separated by noncoding introns. mRNA is synthesized as a precursor known as **pre-mRNA** by transcription of the exon and intron sequences. Before acting as a template for protein synthesis, the pre-mRNA undergoes a series of processing events to produce mature mRNA. Noncoding intron sequences are removed by a process called **splicing** which makes the coding sequences continuous and ensures that the mRNA is an accurate template for protein synthesis. In addition, the 5′ end of the RNA is altered by the addition of a modified nucleotide in a process known as **capping** and the 3′ end is modified by the addition of a tail of up to 250 adenines in a process called **polyadenylation**. RNA transcribed by RNA polymerase II exists in the nucleus as a population of molecules of different lengths (reflecting variations in gene size) and at different stages of processing and is known collectively as **heterogeneous nuclear RNA (hnRNA)**.

In prokaryotes, mRNA is not processed and translation of the message begins even before transcription is complete. Prokaryotic genes do not normally contain introns and so splicing is unnecessary.

Splicing

This process takes place in the nucleus and involves the removal of noncoding intron sequences from pre-mRNAs to produce mature mRNAs in which the coding sequences, corresponding to the exons, are continuous. The mature spliced mRNA is then exported to the cytoplasm where it acts as a template for protein synthesis.

Splicing depends on the presence of signal sequences in the pre-mRNA. In almost all genes the first two nucleotides at the 5′ end of an intron are GT and the last two at the 3′ end are AG. These are part of larger signal sequences present at the 5′ and 3′ ends of the introns. The complete 5′ signal sequence is **5′ AGGTAAGT 3′** and the 3′ sequence is **5′ YYYYYYNCAG 3′** (Y = pyrimidine; N = any nucleotide). In addition, in vertebrates the sequence, **5′ CURAY 3′** (R = purine), which is called the **branchpoint sequence**, is present in the intron 10–40 bases upstream of the 3′ signal sequence. A more specific sequence, **5′ UACUAAC 3′**, occurs in introns of yeast. Splicing occurs in two steps (*Fig. 1*). In the first step the 2′ hydroxyl group of the adenine of the branchpoint sequence attacks the phosphodiester bond 5′ to the G of the GT (**5′ splice site**). The bond is broken releasing the 5′ end of the intron and attaching it to the branchpoint sequence. The intron now forms a tailed loop structure called a **lariat**. In the second step the 3′ end of the intron is cleaved after the G of the AG (**3′ splice site**), the intron is released and the two exon sequences are joined together.

Splicing is catalyzed by a group of molecules called **small nuclear ribonucleoproteins** (snRNPs, pronounced *snurps*). These are composed of small RNA molecules rich in uracil called **U RNAs** or **small nuclear RNAs** (snRNAs) that exist complexed with proteins. Many different snRNPs exist but the most

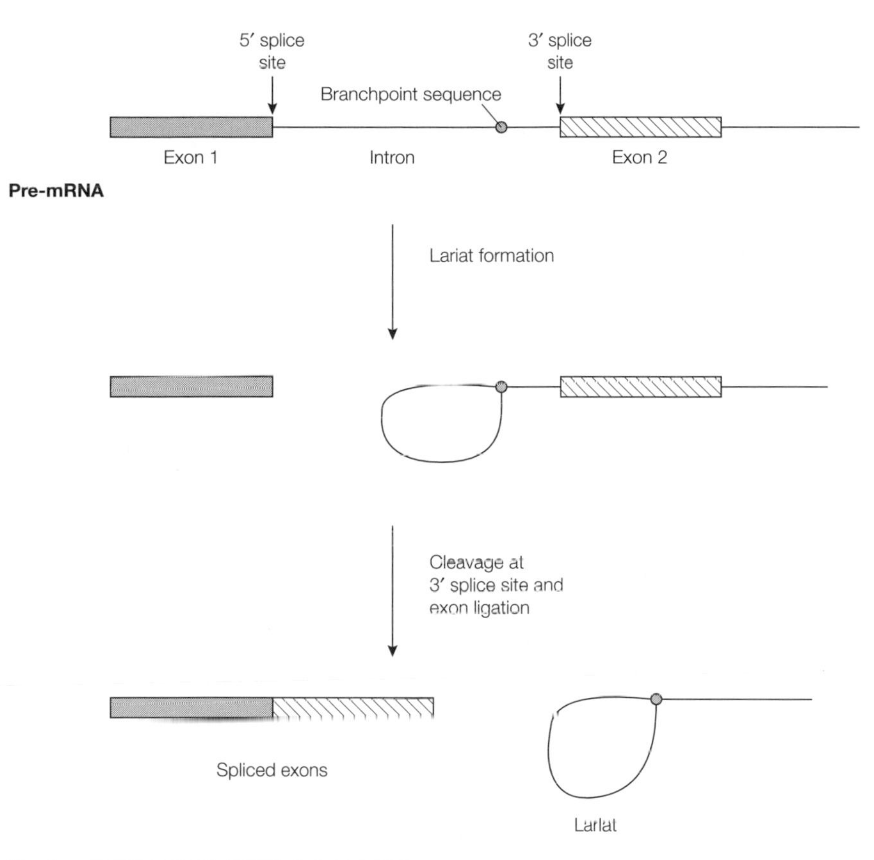

Fig. 1. Splicing of pre-mRNA in eukaryotes.

abundant are U1, U2, U4, U5, and U6 which catalyze the splicing reaction. Each snRNP contains a single U RNA molecule except U4 and U6 which exist base-paired to each other in the same snRNP. The RNA components of the snRNPs interact by base-pairing with the splicing signal sequences. The U1 snRNP binds to the 5′ splice site and the U2 snRNP binds to the branchpoint sequence. The remaining snRNPs, U5 and U4/U6, then form a complex with U1 and U2 causing the intron to loop out and the exons to be brought together. The combination of the pre-mRNA and the snRNPs is called the **spliceosome** and this is responsible for folding the pre-mRNA into the correct conformation for splicing (*Fig. 2*). The spliceosome also catalyzes the cutting and joining reactions that excise the intron and ligate the exons. Once splicing is completed the spliceosome dissociates. The functioning of the spliceosome is not fully understood and the components responsible for all of its enzymatic activities have not been identified.

Although almost all introns are spliced by a spliceosome, there are some examples of intron splicing which occur by different mechanisms. Introns in ribosomal RNA genes in some unicellular organisms can adopt a three-dimensional shape by base-pairing; this then acts as an RNA cutting enzyme, known as a **ribozyme** that contributes to its own splicing. Introns also occur in transfer RNA genes which are removed by the action of ribonucleases in a similar way to the processing of transfer RNA molecules.

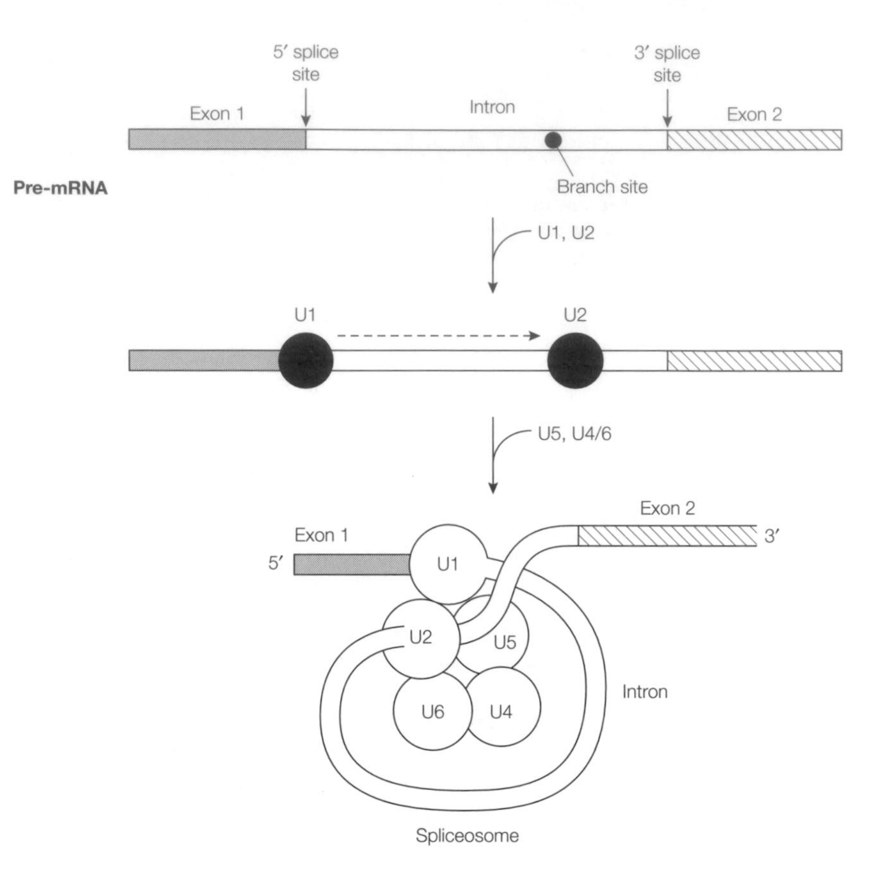

Fig. 2. Spliceosome formation.

Capping

Eukaryotic mRNAs are altered at their 5′ end by a modification known as capping which involves addition of the modified nucleotide, **7-methylguanosine**. The cap is added by the enzyme guanyltransferase which joins GTP by an unusual 5′→5′ triphosphate linkage to the first nucleotide of the mRNA. Methyl transferase enzymes then add a $-CH_3$ group to the 7-nitrogen of the guanine ring and, usually, to the 2′ hydroxyl group on the ribose sugar of the next two nucleotides. Capping protects the mRNA from being degraded from the 5′ end by exonucleases in the cytoplasm and is also a signal allowing the ribosome to recognize the start of a mRNA molecule.

Polyadenylation

Most eukaryotic pre-mRNAs are modified at their 3′ ends by the addition of a sequence of up to 250 adenines known as a **poly A tail**. This modification is called polyadenylation and requires the presence of signal sequences in the pre-mRNA. These consist of the **polyadenylation signal sequence**, 5′ AAUAAA 3′, which occurs near the 3′ end of the pre-mRNA. The sequence YA (Y = pyrimidine) occurs in the next 11–20 bases and a GU rich sequence is often present further downstream. A number of specific proteins recognize and bind these signal sequences forming a complex which cleaves the mRNA about 20 nucleotides downstream of the 5′ AAUAAA 3′ sequence. The enzyme **poly(A) polymerase** then adds adenines to the 3′ end of the molecule. The purpose of the poly A tail is uncertain but it may serve to protect the mRNA from degradation of the coding sequence at the 3′ end by exonucleases. However, some mRNAs, notably those encoding histone proteins, have no poly A tail.

mRNA stability

Unlike ribosomal and transfer RNAs which are stable within cells, mRNA is relatively short-lived. This is because cells regulate protein levels in the cytoplasm primarily by changing the rate of gene transcription. Because mRNAs are short-lived, changes in the rate of transcription of genes are reflected by changes in the amount of mRNA available for protein synthesis. In bacterial cells the half-life for a mRNA is just a few min. In eukaryotic cells a typical half-life might be as much as 6 h, although some mRNAs, such as those encoding the globin polypeptides that make up hemoglobin, are very long-lasting.

Alternative processing of mRNA

Variations can occur in the way pre-mRNAs are processed which generate different mRNAs and hence different proteins from a single gene sequence. This occurs by **alternative splicing** of the pre-mRNA in which cells vary the splice sites they use such that particular exons may be removed or retained during splicing. In addition, **alternative polyadenylation** signals present in the pre-mRNA can lead to the production of mRNAs with different sequences at the 3′ end (*Fig. 3*). For example, the use of an upstream version of alternative polyadenylation sites may exclude exons downstream of it producing an mRNA encoding a truncated protein.

The same pre-mRNA may be alternatively processed within a single cell type, between different cell types and in the same cell type at different stages of development. The proteins produced following alternative processing of a pre-mRNA are related to each other but may have some different functions or characteristics. For example, alternative processing of immunoglobulin pre-mRNAs leads to the synthesis of proteins that may or may not contain

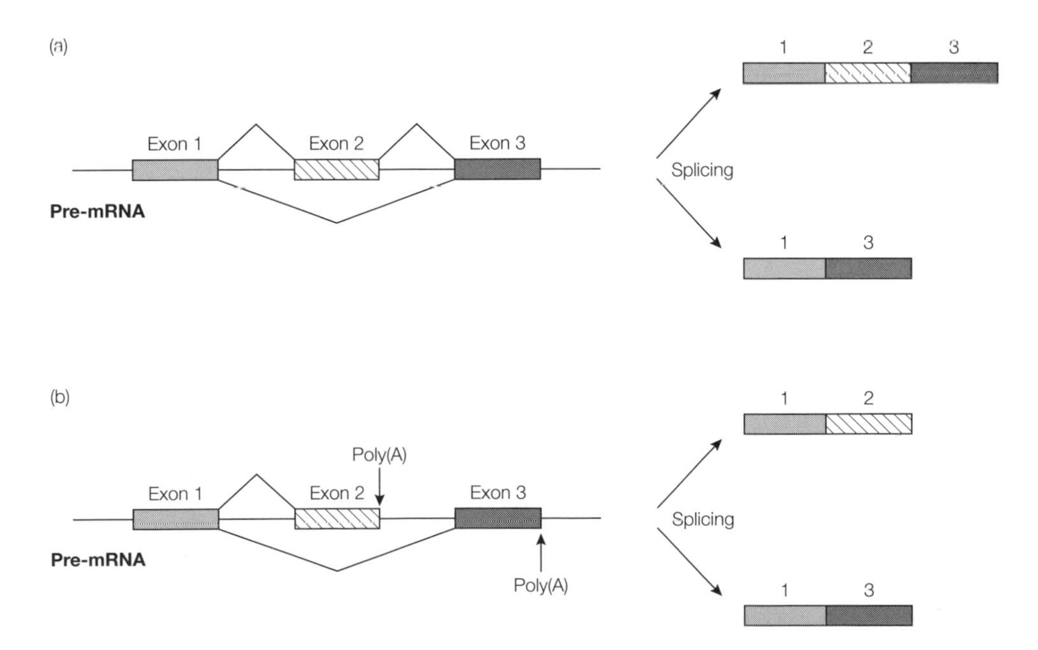

Fig. 3. Alternative splicing of pre-mRNA. (a) Exon skipping; (b) use of alternative polyadenylation sites.

hydrophobic amino acid sequences that allow them bind to cell membranes; this leads to the production of alternative membrane-bound and secreted forms of immunoglobulin proteins.

Pre-mRNAs may also undergo alternative processing by **RNA editing**. In this process the sequence of the pre-mRNA is altered by the insertion, deletion or substitution of bases. RNA editing was first identified in association with some parasitic protozoa in which the transcripts of many of the mitochondrial genes were found to be extensively modified by the insertion of uracil residues. Examples of RNA editing have also been identified in vertebrates, although the modifications found were much less extensive. In humans, the pre-mRNA of the apolipoprotein B gene is edited in intestinal cells by a C to U substitution that produces a stop codon; this leads to the synthesis of a shortened form of the protein. In liver cells, where editing does not occur, a full-length version of the protein is produced.

A8 TRANSLATION

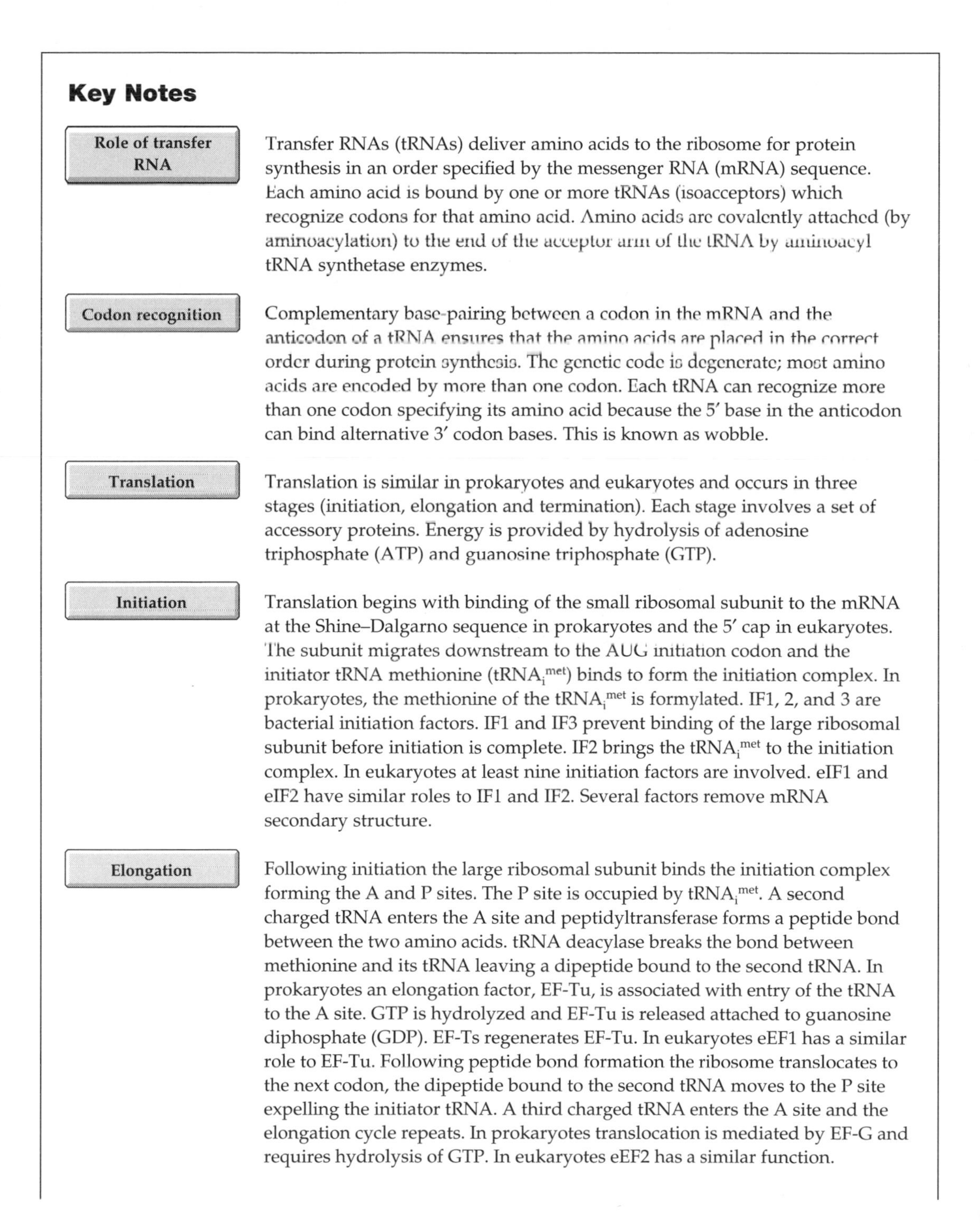

Key Notes

Role of transfer RNA

Transfer RNAs (tRNAs) deliver amino acids to the ribosome for protein synthesis in an order specified by the messenger RNA (mRNA) sequence. Each amino acid is bound by one or more tRNAs (isoacceptors) which recognize codons for that amino acid. Amino acids are covalently attached (by aminoacylation) to the end of the acceptor arm of the tRNA by aminoacyl tRNA synthetase enzymes.

Codon recognition

Complementary base-pairing between a codon in the mRNA and the anticodon of a tRNA ensures that the amino acids are placed in the correct order during protein synthesis. The genetic code is degenerate; most amino acids are encoded by more than one codon. Each tRNA can recognize more than one codon specifying its amino acid because the 5′ base in the anticodon can bind alternative 3′ codon bases. This is known as wobble.

Translation

Translation is similar in prokaryotes and eukaryotes and occurs in three stages (initiation, elongation and termination). Each stage involves a set of accessory proteins. Energy is provided by hydrolysis of adenosine triphosphate (ATP) and guanosine triphosphate (GTP).

Initiation

Translation begins with binding of the small ribosomal subunit to the mRNA at the Shine–Dalgarno sequence in prokaryotes and the 5′ cap in eukaryotes. The subunit migrates downstream to the AUG initiation codon and the initiator tRNA methionine ($tRNA_i^{met}$) binds to form the initiation complex. In prokaryotes, the methionine of the $tRNA_i^{met}$ is formylated. IF1, 2, and 3 are bacterial initiation factors. IF1 and IF3 prevent binding of the large ribosomal subunit before initiation is complete. IF2 brings the $tRNA_i^{met}$ to the initiation complex. In eukaryotes at least nine initiation factors are involved. eIF1 and eIF2 have similar roles to IF1 and IF2. Several factors remove mRNA secondary structure.

Elongation

Following initiation the large ribosomal subunit binds the initiation complex forming the A and P sites. The P site is occupied by $tRNA_i^{met}$. A second charged tRNA enters the A site and peptidyltransferase forms a peptide bond between the two amino acids. tRNA deacylase breaks the bond between methionine and its tRNA leaving a dipeptide bound to the second tRNA. In prokaryotes an elongation factor, EF-Tu, is associated with entry of the tRNA to the A site. GTP is hydrolyzed and EF-Tu is released attached to guanosine diphosphate (GDP). EF-Ts regenerates EF-Tu. In eukaryotes eEF1 has a similar role to EF-Tu. Following peptide bond formation the ribosome translocates to the next codon, the dipeptide bound to the second tRNA moves to the P site expelling the initiator tRNA. A third charged tRNA enters the A site and the elongation cycle repeats. In prokaryotes translocation is mediated by EF-G and requires hydrolysis of GTP. In eukaryotes eEF2 has a similar function.

Termination

Translation ends when a termination codon enters the A site. Release factors enter the A site and cause release of the polypeptide. In *E. coli* RF 1, 2 and 3 cause termination. In eukaryotes a single protein, eRF, is involved. Following termination the ribosome dissociates and the mRNA is released.

Post-translational modifications

Following translation, polypeptides may be modified by the addition of chemical groups to amino acid side chains and the N and C termini or by proteolytic cleavage. Modifications may be required for full functional activity.

Related topics

The genetic code (A3)
Transfer RNA (A5)
Ribosomal RNA (A6)

Role of transfer RNA

Translation is the process by which cells synthesize proteins. During translation information encoded in mRNA molecules is used to specify the amino acid sequence of a protein. Transfer RNA molecules play a key role in this process by delivering amino acids to the ribosome in an order specified by the mRNA sequence; this ensures that the amino acids are joined together in the correct order (*Fig. 1*). Cells usually contain between 31 and 40 individual species of tRNA, each of which binds specifically to one of the 20 amino acids. Consequently, there may be more than one tRNA for each amino acid. Transfer RNAs that bind the same amino acid are called **isoacceptors**. Before translation begins, amino acids become covalently linked to their tRNAs which then recognize codons in the mRNA specifying that amino acid. The attachment of an amino acid to its tRNA is called **aminoacylation** or **charging**. The amino acid is covalently attached to the end of the acceptor arm of the tRNA which always

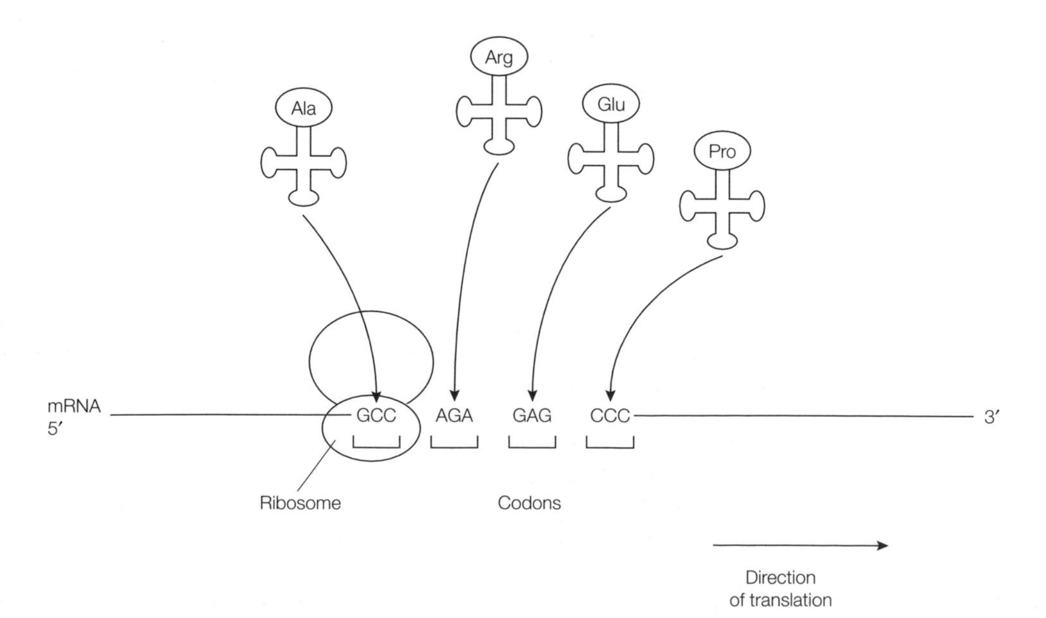

Fig. 1. Role of tRNAs in translation.

ends with the base sequence 5′ CCA 3′. A bond forms between the carboxyl group of the amino acid and the 3′-hydroxyl of the terminal adenine of the acceptor arm. Charging is catalyzed by enzymes called **aminoacyl tRNA synthetases** in a reaction requiring the hydrolysis of ATP. A separate enzyme exists for each amino acid and each enzyme can charge all the isoacceptor tRNAs for that amino acid. The aminoacyl tRNA synthetase recognizes both its appropriate amino acid and the corresponding tRNA. The amino acid is recognized primarily by its side chain. It is not clear exactly how the tRNA is recognized but variant nucleotides that are specific for individual tRNAs may be involved.

Codon recognition

When the correct amino acid has been attached to the tRNA it recognizes the codon for that amino acid in the mRNA allowing it to place the amino acid in the correct position, as specified by the sequence of the mRNA. This ensures that the amino acid sequence encoded by the mRNA is translated faithfully. Codon recognition takes place via the anticodon loop of the tRNA and specifically by three nucleotides in the loop known as the **anticodon** which binds to the codon by complementary base-pairing. The four bases present in DNA can combine as 64 codons. Three codons act as signals for translation to stop and the remaining 61 encode the 20 amino acids present in proteins. Consequently, most amino acids are represented by more than one codon, a feature referred to as the **degeneracy of the genetic code**. To deal with this, individual tRNAs can recognize more than one of the codons for their amino acid. They can achieve this because the anticodon is capable of binding to alternative bases present at the third position of the codon, a feature known as **third base degeneracy** or **wobble**. Binding between the codon and the anticodon can tolerate variation at the third base because the anticodon loop is not linear and when the anticodon binds to the codon it does not form a perfect RNA double helix. This permits the formation of a few nonstandard base pairs. For example, G can base-pair with U as well as with C at the wobble position, and inosine, a deaminated form of guanine sometimes present in the wobble position of the anticodon loop, can base-pair with C but also with A and U. Wobble allows a single tRNA to decode more than one member of a codon family decreasing the number of tRNAs required by the cell. The rules of the genetic code however are not violated and polypeptides are synthesized strictly in accordance with the nucleotide sequence of the mRNA. An important consequence of wobble is that it serves to minimize the effects of mutations.

Translation

Translation occurs by similar mechanisms in prokaryotes and eukaryotes and is conveniently described as occurring in three stages: **initiation**, **elongation** and **termination**. Initiation involves binding of a ribosome to mRNA. Elongation involves repeated addition of amino acids and termination involves release of the new polypeptide chain. Sets of **accessory proteins** assist the ribosome in each of the three stages. Translation requires the use of energy by the cell which is provided by the hydrolysis of guanosine triphosphate (GTP) and adenosine triphosphate (ATP). GTP is used for ribosome movement and in binding of accessory factors. ATP is used to charge tRNAs and in removing secondary structure from mRNA. Up to 90% of ATP produced in a bacterium is used for translation.

Initiation

When they are not actively involved in translation ribosomes exist as separate large and small subunits. The first step in translation involves the binding of

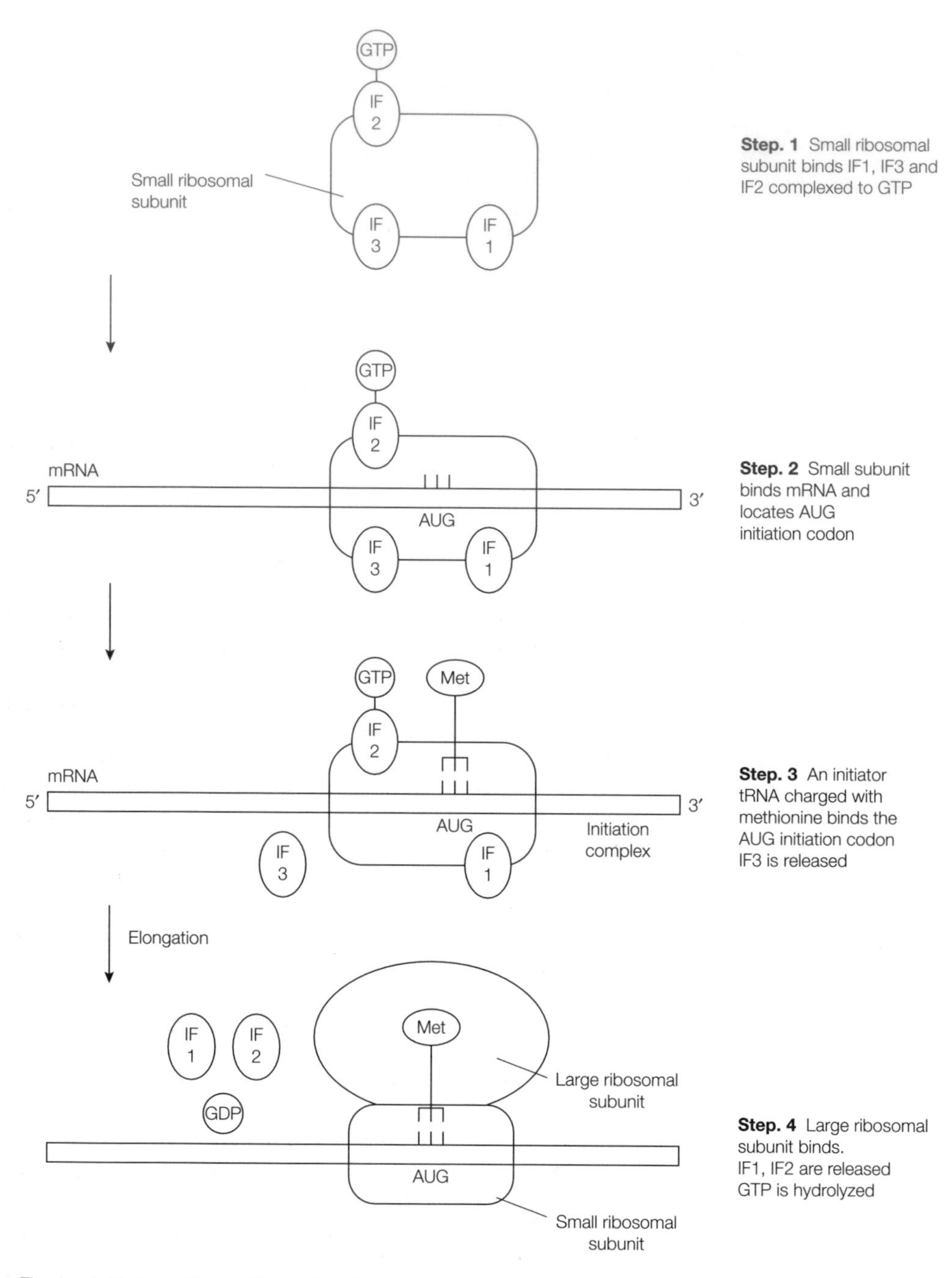

Fig. 2. Initiation of translation in E. coli.

the small ribosomal subunit to the mRNA (*Fig. 2*). Translation usually begins at the sequence **AUG** (bacteria sometimes use GUG or UUG) which encodes methionine and is known as the **translation initiation codon**. The small subunit binds to the mRNA at a specific point upstream of the AUG. In prokaryotes, this is the **Shine–Dalgarno sequence** (5′ AGGAGGU 3′) found near the start of the mRNA. Once bound, the small subunit migrates in a 3′ direction along the mRNA until it finds the AUG, usually about 10 nucleotides downstream. In eukaryotes, the small ribosomal subunit recognizes the cap structure at the 5′ end of the mRNA. It then moves downstream until it encounters the first AUG, although sometimes other AUGs are recognized. A tRNA charged with methionine binds to the AUG located by the small ribosomal subunit. In bacteria, the methionine is modified by the addition of a formyl group (–CHO) to one of the hydrogens of the amino group ($tRNA^{fmet}$). This blocks the amino group, thus preventing it from forming a peptide bond and ensuring that polymerization of the polypeptide can only occur in the amino to carboxy direction. The combination of the mRNA, the small ribosomal subunit and the $tRNA^{fmet}$ is called the **initiation complex**. Two tRNAs recognize AUG and carry methionine. One is used for initiation ($tRNA_i^{met}$) and the other recognizes internal AUGs. Only the initiator tRNA is capable of binding to the initiation complex.

A number of accessory proteins called **initiation factors** are required for initiation. Bacteria have three, known as IF1, IF2, and IF3. Initiation begins with binding of IF1 and IF3 to the small ribosomal subunit. This helps to prevent binding of a large subunit before the mRNA has bound. Next, IF2 complexed with GTP binds the small subunit. Its purpose is to assist binding of the initiator tRNA. The small subunit then binds the mRNA and locates the AUG initiation codon. The initiator tRNA charged with methionine binds to the complex and IF3 is released. This marks the end of initiation. Elongation begins with binding of a large ribosomal subunit to the initiation complex to from a complete ribosome. This is accompanied by the release of IF1 and IF2 and hydrolysis of GTP (*Fig. 2*).

Initiation in eukaryotes is similar to that in prokaryotes; however, the initiation codon is almost always AUG and the methionine is not formylated. Many more initiation factors (at least nine) are involved, some have roles analogous to those of their bacterial counterparts. Two eukaryotic factors, eIF2 and eIF3 have roles similar to bacterial IF2 and IF3. Several of the eukaryotic initiation factors are involved in removing secondary structure from the mRNA before it is translated. Energy for this is provided by hydrolysis of ATP.

Elongation

As soon as the initiation complex has formed, a large ribosomal subunit binds to it. The complete ribosome contains two binding sites for tRNA molecules (*Fig. 3*). The first site is the **P** or **peptidyl site** and is occupied by the $tRNA_i^{met}$ base-paired to the AUG. The second site is the **A** or **aminoacyl site** and is positioned over the second codon. Elongation begins when a tRNA enters the A site and base pairs with the second codon. With both sites occupied by charged tRNAs, the attached amino acids are placed in close contact and a peptide bond can form between the carboxyl group of the methionine and the amino group of the second amino acid. The reaction is catalyzed by a complex enzyme called **peptidyl transferase** which probably contains several different ribosomal proteins. Peptidyl transferase works in conjunction with another enzyme, **tRNA deacylase**, which breaks the link between methionine and its tRNA after formation of the peptide bond.

A number of accessory proteins are required for elongation. In prokaryotes two **elongation factors**, EF-Tu and EF-Ts, are involved. EF-Tu is associated with entry of a tRNA into the A site. EF-Tu binds charged tRNAs in association with GTP. Following entry to the A site, the GTP is hydrolyzed and EF-Tu is released bound to guanosine diphosphate (GDP). Before another tRNA can bind, EF-Tu

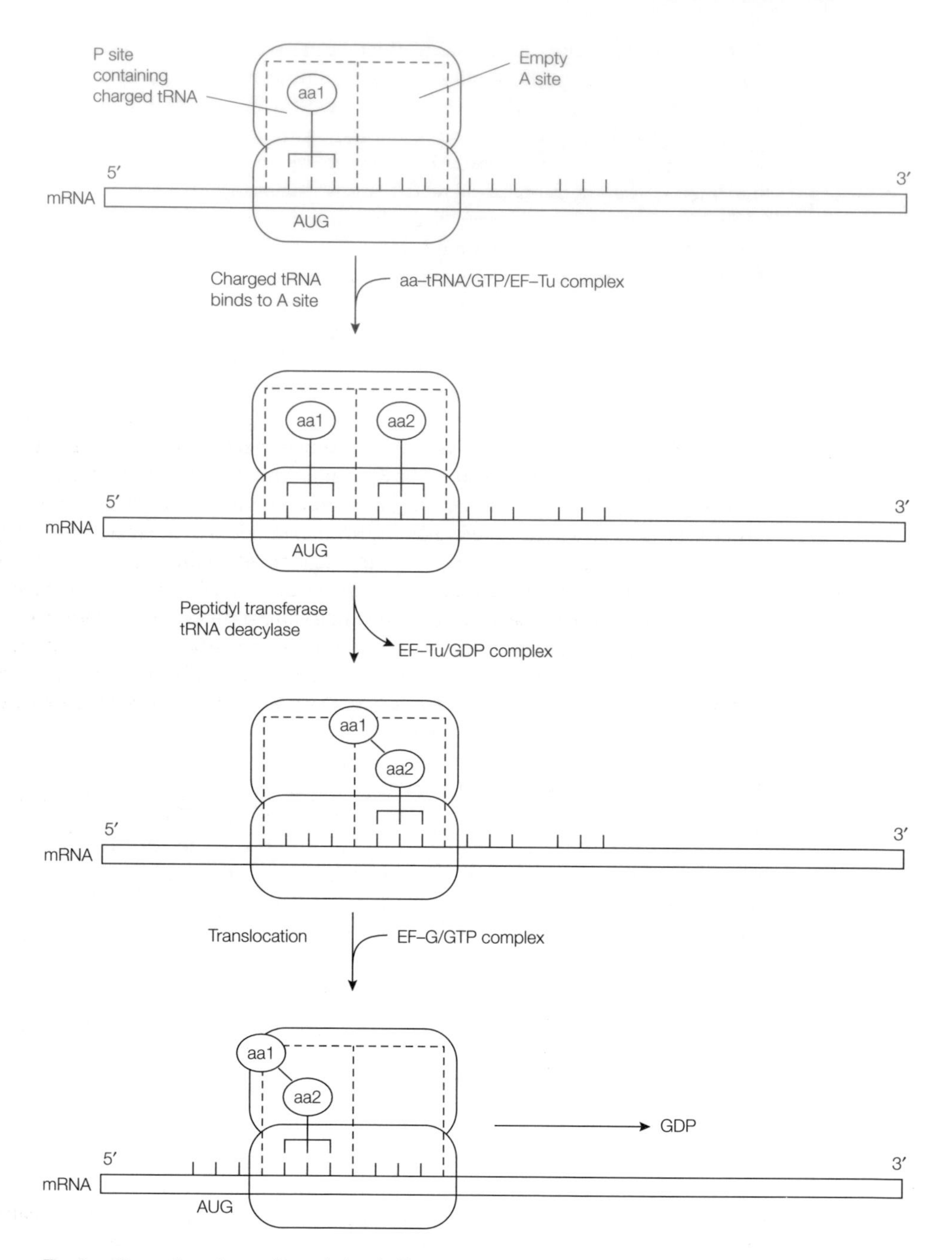

Fig. 3. Elongation stage of translation in E. coli.

must be regenerated with the help of EF-Ts. First, EF-Ts displaces GDP by binding to EF-Tu; a new molecule of GTP then replaces EF-Ts (*Fig. 4*). In eukaryotes, a complex protein called eEF-1 brings the tRNA to the A site. Again, the reaction is associated with hydrolysis of GTP. The details of how eEF-1 is regenerated are unknown but components equivalent to EF-Tu and EF-Ts may be present.

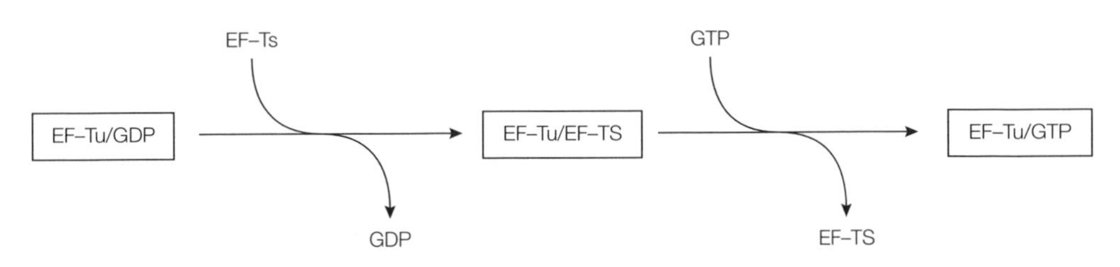

Fig. 4. Regeneration of EF-Tu/GTP complex.

Following peptide bond formation translocation occurs and the ribosome moves on to the next codon. The newly formed dipeptide which is bound to the second tRNA (aa–aa–tRNA) moves into the P site expelling the uncharged tRNA and the A site becomes vacant. A third charged tRNA enters the A site and the elongation cycle is repeated. Following each addition of an amino acid to the growing polypeptide chain the ribosome translocates to the next codon. In bacteria, translocation is mediated by the elongation factor EF-G which binds to the ribosome in a complex with GTP which is then hydrolyzed providing energy for translocation. Binding of EF-G and EF-Tu to the ribosome is mutually exclusive. This ensures that translocation is completed before the next round of elongation begins. The eukaryotic equivalent of EF-G is eEF-2 which also requires GTP and functions in a similar way. As translation proceeds the ribosome moves along the mRNA away from the initiation site which is then free to bind a new ribosome. Thus mRNAs can be translated by several ribosomes at once forming a structure known as a **polysome**.

Termination

Translation ends when a termination codon enters the A site (*Fig. 5*). There are no tRNAs able to bind to the termination codons. Instead, proteins known as **release factors** enter the A site and cause the completed polypeptide to be released. In *E. coli* two release factors, RF1 and RF2, perform this function. RF1 recognizes UAA and UAG stop codons and RF2 recognizes UAA and UGA. A third release factor, RF3, plays an ancillary role in the process. The release factors cause peptidyl transferase to transfer the polypeptide to a water molecule rather than to another aminoacyl tRNA.

In eukaryotes a single protein, eRF, is involved which requires GTP for ribosome binding. This is subsequently hydrolyzed and eRF is released from the ribosome. Following release of the polypeptide, the ribosome releases the mRNA and dissociates; it then joins the pool of ribosomal subunits before becoming involved in further translation.

Post-translational modifications

Following translation, newly synthesized polypeptides may undergo a range of modifications before becoming functional proteins. These involve mainly the covalent attachment of chemical groups and cleavage of the polypeptide chain. Many different chemical modifications of the side chains of amino acids or the amino and carboxyl termini of proteins are found. Modifications may involve addition of small groups, such as methylation, phosphorylation, acetylation, and hydroxylation as well as the addition of larger molecular structures such as lipids and oligosaccharides (glycosylation). Some modifications such as phosphorylation regulate enzyme activity. Cleavage of polypeptide chains is a very common modification; this may involve trimming of amino acids from the termini of the

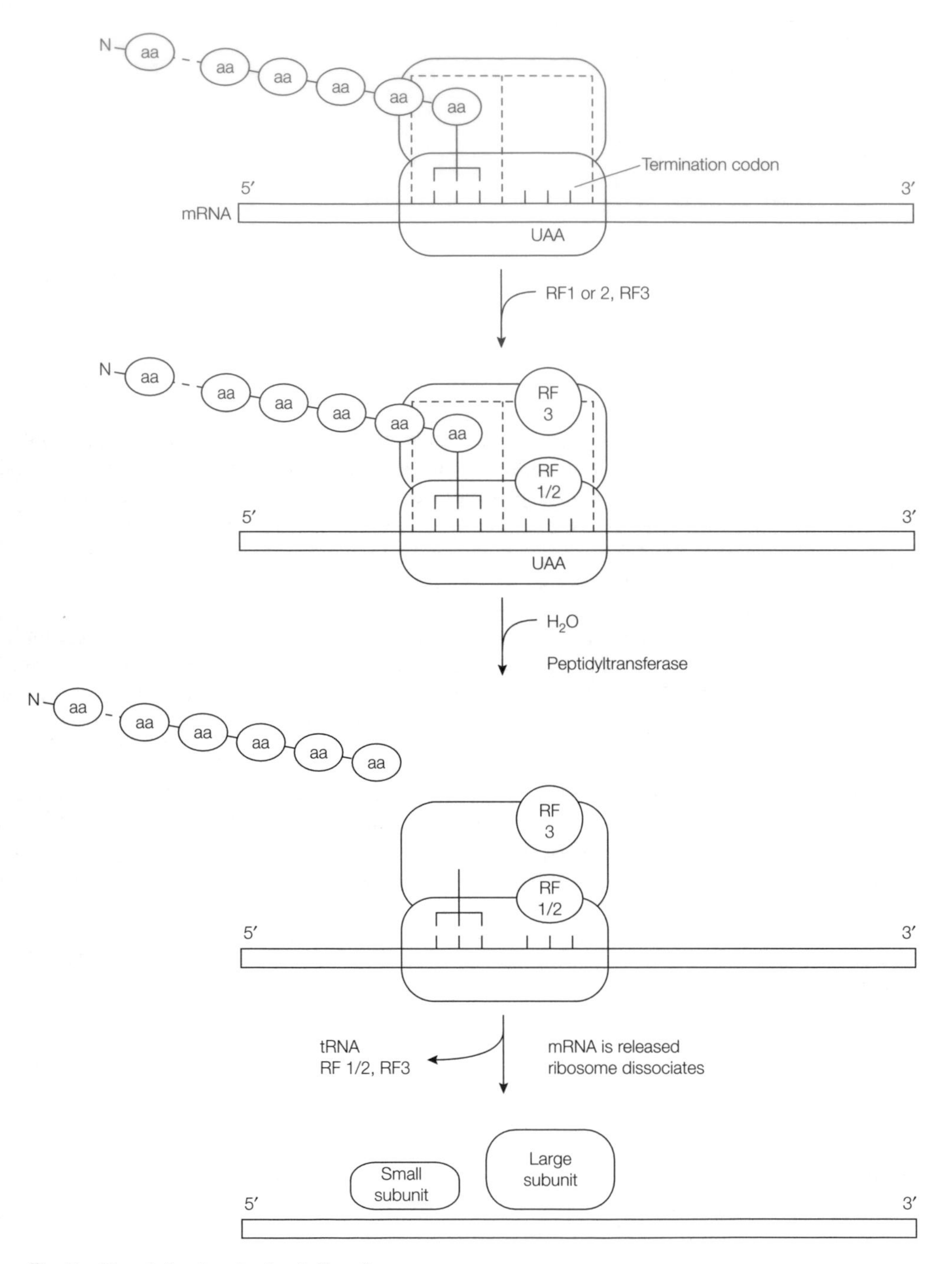

Fig. 5. Translation termination in E. coli.

proteins, removal of internal peptides, removal of amino-terminal signal peptide sequences from secreted proteins and the cleavage of polyproteins into smaller peptides. Cleavage of proteins is sometimes associated with the activation of inactive precursor proteins.

A9 DNA REPLICATION

Key Notes

DNA replication

In this process a cell copies its DNA prior to dividing. The DNA is copied 5′→3′ by DNA polymerases using single-stranded DNA as a template. Replication is semi-conservative. In *E. coli*, DNA polymerases I and III have 3′→5′ exonuclease activity that allows them to proofread sequences ensuring a very low error rate.

The replication fork

DNA synthesis occurs at the replication fork. A helicase separates the double helix and single-strand binding (SSB) protein keeps the strands separate. DNA is synthesized continuously on the leading strand and discontinuously as segments (Okazaki fragments) on the lagging strand. In *E. coli*, DNA synthesis by DNA polymerase III and is initiated at a short RNA primer synthesized by primase RNA polymerase. Primer sequence is replaced with DNA later by DNA polymerase I. In eukaryotes, DNA polymerase α initiates DNA synthesis by its integral primase activity. The leading and lagging strands are synthesized respectively by DNA polymerases δ and α. DNA ligase joins the Okazaki fragments by a phosphodiester bond.

Bacterial DNA replication

Circular bacterial DNA molecules and linear eukaryotic chromosomes are replicated differently. Circular DNA molecules are replicated from a single origin. Replication forks progress in both directions eventually meeting and merging. Unwinding of the double helix produces supercoiling of circular DNA molecules which is removed by topoisomerase I. Replication of circular DNA produces interlocked daughter molecules which are separated by topoisomerase II.

DNA replication in eukaryotes

Cell division in eukaryotes occurs as a series of phases known as the cell cycle. Eukaryotic chromosomes are replicated from multiple origins. Replication bubbles form and these eventually meet and merge. Transcriptionally active regions are replicated first. Replication requires DNA to be unwound from nucleosomes. Special strategies are required to replicate the ends of chromosomes. Telomerase adds noncoding sequences that allow replication of chromosome ends.

Related topics

DNA structure (A1)
Chromosomes (B1)
Prokaryotic genomes (B3)
DNA mutation (B5)
Mutagens and DNA repair (B6)

DNA replication

This is the process by which a cell copies its DNA. Replication is necessary so that the genetic information present in cells can be passed on to daughter cells following cell division. The DNA is copied by enzymes called DNA polymerases. These act on single-stranded DNA synthesizing a new strand complementary to the original strand. DNA synthesis always occurs in the 5′→3′ direction.

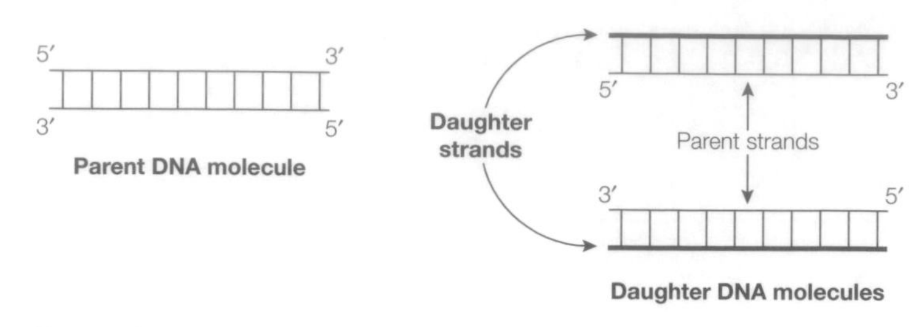

Fig. 1. Semi-conservative replication.

Replication is said to be **semi-conservative** (*Fig. 1*). This means that each copied DNA molecule contains one strand derived from the parent molecule and one newly synthesized strand.

The mechanism of DNA replication is very similar in most organisms. Differences exist only with respect to the enzymes and proteins involved. In prokaryotes such as *E. coli*, two enzymes, DNA polymerases I and III, are responsible for DNA synthesis. In eukaryotes, DNA is replicated by five DNA polymerases ($\alpha,\beta,\gamma,\delta,\varepsilon$). Replication needs to be very accurate because even a small error rate would result in the loss of important genetic information after just a few cell divisions. Accuracy is ensured by the ability of the DNA polymerases to check that the correct bases have been inserted in the newly synthesized strand. This is achieved through the reverse (3′→5′) exonuclease activity of the enzymes which allows them to remove incorrectly inserted bases from newly synthesized DNA and replace them with the correct base. This is referred to as **proofreading** ability. It is estimated that just one base in five billion is inserted incorrectly.

The replication fork

During DNA replication the double helix of a cell's entire DNA is progressively unwound producing segments of single-stranded DNA which can be copied by DNA polymerases. Unwinding of the double helix begins at a distinct position called the **replication origin** and gradually progresses along the molecule, usually in both directions. Replication origins usually contain sequences rich in weak A–T base pairs. The region where the helix unwinds and new DNA is synthesized is called the **replication fork** (*Fig. 2*). At the replication fork a number of distinct events occur:

- **Separation of the double helix**. This is achieved by the action of a **helicase** enzyme. Following separation of the strands, **single-strand binding (SSB) protein** attaches to the DNA and prevents the double helix from reforming.
- **Synthesis of leading and lagging strands**. Synthesis of DNA by DNA polymerases occurs only in the 5′→3′ direction. As the two strands of the double helix run in opposite directions (one strand runs 5′→3′ and the other 3′→5′) slightly different mechanisms are required to replicate each. One strand, called the **leading strand**, is copied in the same direction as the unwinding helix and so can be synthesized continuously (*Fig. 3a*). The other strand, known as the **lagging strand**, is synthesized in the opposite direction and must be copied discontinuously. The lagging strand is synthesized as a series of segments known as **Okazaki fragments** (*Fig. 3b*).
- **Priming**. DNA polymerases require a short double-stranded region to initiate or **prime** DNA synthesis. This is produced by an RNA polymerase, called **primase**, which is able to initiate synthesis on single-stranded DNA. The

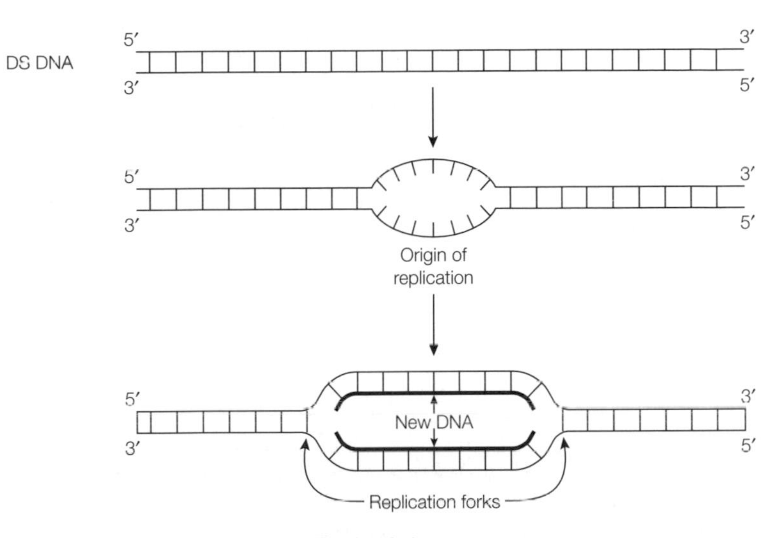

Fig. 2. Replication origin and replication forks.

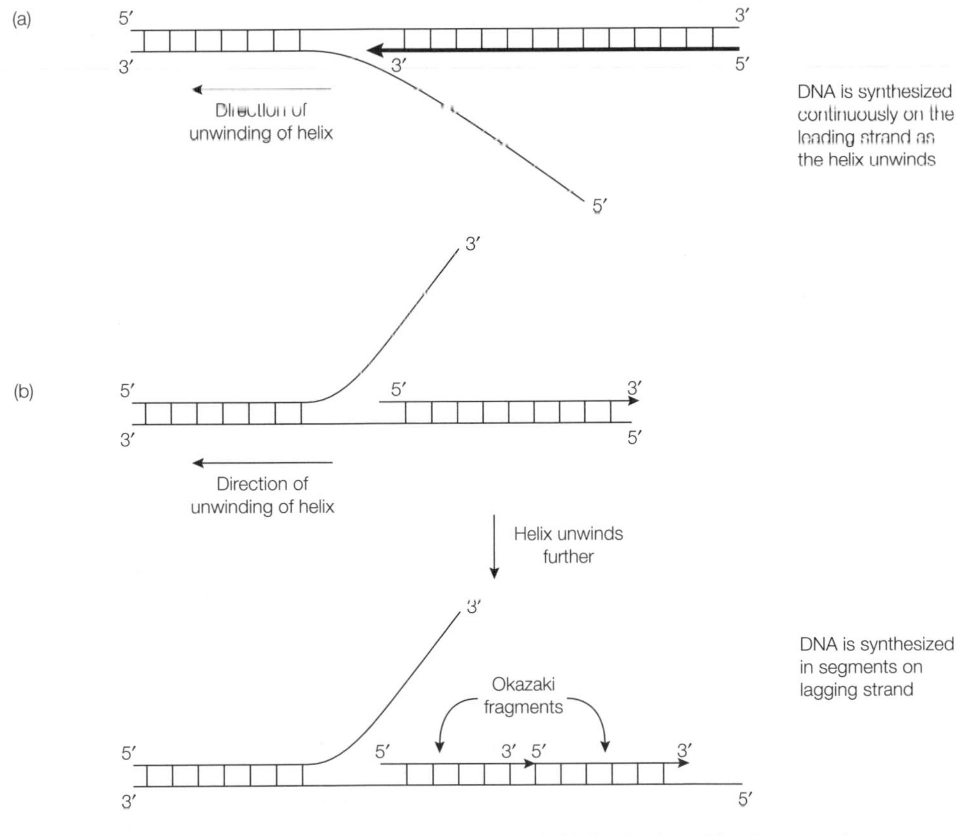

Fig. 3. (a) Replication of leading strand. (b) Replication of lagging strand.

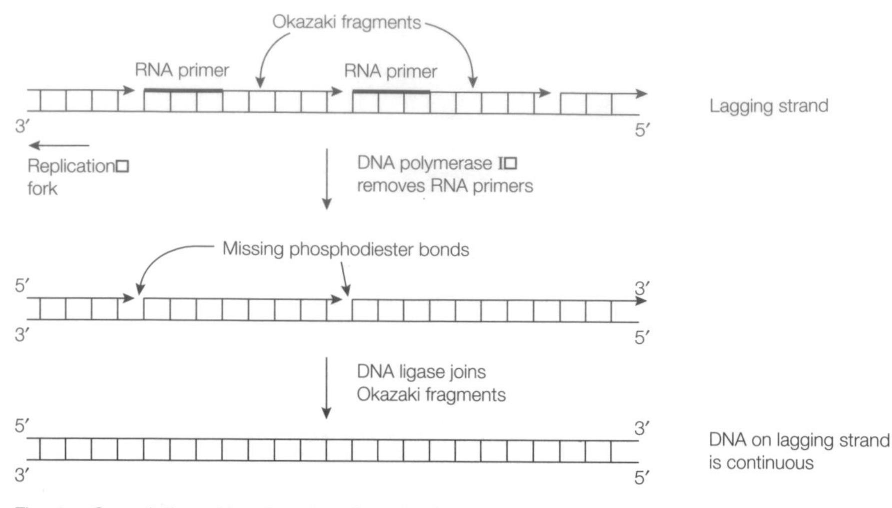

Fig. 4. Completion of lagging strand synthesis.

primase synthesizes a short RNA primer sequence on the DNA template creating a short double-stranded region. In *E. coli*, DNA polymerase III then synthesizes DNA beginning at the RNA primer. On the lagging strand, synthesis ends when the next RNA primer is encountered. At this point DNA polymerase I takes over and removes the RNA primer replacing it with DNA (*Fig. 4*). In eukaryotes the situation is different. DNA polymerase α, which has integral primase activity, is responsible for initiating DNA synthesis. DNA is replicated by DNA polymerases α and δ with α synthesizing the lagging strand and δ synthesizing the leading strand. The other polymerases have ancillary roles. DNA polymerase ε is involved in DNA repair and DNA polymerase γ replicates mitochondrial DNA.

- **Ligation**. The final step required to complete synthesis of the lagging strand is for the Okazaki fragments to be joined together by phosphodiester bonds. This is carried out by a **DNA ligase** enzyme (*Fig. 4*).

Bacterial DNA replication

Although the mechanism of DNA replication is similar in all organisms, the overall process varies depending on the nature of the DNA molecule being copied. Different strategies are required for replication of the circular DNA molecules which occur in bacteria and for the linear chromosomal DNA molecules present in eukaryotes. The simplest and most common form of replication for circular DNA involves a single origin of replication from which two replication forks progress in opposite directions. This results in an intermediate θ form (*Fig. 5*). The replication forks eventually meet and fuse and replication terminates.

The replication of DNA molecules requires unwinding of the DNA double helix. This causes the helix ahead of the replication fork to rotate. For circular DNA molecules that do not have free ends, this produces supercoiling of the DNA preventing the replication fork from progressing. This problem is overcome by the action of enzymes called **topoisomerases** of which there are two types. DNA topoisomerase I produces a transient break in the polynucleotide backbone of one of the DNA strands a short distance ahead of the replication fork allowing the DNA to rotate freely around the other intact strand removing the supercoiling. The enzyme then rejoins the ends of the broken strand. When replication of a bacterial chromo-

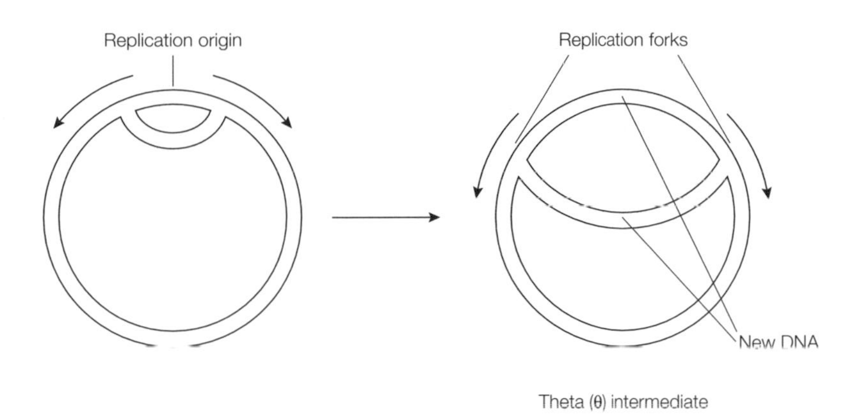

Fig. 5. Replication of circular DNA molecules via theta intermediate.

somes is completed, two circular daughter molecules that are interlocked are produced. These are separated by the action of DNA topoisomerase II which acts by causing transient breaks in both strands of one of the DNA molecules allowing the other DNA molecule to pass through, thus separating the two daughter molecules. The topoisomerase II enzyme then rejoins the broken strands.

DNA replication in eukaryotes

Before a cell can divide it must replicate its DNA. Cell division in eukaryotes is highly regulated and occurs as a series of phases known as the **cell cycle** (*Fig. 6*). The length of the cell cycle varies but is typically several hours. The longest phase is G_1 during which the cells prepare for division. G_1 is followed by the S phase, in which replication of the DNA occurs. A second short gap, G_2, is next and is followed by the M phase during which the cells undergo mitosis involving separation of the chromosomes and cell division. After M phase, proliferating cells enter the G_1 phase of the next cell cycle. Alternatively, cells may exit the cell cycle by entering the G_0 phase where they remain quiescent for extended periods. Some cells such as neurons stop dividing completely and are permanently in G_0 phase.

Due to the extreme length of eukaryotic chromosomes, DNA replication must be initiated at multiple origins to ensure that the process is completed within a reasonable time span. Replication forks proceed in either direction from each replication origin forming replication bubbles which eventually meet and merge. DNA replicated from a single origin is called a **replicon**. A typical mammalian cell has 50–100 000 replicons, each of which replicates 40–200 kb of DNA. Not all the DNA is replicated at once. Clusters of about 50 replicons initiate

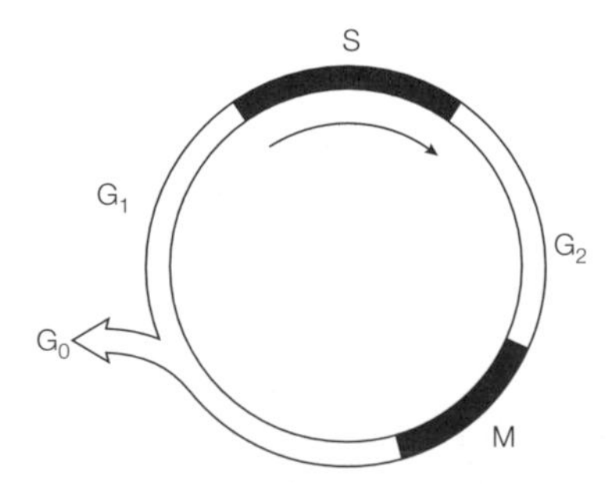

Fig. 6. The eukaryotic cell cycle.

simultaneously at defined points during S phase. Areas containing transcriptionally active genes are replicated first with nonactive regions replicated later.

The DNA in eukaryotic chromosomes is packaged as DNA–protein complexes called **nucleosomes**. As the replication fork progresses DNA must unwind from the nucleosome for replication to occur. This slows the progress of the replication forks and may explain the short length of the Okazaki fragments on the lagging strand in eukaryotes (100–200 bases) compared with prokaryotes (1000–2000 bases). After the replication fork has passed the nucleosomes reform.

Replication of linear eukaryotic chromosomes poses a problem not encountered with circular bacterial chromosomes in that the extreme 5′ end of the lagging strand cannot be replicated because there is no room for an RNA primer to initiate replication. This creates the potential for chromosomes to shorten after each round of replication leading to a loss of genetic information. The problem is overcome by a specialized structure at the end of the chromosome known as the **telomere** which contains tandem (side-by-side) repeats of a simple noncoding sequence. In humans this is 5′ TTAGGG 3′. In addition, the 3′ end of the leading strand extends beyond the 5′ end of the lagging strand. The enzyme **telomerase** contains an RNA molecule which partly overlaps with and binds to the repeat sequence on the leading strand. The enzyme then extends the leading strand using the RNA as a template. The telomerase then dissociates and binds to the new telomere end so that the leading strand is extended again. This process of extension may occur hundreds of times before the telomerase finally dissociates. The newly extended leading strand then acts as a template for replication of the 5′ end of the lagging strand (*Fig. 7*). The two processes whereby the DNA is shortened during normal replication and lengthened by the action of the telomerase are roughly balanced so the overall length of the chromosome remains approximately the same.

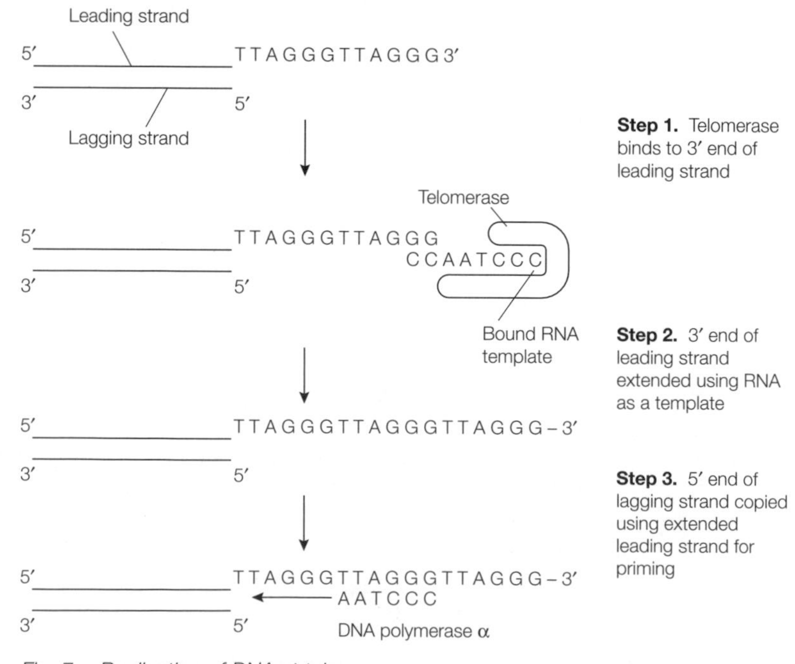

Fig. 7. Replication of DNA at telomeres.

A10 Regulation of Gene Expression in Prokaryotes

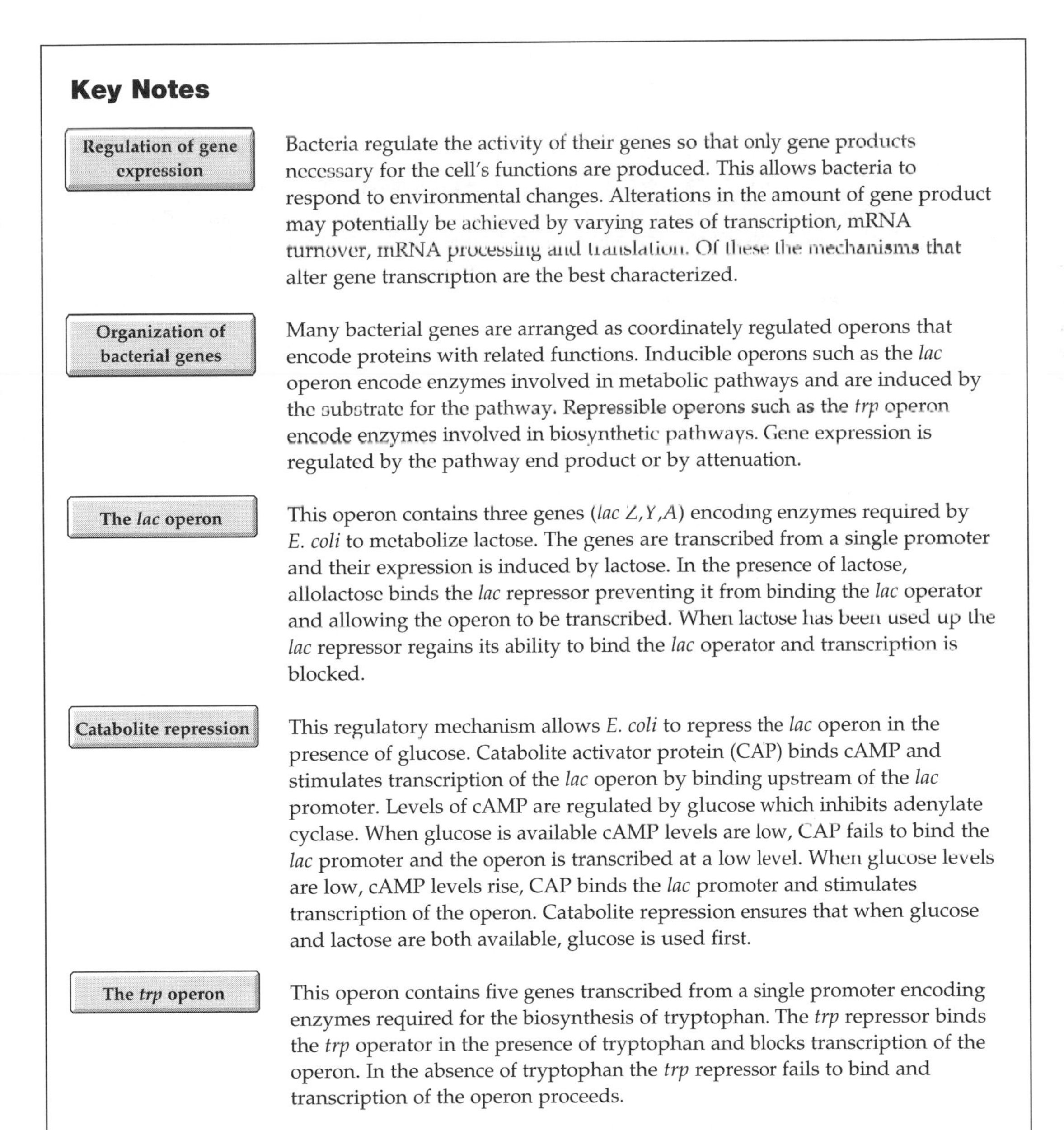

Key Notes

Regulation of gene expression

Bacteria regulate the activity of their genes so that only gene products necessary for the cell's functions are produced. This allows bacteria to respond to environmental changes. Alterations in the amount of gene product may potentially be achieved by varying rates of transcription, mRNA turnover, mRNA processing and translation. Of these the mechanisms that alter gene transcription are the best characterized.

Organization of bacterial genes

Many bacterial genes are arranged as coordinately regulated operons that encode proteins with related functions. Inducible operons such as the *lac* operon encode enzymes involved in metabolic pathways and are induced by the substrate for the pathway. Repressible operons such as the *trp* operon encode enzymes involved in biosynthetic pathways. Gene expression is regulated by the pathway end product or by attenuation.

The *lac* operon

This operon contains three genes (*lac Z,Y,A*) encoding enzymes required by *E. coli* to metabolize lactose. The genes are transcribed from a single promoter and their expression is induced by lactose. In the presence of lactose, allolactose binds the *lac* repressor preventing it from binding the *lac* operator and allowing the operon to be transcribed. When lactose has been used up the *lac* repressor regains its ability to bind the *lac* operator and transcription is blocked.

Catabolite repression

This regulatory mechanism allows *E. coli* to repress the *lac* operon in the presence of glucose. Catabolite activator protein (CAP) binds cAMP and stimulates transcription of the *lac* operon by binding upstream of the *lac* promoter. Levels of cAMP are regulated by glucose which inhibits adenylate cyclase. When glucose is available cAMP levels are low, CAP fails to bind the *lac* promoter and the operon is transcribed at a low level. When glucose levels are low, cAMP levels rise, CAP binds the *lac* promoter and stimulates transcription of the operon. Catabolite repression ensures that when glucose and lactose are both available, glucose is used first.

The *trp* operon

This operon contains five genes transcribed from a single promoter encoding enzymes required for the biosynthesis of tryptophan. The *trp* repressor binds the *trp* operator in the presence of tryptophan and blocks transcription of the operon. In the absence of tryptophan the *trp* repressor fails to bind and transcription of the operon proceeds.

Attenuation

This regulatory mechanism allows fine adjustment of expression of the *trp* operon and other operons. DNA sequences between the *trp* promoter and the first *trp* operon gene are capable of forming either a large stem–loop structure that does not influence transcription or a smaller terminator loop. A short coding region upstream contains tryptophan codons. When tryptophan levels are adequate RNA polymerase transcribes the region closely followed by a ribosome which prevents formation of the larger stem–loop, allowing the terminator loop to form ending transcription. If tryptophan is lacking, the ribosome is stalled, the RNA polymerase moves ahead and the large stem–loop forms. Formation of the terminator loop is blocked and transcription of the operon proceeds.

Regulation by alternative sigma factors

This mechanism is used to make major alterations to gene expression in response to environmental changes. Alternative σ factors alter the specificity of bacterial RNA polymerase allowing it to recognize different gene promoters. Alternative σ factors activate gene transcription in *E. coli* in response to heat shock and in *Bacillus subtilis* during sporulation. Bacteriophages synthesize σ factors that direct the transcription of bacteriophage genes.

Related topics

Genes (A2)
Prokaryotic genomes (B3)
Bacteriophages (B8)

Regulation of gene expression

For a bacterium to function it is not necessary that all of its genes are transcribed at all times. To conserve energy and resources bacteria regulate the activity of their genes so that only those gene products necessary for the cell's functions are produced. For example, it would be wasteful for a bacterium to produce enzymes required to synthesize an amino acid that was already available to it from its environment. Regulation of gene expression allows bacteria to respond to changes in their environment, typically to the presence or absence of nutrients.

Bacteria regulate expression of their genes in order to control the amount of gene product present. The steady state concentration of a gene product is determined by the balance between the rate of synthesis and the rate of degradation of the expressed protein. In practice, changes in the rate of synthesis are what alter the amount of gene product. The rate of synthesis could potentially be altered by a number of factors:

- changes in the rate of gene transcription;
- changes in mRNA turnover time;
- changes in the rate of translation.

In practice, all three mechanisms probably influence gene expression but the best understood examples are those involving the regulation of gene transcription.

Organization of bacterial genes

An important feature which determines how gene transcription is regulated in bacteria is the organization of the genes as **operons**. These are transcriptional units in which several genes, usually encoding proteins with related functions, are regulated together. Other genes also occur which encode regulatory proteins that control gene expression in operons. Many different operons have been identified in *E. coli*. Most contain genes that encode proteins involved in the biosynthesis of amino acids or the metabolism of nutrients. Operons are classified as **inducible** or **repressible**. Inducible operons contain genes that encode enzymes involved in metabolic pathways. Expression of the genes is controlled by a substrate of the pathway. An example of an inducible operon is the ***lac* operon** which encodes enzymes required for the metabolism of lactose. Repressible operons contain genes that encode enzymes involved in biosynthetic pathways and gene expression is controlled by the end product of the pathway which may repress expression of the operon or control it by an alternative mechanism called **attenuation**. An example of a repressible operon is the ***trp* operon** which encodes enzymes involved in the biosynthesis of tryptophan.

The *lac* operon

This operon contains three genes encoding enzymes required by the *E. coli* bacterium for the utilization of the disaccharide sugar lactose. These are lactose permease, which transports lactose into the cell, β-galactosidase which hydrolyzes lactose into its component sugars (glucose and galactose) and β-galactoside transacetylase which is also involved in the hydrolysis of lactose. These enzymes are normally present in *E. coli* at very low levels but in the presence of lactose their levels rise rapidly. The three genes in the *lac* operon are known as *lac Z, Y* and *A* and encode β-galactosidase, lactose permease and β-galactoside transacetylase, respectively. The genes are sequential and are transcribed as a single mRNA from a single promoter. Another regulatory gene, *lac I*, which is expressed separately, lies upstream of the operon and encodes a protein called the ***lac* repressor** which regulates the expression of the *lac Z, Y* and *A* genes. In the absence of lactose the *lac* repressor binds to a DNA sequence called the **operator** positioned between the *lac* promoter and the beginning of the *lac Z* gene. When bound to the operator, the *lac* repressor blocks the path of the RNA polymerase bound to the *lac* promoter upstream of it and prevents transcription of the *lac* genes. When the cell encounters lactose a few molecules of the *lac* enzymes present in the cell allow lactose to be taken up and metabolized. **Allolactose**, an isomer of lactose produced as an intermediate during the metabolism of lactose, acts as an inducer. It binds to the lactose repressor and changes its conformation such that it can no longer bind to the operator. The path of the RNA polymerase is no longer blocked and the operon is transcribed. Large numbers of enzyme molecules are produced which take up lactose and metabolize it. The presence of lactose thus induces the expression of the enzymes needed to metabolize it. When the lactose is used up the *lac* repressor returns to its original conformation and again, binds the *lac* operator preventing transcription and switching off the operon (*Fig. 1*).

Catabolite repression

This term describes an additional regulatory mechanism which allows the *lac* operon to sense the presence of glucose, an alternative and preferred energy source to lactose. If glucose and lactose are both present, cells will use up the glucose first and will not expend energy splitting lactose into its component sugars. The presence of glucose in the cell switches off the *lac* operon by a mechanism called **catabolite repression** which involves a regulatory protein called the **catabolite activator protein** (CAP) (*Fig. 2*). CAP binds to a DNA sequence upstream of the *lac* promoter and enhances binding of the RNA polymerase leading to enhanced transcription of the operon. However CAP only binds in the presence of a derivative of ATP called cyclic adenosine monophosphate (cAMP)

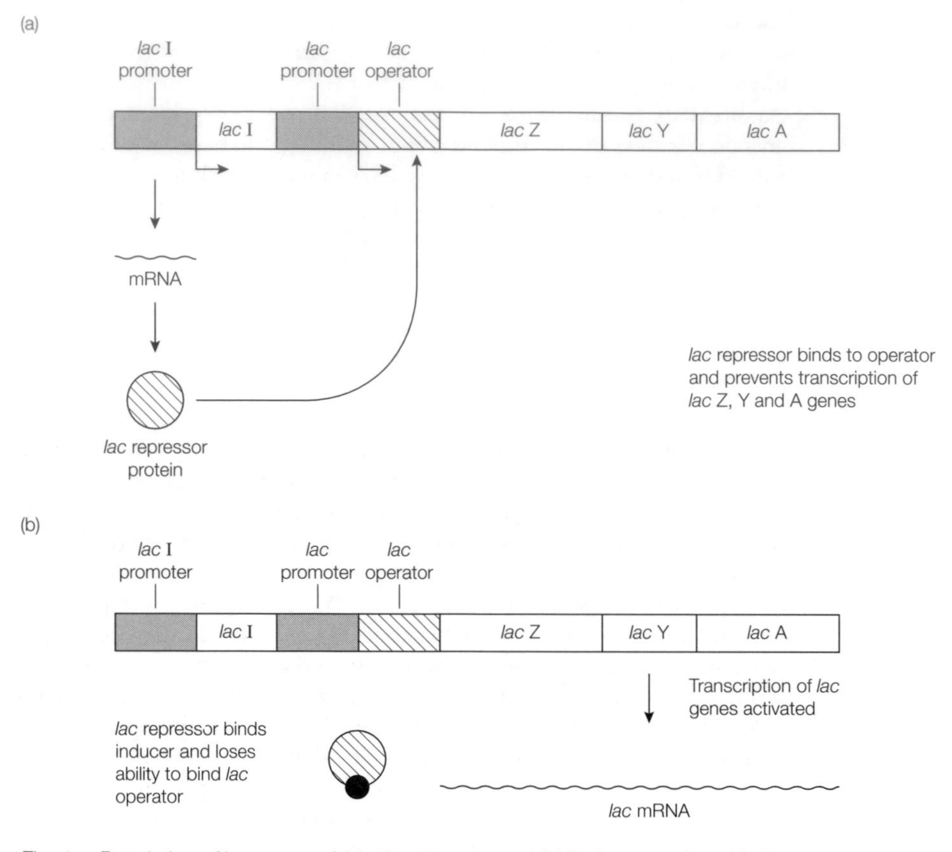

Fig. 1. Regulation of lac *operon (a) in the absence and (b) in the presence of inducer.*

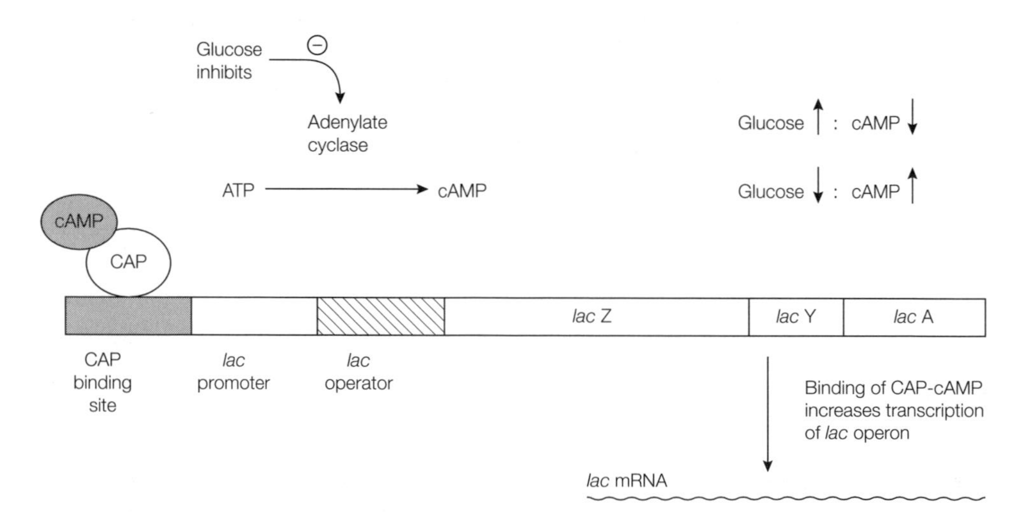

Fig. 2. Catabolite repression of lac *operon.*

whose levels are influenced by glucose. The enzyme **adenylate cyclase** catalyzes the formation of cAMP and is inhibited by glucose. When glucose is available to the cell adenylate cyclase is inhibited and cAMP levels are low. Under these conditions CAP does not bind upstream of the promoter and the *lac* operon is transcribed at a very low level. Conversely, when glucose is low adenylate cyclase is not inhibited, cAMP is higher and CAP binds increasing the level of transcription from the operon. If glucose and lactose are present together the *lac* operon will only be transcribed at a low level. However when the glucose is used up catabolite repression will end and transcription from the *lac* operon increases allowing the available lactose to be used up.

The *trp* operon

This operon contains five genes encoding enzymes involved in biosynthesis of the amino acid tryptophan. The genes are expressed as a single mRNA transcribed from an upstream promoter. Expression of the operon is regulated by the level of tryptophan in the cell (*Fig. 3*). A regulatory gene upstream of the *trp* operon encodes a protein called the ***trp* repressor**. This protein binds a DNA sequence called the ***trp* operator** which lies just downstream of the *trp* promoter partly overlapping it. When tryptophan is present in the cell it binds to the *trp* repressor protein enabling it to bind the *trp* operator sequence, obstructing binding of the RNA polymerase to the *trp* promoter and preventing transcription of the operon. In the absence of tryptophan the *trp* repressor is incapable of binding the *trp* operator and transcription of the operon proceeds. Tryptophan, the end product of the enzymes encoded by the *trp* operon, thus acts as a corepressor with the *trp* repressor protein and inhibits its own synthesis by end product inhibition.

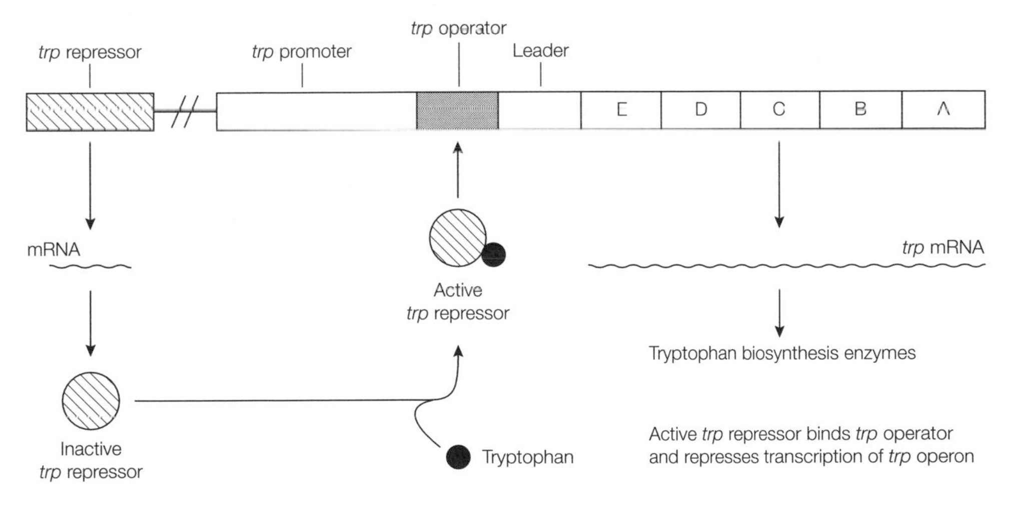

Fig. 3. The trp operon.

Attenuation

The *trp* operon makes use of an alternative strategy for controlling transcription called attenuation which can finely tune expression levels (*Fig. 4*). The transcribed mRNA sequence between the *trp* promoter and the first *trp* gene is capable of forming two stem–loop structures. The relative positions of the sequences mean that both stem–loops cannot form at once: just one or the other may be present at any time. The larger, more stable structure does not influence transcription and occurs upstream of the smaller stem–loop which acts as

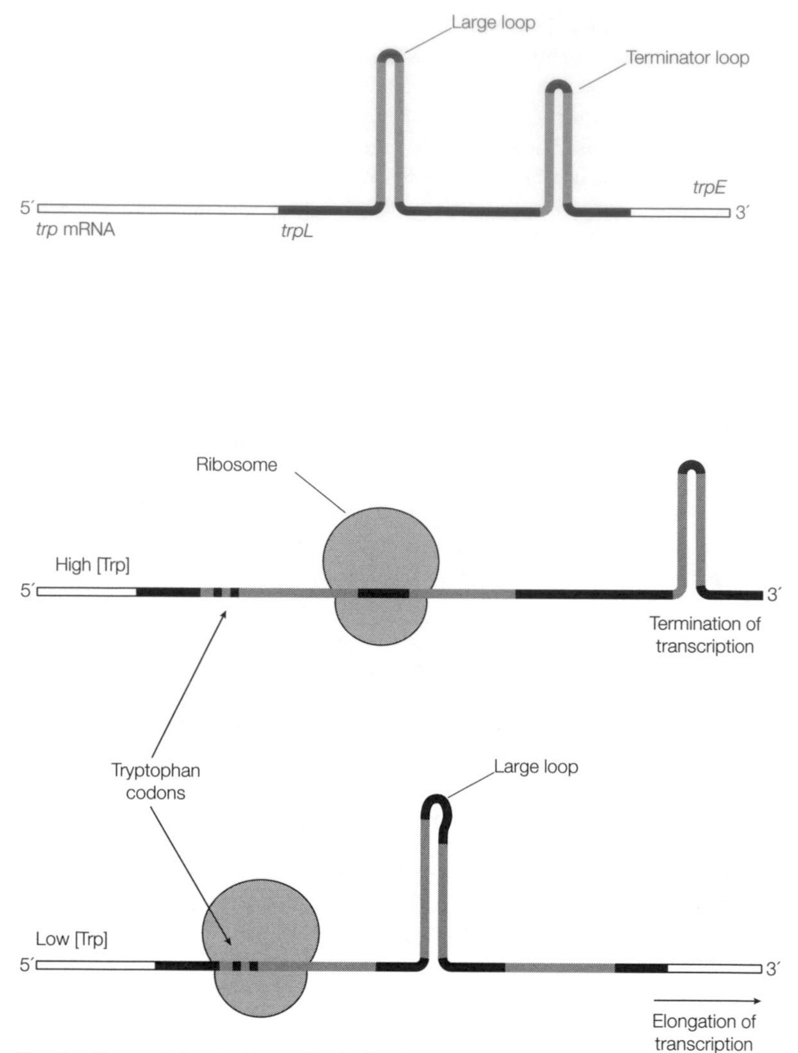

Fig. 4. Transcriptional attenuation in the trp operon.

a transcription terminator. If this structure forms it will terminate transcription before the first gene is reached eliminating gene expression. Attenuation depends on the fact that transcription and translation are linked in bacteria: ribosomes attach to mRNAs as they are being synthesized and begin translating them into protein. An mRNA that is being transcribed may already have one or more ribosomes attached to it. Binding of ribosomes to the *trp* mRNA influences which of the two stem–loop structures can form and so determines whether or not termination occurs. Immediately upstream of the stem–loop region is a short open reading frame containing 14 codons followed by a stop codon which is translated before the structural genes; two out of these 14 codons are for tryptophan. If levels of tryptophan are adequate the ribosome will translate the coding region following closely behind the RNA polymerase. In these circumstances the presence of the ribosome prevents formation of the larger stem–loop allowing the terminator loop to form and transcription ends. If tryptophan is lacking, the ribosome will be stalled as it translates the coding region.

The RNA polymerase will move ahead and the first stem–loop will be free to form. Formation of the terminator loop is then blocked and transcription of the operon can proceed. The speed at which the ribosome translates the coding region will not be the same for each transcript. When tryptophan is present in the medium at intermediate levels some transcripts will terminate and others will not, thus allowing fine adjustments in the levels of transcription of the operon. Overall, the *trp* repressor determines whether the operon is switched on or off and attenuation determines how efficiently it is transcribed, both depend on the level of tryptophan in the cell. Attenuation allows the cell to synthesize tryptophan according to its exact requirements. Attenuation is not restricted to the *trp* operon and occurs in at least six other operons that encode amino acid biosynthetic enzymes. Some operons such as the *trp* and *phe* operons are regulated by repressors and attenuation and others such as the *his*, *leu* and *thr* operons rely only on attenuation.

Regulation by alternative sigma factors

Bacterial RNA polymerase is composed of five individual polypeptide subunits (α_2, β, β', σ) One of the subunits called **sigma (σ) factor** is responsible for initiating transcription by recognizing bacterial promoter DNA sequences. Bacteria, including *E. coli*, make alternative sigma factors that recognize different sets of promoters and cause the RNA polymerase to transcribe different sets of genes. This is used as a way of regulating gene expression where environmental conditions dictate major alterations in the pattern of gene expression. The σ^{70} factor is the most common factor used by *E. coli*. Alternative σ factors come into play in a variety of situations:

- **Heat shock**. When exposed to increased temperature *E. coli* transcribes a set of 17 proteins which help the cell to adapt to the altered environmental conditions. An alternative factor σ^{32} is produced that recognizes promoters of heat shock genes.
- **Sporulation in *Bacillus subtilis***. This bacterium undergoes sporulation in response to adverse environmental conditions. Sporulation requires drastic changes in gene expression involving a shut down of most protein synthesis and the production of proteins needed for resumption of protein synthesis when the spore germinates. *B. subtilis* achieves this change by the use of alternative σ factors.
- **Bacteriophage σ factors**. Some phages make use of the host cell RNA polymerase and supply it with σ factors to instruct it to transcribe phage genes preferentially (see Topic B8). Other phages produce a series of σ factors that allow temporal control of their own gene expression with early genes transcribed by the host polymerase containing a σ factor that directs transcription of later genes.

A11 REGULATION OF GENE EXPRESSION IN EUKARYOTES

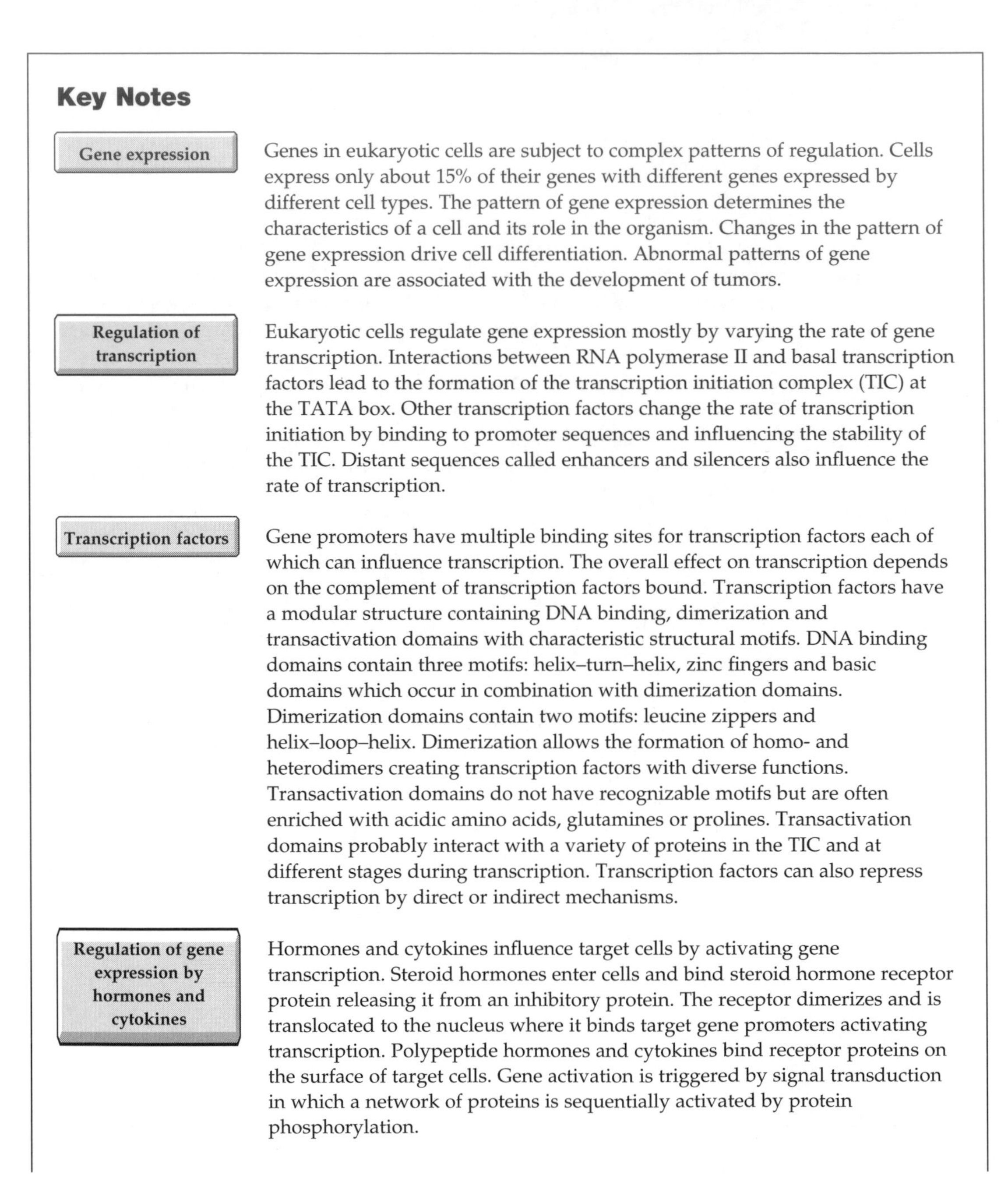

Key Notes

Gene expression

Genes in eukaryotic cells are subject to complex patterns of regulation. Cells express only about 15% of their genes with different genes expressed by different cell types. The pattern of gene expression determines the characteristics of a cell and its role in the organism. Changes in the pattern of gene expression drive cell differentiation. Abnormal patterns of gene expression are associated with the development of tumors.

Regulation of transcription

Eukaryotic cells regulate gene expression mostly by varying the rate of gene transcription. Interactions between RNA polymerase II and basal transcription factors lead to the formation of the transcription initiation complex (TIC) at the TATA box. Other transcription factors change the rate of transcription initiation by binding to promoter sequences and influencing the stability of the TIC. Distant sequences called enhancers and silencers also influence the rate of transcription.

Transcription factors

Gene promoters have multiple binding sites for transcription factors each of which can influence transcription. The overall effect on transcription depends on the complement of transcription factors bound. Transcription factors have a modular structure containing DNA binding, dimerization and transactivation domains with characteristic structural motifs. DNA binding domains contain three motifs: helix–turn–helix, zinc fingers and basic domains which occur in combination with dimerization domains. Dimerization domains contain two motifs: leucine zippers and helix–loop–helix. Dimerization allows the formation of homo- and heterodimers creating transcription factors with diverse functions. Transactivation domains do not have recognizable motifs but are often enriched with acidic amino acids, glutamines or prolines. Transactivation domains probably interact with a variety of proteins in the TIC and at different stages during transcription. Transcription factors can also repress transcription by direct or indirect mechanisms.

Regulation of gene expression by hormones and cytokines

Hormones and cytokines influence target cells by activating gene transcription. Steroid hormones enter cells and bind steroid hormone receptor protein releasing it from an inhibitory protein. The receptor dimerizes and is translocated to the nucleus where it binds target gene promoters activating transcription. Polypeptide hormones and cytokines bind receptor proteins on the surface of target cells. Gene activation is triggered by signal transduction in which a network of proteins is sequentially activated by protein phosphorylation.

Related topics Genes (A2) The human genome (B4)
Gene transcription (A4)

Gene expression

The human genome is estimated to contain as many as 100 000 genes which are subject to complex patterns of regulation that are, as yet, incompletely understood. In eukaryotic cells not all of the genes present in the genome are active. Cells express about 15% of their genes; the rest remain inactive. In multicellular organisms the genes that are active vary between cell types. The genes that are active in a particular cell type may be very different from those in another type of cell. The active genes determine which proteins and enzymes are present in a cell and are responsible for determining the characteristics of the cell and its role in the organism. For example, in lymphocytes which produce antibodies to fight infection the genes that encode the polypeptides that make up the antibodies are expressed at a high level. The pattern of genes expressed can change during the lifetime of a cell. For example, blood cells develop by differentiation from primitive progenitor cells. The changes that occur in the cell's characteristics as it differentiates are a result of changes in the pattern of gene expression. Understanding how gene expression is regulated is important for understanding diseases such as cancer in which abnormal expression of genes leads to uncontrolled cell division and formation of a tumor.

Regulation of transcription

Eukaryotic cells regulate the expression of their genes largely by determining the rate at which they are transcribed into mRNA. This can vary greatly with abundant proteins transcribed at very high rates and rare proteins at much lower rates. Regulation of gene expression is achieved by the interaction of gene promoters and DNA binding proteins called **transcription factors** (*Fig. 1*). Transcription of the gene by the RNA polymerase is initiated at the promoter and the efficiency of transcription initiation can be varied by the interaction between short regulatory DNA sequences present in the promoter that are recognized and bound by transcription factor proteins. The regulatory sequences present in the promoter are parallel with the coding sequence and are said to be ***cis*-acting**.

In eukaryotic cells protein coding genes are all transcribed by RNA polymerase II. Transcription is initiated by the formation of the transcription initiation complex (TIC) which involves binding of RNA polymerase II and a number of associated proteins called **basal transcription factors** to the DNA

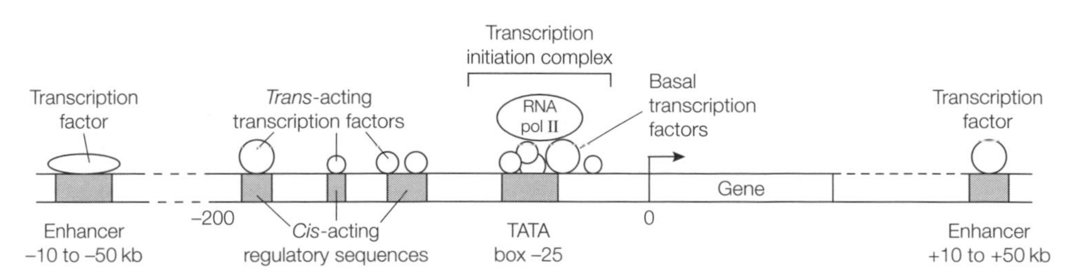

Fig. 1. Regulation of gene expression in eukaryotes.

of the promoter at a characteristic sequence known as the **TATA box** (see Topic A4). This has the sequence 5′ TATA(A/T)A(A/T) 3′ and is present in most but not all eukaryotic genes located approximately 25 bp upstream of the transcription initiation site. Its function is to locate the RNA polymerase in the correct position to initiate transcription. Some genes, especially those expressed only in specific tissues or cells, do not have a TATA box but instead have an initiator sequence usually located over the transcription start site. Other genes, usually those expressed at low levels, have neither a TATA box nor an initiator element.

The efficiency of transcription initiation and hence the amount of mRNA produced is influenced by additional transcription factors that bind other DNA sequences present in the promoter and can interact with the proteins of the TIC affecting its stability. Transcription factors can increase or decrease the rate of transcription. Many different transcription factors exist each of which recognizes and binds a DNA sequence in the promoter. The sequence recognized can vary between promoters and the binding site is usually described as a **consensus sequence** which incorporates possible variations. The transcription factors are synthesized in the cytoplasm of the cell but exert their effects in the nucleus. As such, they are often referred to as ***trans*-acting** factors.

The rate of transcription of a gene can also be influenced by sequence elements called **enhancers** that may be located thousands of base pairs distant from the transcription start site. Enhancers are typically 100–200 bp long and contain sequences that bind transcription factors and can stimulate transcription of the linked gene. The position of the enhancer relative to the gene it influences can vary, and may be upstream or downstream. Enhancers work independently of their orientation and are equally effective facing either forward or reverse. Interaction between the enhancer and its promoter occurs by looping of the intervening DNA to bring the two into close proximity. Some enhancers contain sequences that bind transcription factors that influence transcription negatively. These are known as **silencers** and they may be responsible for restricting expression to specific cell types. Other distant sequences called **locus control elements** exist which influence expression of entire families of genes by controlling access of transcription proteins to the DNA. An example is the locus control regions that regulate expression of the globin gene family.

Transcription factors

These are a large family of proteins that regulate the expression of protein coding genes. They are distinct from the basal transcription factors that interact with the RNA polymerase II to form the TIC. Transcription factors have varied patterns of expression: some occur only in specific cell types whereas others occur in all cell types. Each transcription factor bound by a gene promoter can regulate the transcription of the gene either positively or negatively. The overall effect on transcription depends on the complement of transcription factors bound. The ability of transcription factors to form dimers with themselves and other transcription factors adds further to the possibilities for regulation of gene expression.

Transcription factors have a modular structure composed of discrete protein domains with specific functions. Three types of domain occur commonly:

- DNA binding domains;
- dimerization domains;
- transactivation domains.

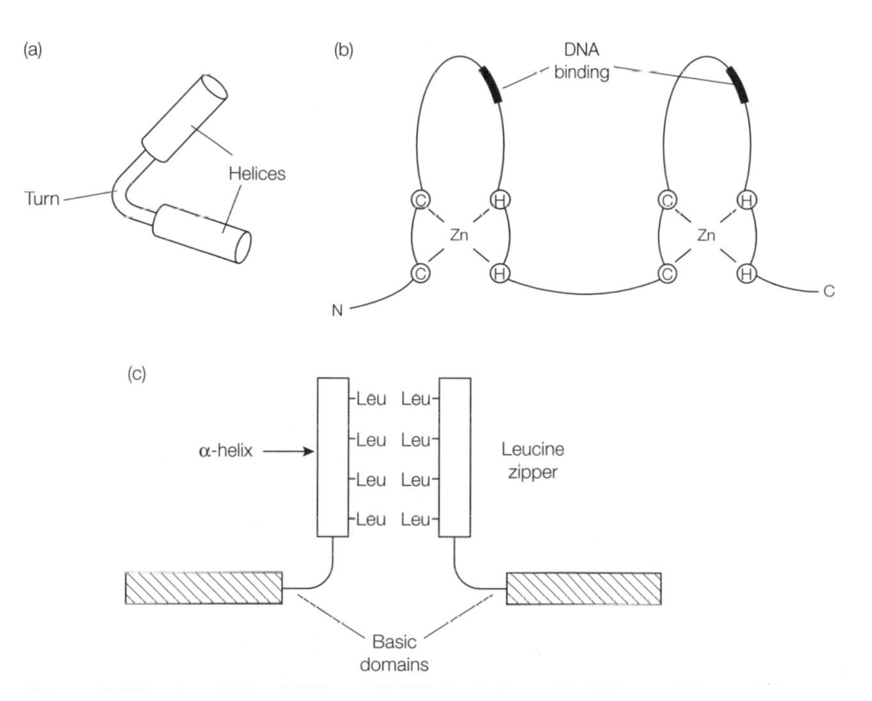

Fig. 2. Transcription factor motifs. (a) Helix–turn–helix motif. (b) Two linked zinc finger motifs. (c) Leucine zipper and basic DNA binding domains of bZip transcription factor.

The DNA binding and dimerization domains contain characteristic protein structures called **motifs** that identify them (*Fig. 2*).

DNA binding domains

Three types of DNA binding domain have been identified on the basis of the motifs present:

- **Helix–turn–helix**. This motif is composed of two α helices separated by a β turn. One of the helices, called the recognition helix, binds to the DNA by making contact with the major groove of the double helix. An example of proteins containing this motif are the homeodomain family of transcription factors encoded by the highly conserved homeobox genes that play an important role in embryonic development.
- **Zinc fingers**. This DNA binding motif occurs in two forms called C_2H_2 and C_4. The C_2H_2 form has a loop of 12 amino acids anchored at the base by two cysteines and two histidines that tetrahedrally coordinate a zinc ion. The motif forms a compact structure composed of two β strands and an α helix that contains basic amino acids that interact with the DNA via the major groove of the double helix. The zinc finger motif is repeated multiple times in the DNA binding domain and usually three or more are required for binding. An example of this motif is found in the ubiquitously expressed transcription factor Sp1. The C_4 motif has a similar structure but has four cysteines coordinated to zinc. An example of this type of domain is found in steroid hormone receptor transcription factors.
- **Basic domains**. This DNA binding domain is usually found in association with one of two dimerization domains called the leucine zipper or the

helix–loop–helix (HLH) which give rise to basic leucine zipper (bZIP) and basic HLH proteins. Binding to DNA requires the presence of two basic domains which are brought together by dimerization.

Dimerization domains

Two types of motif are found in dimerization domains:

- The **leucine zipper** motif is usually present on the carboxyl terminal side of a DNA binding domain. It contains an α helix in which every seventh amino acid is leucine such that a leucine is present on the same side of the helix every second turn creating a hydrophobic face. Dimerization is achieved by the interaction between the hydrophobic faces of two leucine zippers and results in the basic DNA binding domains of the two proteins being brought into close proximity. The two DNA binding domains face in opposite directions allowing them to bind a DNA sequence which has inverted symmetry.
- The **HLH** dimerization domain contains two α helices separated by a nonhelical loop. Dimerization is achieved by interaction between hydrophobic amino acids present on one side of the carboxyl terminal helix. Dimerization can occur not only between two molecules of the same transcription factor (homodimers) but between different transcription factors with the same dimerization domain (heterodimers). The formation of heterodimers creates transcription factors with new functions and increases the possibilities for regulating expression of target genes. An example of this is the MyoD family of transcription factors which form homo- and heterodimers that regulate gene expression in developing muscle cells.

Activation domains

Unlike DNA binding and dimerization domains, no motifs have been identified which characterize activation domains. Analysis of amino acid sequences has shown only that activation domains are often enriched for certain amino acids. Specifically, activation domains have been identified that are rich in **acidic amino acids** (for example, in yeast Gal4 transcription factor), **glutamines** (for example, in Sp1 transcription factor), or **prolines** (for example, in c-Jun Ap2 and Oct-2 transcription factors).

Transcription factors regulate the expression of target genes by binding to the TIC and changing its stability. It is probable that activation domains interact with different proteins in the TIC and at a variety of stages during transcription initiation and elongation.

Some transcription factors can repress transcription. This may be achieved in a number of ways. Some may repress transcription directly by interacting with the transcription initiation complex. Other may act indirectly in a number of ways including: (i) blocking the DNA binding site of an activating transcription factor; (ii) formation of a dimer that lacks a DNA binding domain; or (iii) binding of a repressor protein to the activation domain of another transcription factor.

Regulation of gene expression by hormones and cytokines

Hormones are agents produced by cells that act on other cells influencing their characteristics and functions by activating transcription of specific genes. Hormones may be small molecules, often steroids such as estrogens and glucocorticoids, or polypeptides such as insulin. Cytokines are proteins that act in a similar way to hormones, often with blood cells as their targets. Hormones and cytokines modulate gene expression in target cells in a number of ways. Steroid hormones are lipid-soluble and so can pass through the cell membrane into the cytoplasm where they bind to a transcription factor called the **steroid hormone receptor**. Binding causes the steroid hormone receptor to be released from an inhibitory protein. It then dimerizes and is translocated to the nucleus where it activates transcription of target genes by binding to promoter sequences. Polypeptide hormones and cytokines act in a different way to steroid hormones by binding to receptor proteins on the surface of the target cell. Gene activation is triggered by a process called **signal transduction** in which a network of proteins is sequentially activated by protein phosphorylation. Ultimately this leads to stimulation of transcription of target genes by binding of transcription factors to gene promoter sequences.

B1 CHROMOSOMES

Key Notes

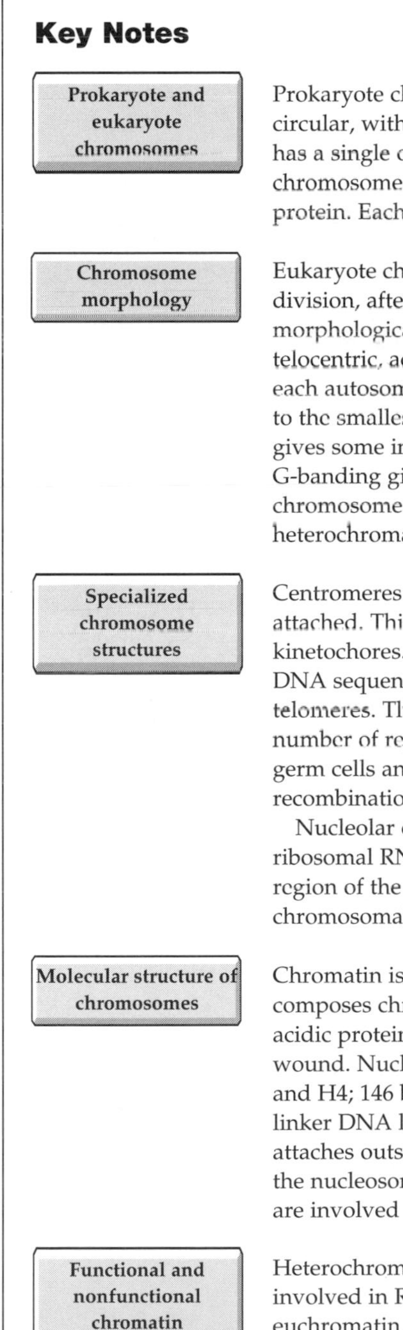

Prokaryote and eukaryote chromosomes

Prokaryote chromosomes consist of a single DNA molecule, that is usually circular, with only a small amount of associated protein. Each chromosome has a single origin of DNA replication. Eukaryotes have several linear chromosomes, and the DNA is tightly associated with large amounts of protein. Each eukaryote chromosome has multiple origins of DNA replication.

Chromosome morphology

Eukaryote chromosomes are visible by light microscopy only during cell division, after they have replicated. Chromosomes are placed into morphological groups, metacentric, submetacentric, acrocentric and telocentric, according to the position of their centromere. Within a species each autosome is given a specific number in ascending order from the largest to the smallest. Chromosome banding aids chromosome identification and gives some information as to the underlying organization of the chromosome. G-banding gives a series of light and dark bands along the length of the chromosome. C-banding produces dark bands in regions of constitutive heterochromatin.

Specialized chromosome structures

Centromeres are points on chromosomes to which the spindle fibers are attached. This is mediated through specialized protein structures known as kinetochores. Centromeres are composed largely of highly repeated satellite DNA sequences. Specialized structures at chromosome ends are known as telomeres. These are also composed of short repeated DNA sequences. The number of repeats decreases with age in somatic cells, but is maintained in germ cells and tumor cells by the enzyme telomerase. Telomeres prevent recombination between the ends of chromosomes.

Nucleolar organizer regions (NORs) contain tandem repeats of the major ribosomal RNA genes and are located in secondary constrictions. When the region of the chromosome distal to the NOR is small it is referred to as a chromosomal satellite.

Molecular structure of chromosomes

Chromatin is the term given to the association of DNA and proteins that composes chromosomes. It contains basic proteins, histones, and nonhistone acidic proteins. The histones form nucleosomes around which the DNA is wound. Nucleosomes consist of two discs containing histones H2a, H2b, H3 and H4; 146 base pairs of DNA are associated with each nucleosome and linker DNA leads to the next nucleosome. A single molecule of histone H1 attaches outside the core. This molecule is responsible for further folding of the nucleosomes into solenoids and more complex structures. Acidic proteins are involved in the chromosomal scaffold and in gene regulation.

Functional and nonfunctional chromatin

Heterochromatin is inactive chromatin, whereas euchromatin is actively involved in RNA transcription. Heterochromatin appears denser than euchromatin under the electron microscope and stains darker under the light microscope. Some chromatin can exist as either hetero- or euchromatin. This is called facultative heterochromatin. One of the two X chromosomes in cells of female mammals is converted to heterochromatin. It forms a small dark body attached to the nuclear membrane. Chromatin that is permanently heterochromatic is called constitutive heterochromatin, and can be identified by C-banding.

Alteration to chromosome numbers

Polyploidy is where the altered chromosome number is a multiple of the haploid chromosome number. This is rare in animals but important in plants. Small changes in chromosome number are classed as aneuploidy. In humans up to 4% of conceptuses are aneuploid, but very few of these survive to birth. Those that do survive tend to involve smaller chromosomes or alterations of the sex chromosomes. Aneuploidy arises by nondisjunction of homologous chromosomes or chromatids. It is more common in older mothers. Translocations can cause inheritance of trisomy 21 within families. Loss of a chromosome has more severe effects than gain of an extra chromosome. Some aneuploidies are often found as mosaics.

Related topics

Cell division (B2)
The human genome (B4)
Meiosis and gametogenesis (C3)
Polyploidy (D8)

Prokaryote and eukaryote chromosomes

All cellular life-forms have structures carrying genes, encoded in DNA, that are referred to as chromosomes. There are, however, major differences between these structures in prokaryotes and even the most simple of the eukaryotes. In prokaryotes the chromosome consists of a single DNA double helix, that is usually circular and has relatively few proteins associated with it. DNA replication proceeds from a single origin of replication (see Topic A9). Eukaryotes have, in almost all cases, a number of different chromosomes which are linear and which are contained within a membrane-bound organelle, the nucleus. The DNA molecules are intimately associated with large amounts of specific proteins. These may have functional or structural roles. The amount of DNA per chromosome is much greater in eukaryotes and because of this there are multiple origins of replication on each chromosome.

Chromosome morphology

Eukaryote chromosomes are usually only visible when a cell is in the process of dividing (see Topics B2 and C3), after the chromosome has been replicated into identical double structures known as **chromatids** (daughter chromosomes). Chromosomes are classified on the basis of their morphology. This is determined by the position of the centromere (primary constriction). *Fig. 1* shows four typical morphologies for chromosomes. In **metacentric** (mediocentric) chromosomes the centromere is close to the midpoint of the chromosome. This divides the chromosome into two roughly equal halves (arms). Where the centromere is sufficiently far away from the midpoint for a long arm (q arm) and a short arm (p arm) to be distinguished the chromosome is referred to as **submetacentric**. The other two morphological classes relate to chromosomes in which the centromere is close to one end of the chromosome. In **telocentric** chromosomes the centromere is at the end of the chromosome and there is only one arm. If the centromere is so close to the end of the chromosome that the

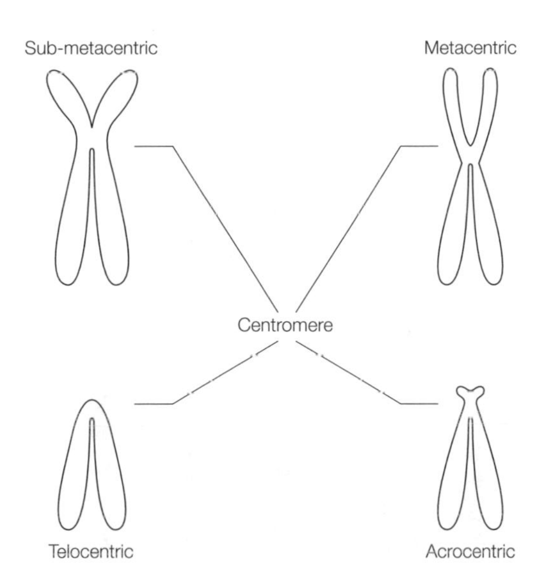

Fig. 1. Chromosome morphological. The morphological class is based on the position of the centromere.

short arm is only just discernible then the chromosome is termed **acrocentric**. Students often confuse submetacentric and acrocentric morphologies. The differences between the two are easily recognized in the chromosomes of humans (see *Fig. 2*).

In any species the complete diploid set of chromosomes is referred to as the **karyotype**. The autosomes are numbered in order of decreasing size, and the sex chromosomes are referred to as X or Y (see Topic C8). A chromosome preparation can be photographed, the homologous chromosomes paired, and set out in order. This is known as an **ideogram**, and is the conventional way to display karyotypes.

By grouping chromosomes by relative size and morphology it is usually possible to individually identify each chromosome in a species. This process was made much simpler by the development of treatments which, when applied to chromosomes prior to staining, produce a pattern of dark and light bands unique to each chromosome. Although there are several different banding techniques available only the two major processes, G-banding and C-banding, will be described here.

G-banding patterns can be produced by a wide range of different treatments but the most commonly used is a mild pretreatment of chromosome preparation slides with protease enzymes such as trypsin. The treated slides are then stained with Giemsa (hence the term G-banding). The process causes chromosomes to stain as a series of dark G-bands and pale interbands. The pattern is unique to each homologous chromosome pair and greatly aids the process of chromosome identification. These patterns have been used to create cytological maps of each chromosome in many species so that subchromosomal regions can be accurately identified. This is important in various processes such as gene mapping (see Topic F5) and medical genetics (see Topic F1).

Fig. 2 sets out the G band pattern for each of the human chromosomes diagrammatically. The pattern of bands allows each individual chromosome to be divided into a series of regions and subregions. Using chromosome 1 as an

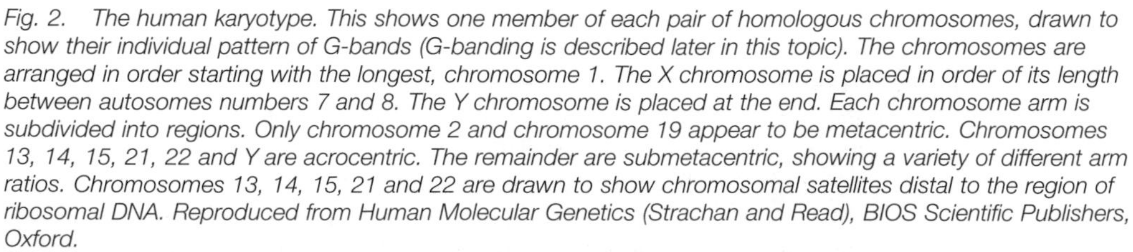

Fig. 2. The human karyotype. This shows one member of each pair of homologous chromosomes, drawn to show their individual pattern of G-bands (G-banding is described later in this topic). The chromosomes are arranged in order starting with the longest, chromosome 1. The X chromosome is placed in order of its length between autosomes numbers 7 and 8. The Y chromosome is placed at the end. Each chromosome arm is subdivided into regions. Only chromosome 2 and chromosome 19 appear to be metacentric. Chromosomes 13, 14, 15, 21, 22 and Y are acrocentric. The remainder are submetacentric, showing a variety of different arm ratios. Chromosomes 13, 14, 15, 21 and 22 are drawn to show chromosomal satellites distal to the region of ribosomal DNA. Reproduced from Human Molecular Genetics (Strachan and Read), BIOS Scientific Publishers, Oxford.

example the first division is between the long (q), and the short (p) arm. *Fig. 2* shows that the q arm is further divided into regions 1, 2, 3 and 4. There are two further levels of subdivision so that a specific region may be accurately defined as 1q42.1. Note that not all chromosomes nor chromosome regions are equally subdivided. These regions are arbitrary, and are decided by international conventions of cytogeneticists. As techniques have allowed the banding of longer, less condensed chromosomes so the number of subregions has increased. G-banding not only provides us with a convenient method for identifying chromosomes, but it also gives us information as to the overall organization of DNA and genes within eukaryote chromosomes. In general the darkly stained G bands are rich in the bases adenine and thymine whereas the pale staining interbands are richer in guanine and cytosine. More genes are located in the interbands than in the G bands. C-banding also gives us an insight into the organization of chromosomes. C-banding produces a number of dark bands. These are largely confined to areas around centromeres. These indicate regions of **constitutive heterochromatin** and are discussed in greater detail later in this topic.

Specialized chromosomal structures

All eukaryote chromosomes contain two different areas which have specific structural importance. These are the **centromeres** and **telomeres**. In addition some chromosomes contain **nucleolar organizer regions** (NORs). Centromeres are the sites at which the spindle attaches during cell division and functional centromeres are essential to this process. Any chromosome fragment which loses its connection to a centromere will not segregate to daughter cells at the end of cell division. The best studied centromeres are those of yeast where some are as short as 200 bp. Most centromeres are much larger than this. Normally the centromere consists of highly repeated satellite DNA (see Topic B4). In humans different chromosomes can be distinguished by the presence of specific alphoid satellite DNAs within their centromeres. Connection of the chromosome to the microtubular spindle fibers is effected by proteins that attach to the centromere forming a multilayered structure known as a **kinetochore**.

Telomeres are not simply the ends of chromosomes and DNA molecules, but are specialized structures. They contain multiple repeats of simple, short DNA sequences. In humans the repeat sequence is TTAGGG, but there is little variation between eukaryotes: similar sequences are found in plant and protist species. Specific proteins bind to the telomere region and the resulting nucleoprotein structures are thought to prevent recombination between the ends of different chromosomes. The number of repeats per telomere is high in germ cells but decreases with age in somatic tissues; this is a molecular marker of the aging process. Telomere length is maintained by the enzyme **telomerase**, a protein that contains RNA complementary to the telomere repeat DNA sequence, that acts as a template for extension of the telomere. Telomerase is absent from somatic cells but reappears in tumor cells, where telomere length is stabilized.

NORs are usually found at secondary constrictions. They consist of tandemly repeated 5.8S, 18S and 28S rRNA genes (see Topic A5). In most species the 5S rRNA genes are clustered elsewhere in the genome. In humans NORs are found on the short arms of all acrocentric chromosomes except the Y chromosome. Each NOR consists of approximately 80–100 repeats. During interphase the NOR decondenses and a nucleolus forms around it; NORs from different chromosomes can be incorporated into a single nucleolus. When the cell enters metaphase of mitosis (see Topic B2) the chromosomes may appear to be still attached at their short arms. This is known as **satellite association**.

The secondary constriction can be so pronounced that the small distal region of the chromosome appears to be unconnected from the body of the chromosome; this has given rise to the term **chromosomal satellite** and in humans these can be seen on chromosomes 13, 14, 15, 21 and 22 (*Fig. 2*). The term chromosomal satellite must not be confused with the term satellite DNA sequences (see Topic B4).

Molecular structure of chromosomes

Chromosomes are composed of DNA and proteins; a small amount of RNA is also present but this is effectively only in transit to the cytoplasm. The mixture of DNA and protein is called **chromatin**. The proteins are divided into two classes, **histones** and **nonhistone** or **acidic** proteins, both of which play important roles in chromatin structure and function. Histones are a group of small proteins with molecular masses of less than 23 kDa. In terms of dry weight they approximately equal DNA in the composition of chromatin. At physiological pH they have a basic charge due to the high frequency of the amino acids lysine and arginine. This basic charge assists their intimate interactions with the polyanion DNA. Five types of histone are found, H1 H2a, H2b, H3, and H5. This is true for all species and tissues with only rare exceptions relating to H1. The amino acid sequence of each of the histones is highly conserved throughout evolution, suggesting that these molecules have an important role, essential for the survival of eukaryotes. This has been elucidated by the identification of **nucleosomes**, the basic building blocks of chromatin structure.

Nucleosomes consist of a core of histones around which DNA is wound. The core consists of two discs arranged in parallel each composed of four histone molecules, one each of H2a, H2b, H3, and H4. The DNA molecule runs along the rim of the discs, and a molecule of histone H1 sits on the outside of the nucleosome complex acting as a seal; 146 bp of DNA are associated with a nucleosome core. The length of the linker between nucleosomes varies between species but in humans it is about 60 bp giving a total length of DNA per nucleosome of 200 bp. This is the basic level of packing of DNA in chromatin. Further packing depends to a great extent on histone H1. H1 molecules can interact to hold the individual nucleosomes in a helical structure giving rise to a **solenoid** of 30 nm diameter. This is the diameter of the fiber most commonly seen in electron micrographs of chromatin, but more densely coiled structures are also found. Simple nucleosome structures are shown in *Fig. 3*. Increasing levels of packing are observed within the nucleus. The highest level of packing is found in chromosomes at the metaphase of cell division. The organization of these structures involves the binding of chromatin fibers on to a chromosomal scaffold. This is made up largely of the acidic (nonhistone) nuclear protein, **topoisomerase II**. Specific regions of the DNA which run for several hundred base pairs and are rich in the bases adenine and thymine, known as **scaffold attachment regions** (SARs), link the DNA molecule to the chromosomal scaffold. The intervening material is arranged as loops of different lengths. The DNA in these is shown when histones are removed from preparations of metaphase chromosomes; electron micrographs show long lengths of DNA spooling out from the chromosomal scaffold. Nonhistone nuclear proteins are involved in a number of aspects of chromatin function, including the regulation of gene expression, where transcription factors are of major importance. These are described in more detail in Topic A11.

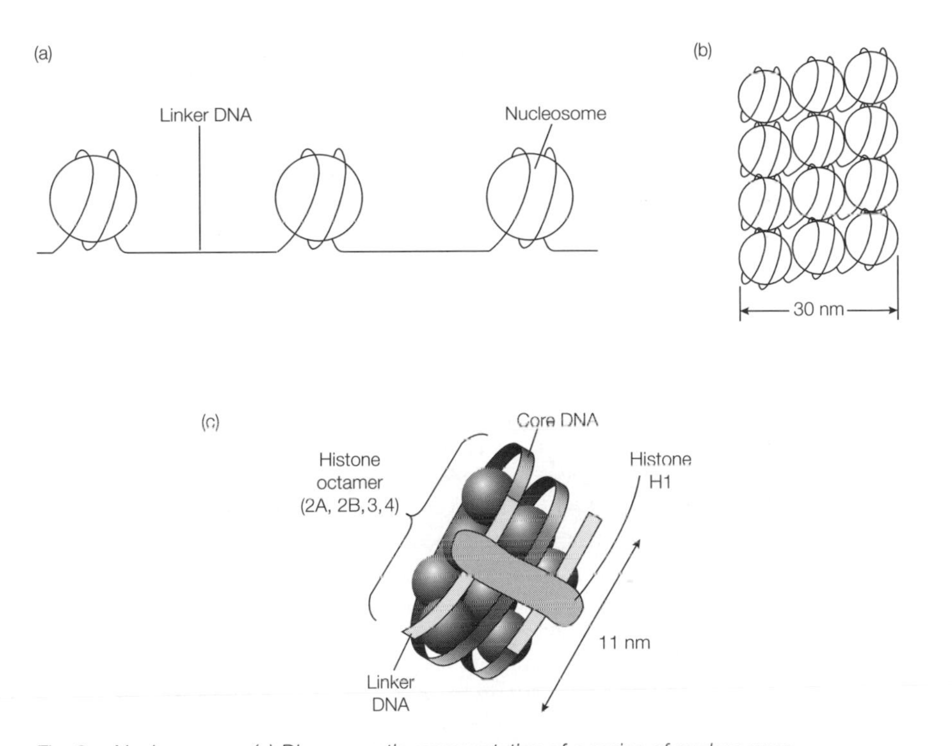

Fig. 3. Nucleosomes. (a) Diagrammatic representation of a series of nucleosomes. (b) Nucleosomes coiled together to form a solenoid. (c) Basic structure of nucleosome showing histones and DNA.

Functional and nonfunctional chromatin

Not all areas of chromatin are equally involved in gene transcription. Some chromatin is effectively inert. This is known as **heterochromatin** as compared with active chromatin or **euchromatin**. As seen by electron microscopy heterochromatin has a denser pattern of chromatin fibrils. It can also be differentiated by staining at the light microscopy level. Some chromatin is heterochromatic in all tissues and at all stages of development; this is **constitutive heterochromatin**, and is detectable by C-banding. As noted above centromeres are often detected by C-banding. There is a strong association between protein and satellite DNA sequences in these regions that prevents loss of DNA during the C-banding procedure; these regions therefore stain strongly. Other regions of constitutive heterochromatin include the long arm of the human Y chromosome and regions of sex chromosomes in other animals. Interstitial C-bands are more frequently encountered in plants, where they can be used to identify chromosomes because plant chromosomes do not G band.

Some regions of chromatin can exist in either the heterochromatic or euchromatic state; these are known as **facultative heterochromatin**. Female mammals have two X chromosomes, one of which is largely inactive as far as transcription is concerned. It is converted into heterochromatin and can be observed as a small dense spot on the side of the interphase nucleus, known as a **Barr body** or **X chromatin**. In this way there is a dosage compensation between males and females because in the male there is only one X chromosome and the Y chromosome is composed largely of constitutive heterochromatin. Which X chromosome is inactivated is a random event; in approximately half of the cells of a female mammal the paternally inherited X chromosome is inactivated and in the other half of the cells the maternally inherited X is inactivated; the inactivated X chromosome is reactivated during gametogenesis.

Alteration to chromosome numbers

Chromosome numbers may be altered by errors that take place in meiosis, the division process associated with gametogenesis (see Topic C4). These result in the formation of gametes with gains or losses of genetic material. If the gametes are viable they will give rise to progeny in which each cell has the same, abnormal, chromosome complement. Errors that occur during division of somatic cells, **mitosis**, (see Topic B12) result in an individual with more than one karyotype. Such an individual is termed a **mosaic.**

Polyploidy occurs when the chromosome number is increased, or decreased by a multiple of the haploid chromosome number. The euploid series describes different ploidy levels. An example using humans is given in *Table 1*. Although triploids and tetraploids make up approximately 10% of spontaneously aborted human fetuses, they are not observed in liveborn infants. Polyploidy is rare in the animal kingdom, but frequently observed in plants, where it is important in evolution (see Topic D8).

Aneuploidy refers to small deviations from the chromosome numbers of the euploid series. **Nullisomy** is the absence of any copy of a specific chromosome in a cell. One copy of a chromosome per cell is referred to as **monosomy,** two as **disomy** and three as **trisomy**. Monosomy is the norm in gametes and disomy in somatic tissue. *Table 2* contains examples of aneuploid syndromes found in humans.

Aneuploids for other chromosomes are also found, but usually only in fetuses that undergo spontaneous abortion. In total, 4% of human conceptuses are thought to be trisomic. In those aneuploid syndromes that survive to term there is a marked reduction in life expectancy. Individuals who have only part of a chromosome present three times express the associated syndrome less severely. In this context it is relevant to note that the autosomal trisomies listed above all involve small chromosomes. The presence of an extra copy of a large chromosome would probably affect the expression of too many genes to allow development of the fetus.

Sex chromosome aneuploids are relatively common, despite the fact that the X is

Table 1. The human euploid series

Chromosome number	Ploidy	Normally found in
23	Haploid	Gametes
46	Diploid	Somatic cells
69	Triploid	
96	Tetraploid	

Table 2. Aneuploidies in human populations

Condition	Chromosome involved	Approximate frequency
Involving gain of an autosomal chromosome		
Edward's Syndrome	18	1 in 5000
Pateau's syndrome	13	1 in 5000
Down's syndrome	21	1 in 750
Sex chromosome aneuploidies		
Turner's syndrome	XO	1 in 10 000
Klinfelter's syndrome	XXY	1 in 2000
Triple X syndrome	XXX	1 in 2000

a large chromosome known to carry many genes. How can this be explained? As noted above only one X chromosome is active in any mammalian cell so the presence of extra copies has less effect than would otherwise be predicted. Turner's syndrome (XO) is rarer than other sex chromosome aneuploidies; it is often found as a mosaic, where it is mixed with (XX) cells, and this may aid survival.

Aneuploid gametes arise from errors at both first and second division of meiosis (see Topic C3) through a process known as **nondisjunction**. This is where either two homologous chromosomes fail to separate at meiosis I or a centromere fails to split at metaphase of meiosis II (see Topic C3). Nondisjunction involving the splitting of centromeres can also arise at mitosis. Nondisjunction is thought to arise as random events similar to mutations in the DNA sequence; however, they are affected by external factors. This is shown clearly for trisomy 21, where the incidence increases with the age of the mother. In some cases there is a familial predisposition to trisomy 21, due to the presence of carriers in the family who have a normal phenotype but carry a translocation involving chromosome 21. They can transmit both a normal copy of chromosome 21 and the translocation chromosome which also bears chromosome 21, to their offspring. An affected child will also inherit a normal copy of chromosome 21 from their other parent and thus will have three copies of the chromosome. This is described in *Fig. 4.*

All of the syndromes mentioned, with the exception of Turner's Syndrome (XO), have involved presence of an extra chromosome. Individuals with loss of a single autosome, monosomy, are very rare and have a very short life-span. In humans nullisomics are never viable.

Carrier	Normal	
21,14	21,14	Viable normal
21/14	21,14	Viable carrier
21, 21/14	21,14	Trisomy 21
14	21, 14	Lethal nullisomic
14, 21/14	21, 14	Nonviable
21	21, 14	Lethal nullisomic

Fig. 4. Trisomy 21 inheritance. If an individual is a carrier for trisomy 21 (Down's syndrome) because of the presence of a translocation between chromosome 21 and chromosome 14 they will produce a number of genetically different gemetes due to irregularities in segregation of the translocation at meiosis. These gametes will be fertilized by normal gametes monosomic for both chromosome 21 and 14. The two sets of gametes and the outcome of their fusion are shown below. The translocation is denoted as 21/14. Note that the gametes will not all be produced at equal frequency.

B2 Cell division

Key Notes

Eukaryote cell cycle

The cell cycle is the period between two divisions. It consists of four stages, G_1, S, (the DNA synthesis phase), G_2 and mitosis. Cells in different phases of the cell cycle can be identified using a fluorescence activated cell sorter. Cells in S phase can be identified by labeling with analogs of thymidine.

Mitosis and cytokinesis

Nondividing cells are in interphase. Mitosis is divided into four phases. In prophase the chromosomes condense. At metaphase they are aligned on the equator of the cell. When the centromeres split, the two chromatids separate towards opposite poles. This is anaphase. As the chromatids near the poles the cell moves into telophase. Nuclear membranes form around the two nuclei and the chromosomes begin to decondense. At the same time cytokinesis is initiated at the equator of the cell.

Regulation of the cell cycle

Cells in multinucleate organisms are often not traversing the cell cycle. Cells are stimulated to grow and divide by growth hormones. Once a cell passes the restriction point it is committed to divide. Growth hormones act through the signal transduction pathway. Progress through the cycle is dependent on complexes between cyclins and cyclin-dependent kinases. Checkpoints exist at several crucial stages in the cycle to ensure that the process of cell division is highly regulated. Mutations in cell cycle genes are usually found in tumor cells.

Related topics

DNA replication (A9)
Meiosis and gametogenesis (C3)
Genes and cancer (F2)

Eukaryote cell cycle

The progression from one cell division to the next can be regarded as a cyclic process, the **cell cycle**. During this time the cell must replicate its contents and then organize the distribution of its components equally between two daughter cells. Except in the production of gametes (see Topic C3) the nuclei of eukaryote cells divide by **mitosis** and in parallel with this the cytoplasm divides by **cytokinesis**. These processes are easily visualized in fixed or living cells. The period between two consecutive divisions is referred to as a **cell cycle**. The replication of DNA is accomplished during a period in the cell cycle known as the **synthetic** or **S phase**. The period preceding S phase is called $\mathbf{G_1}$ (gap 1), and the period between S phase and division is known as $\mathbf{G_2}$ (gap 2).

Where suspensions of single cells can be produced, **fluorescence activated cell sorters** can identify cells as being in G_1, S, or G_2 phases of the cell cycle. Apart from this cells in G_1 and G_2 are not easily identified, but cells in S phase can be detected if cells are allowed to incorporate a labeled precursor of DNA. Analogs of thymidine are best suited to this purpose as this nucleoside is specific to DNA. The proportion of cells labeled after a short exposure to labeled precursor is equal to the average proportion of the cell cycle that the cells spend in S phase.

Mitosis and cytokinesis

Mitosis has many similarities with **meiosis** (see Topic C3), the reduction division in gametogenesis, and care must be taken to avoid confusion. The stages of mitosis are set out in *Fig. 1*. G_1, S and G_2 phases are collectively referred to as **interphase**. As a cell leaves G_2 it enters **prophase** of mitosis. The individual chromosomes gradually become apparent as their chromatin structure begins to condense. They are now double structures, **chromatids**, as DNA has been replicated in S phase. Towards the end of prophase the nuclear membrane and nucleolus begin to break down. Released nucleolar proteins attach to the surfaces of condensing chromosomes. The cell moves from prophase to **metaphase**. At this point the nuclear membrane and nucleolus are absent and the chromosomes are aligned on the equator of the cell. This is achieved by the action of tublin-containing spindle fibers which run from both poles of the cell and attach to the centromeres of the chromosomes (see Topic B1). The subsequent phase, **anaphase**, commences as soon as the centromeres are cleaved allowing the chromatids to separate. The individual chromatids, better described as **daughter chromosomes**, are pulled to opposite poles of the cell by the mitotic spindle. When the two sets of chromosomes are well separated toward the poles of the cell they begin to decondense, nuclear membranes form around the decondensing chromatin and the cell is transiently binucleate; this is **telophase**. At the same time, in animal cells, a **cleavage furrow** forms across the equator of the cell. This structure, which is composed of actin-containing microfilaments, progressively tightens and eventually divides the cell into two daughter cells; this is **cytokinesis**. In plant cells cytokinesis is achieved by the formation of a new cell wall across the equator of the cell.

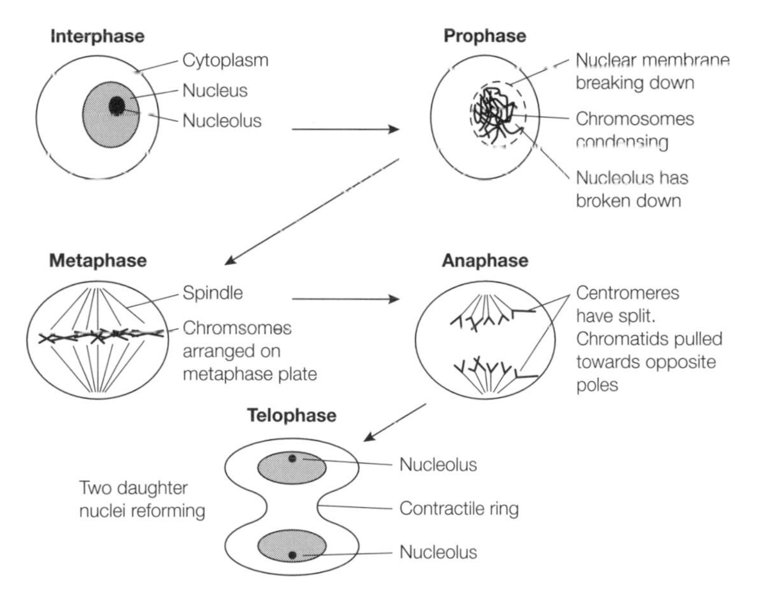

Fig. 1 Schematic representation of mitosis in an animal cell.

Regulation of the cell cycle

This refers to the rate at which cells traverse the cell cycle. The essential point to grasp is that in multicellular organisms most cells do not divide regularly. These quiescent cells require stimulation before they divide and are referred to as being in the $\mathbf{G_0}$ phase. Only specific tissues such as bone marrow in animals or root tips in plants have high proportions of cycling cells. Cells are triggered to leave G_0 by specific hormonal growth factors. They then move through G_1 phase. This can vary in length from cell to cell, but once a cell passes a specific point, the **R** or **restriction point**, it is committed to attempt to complete the cycle, pass through division and reach G_1 again. The subsequent portions of the cycle, S phase, G2 and mitosis have relatively fixed lengths.

Progress through the cell cycle as a response to the effects of growth stimulating hormones, is mediated by a process known as **signal transduction** (see Topic F2). This comprises a series of effector molecules that translocate and amplify the initial signal produced by the growth hormone, eventually causing transcription of genes involved in cell growth and division. At the same time progress through the cell cycle is negatively regulated by a series of **checkpoints** which the cell cannot pass until certain criteria have been fulfilled. Thus growth and division of cells remains a highly regulated process as is necessary for the development of multicellular organisms. Key players in the progression of the cell cycle are a group of kinases, enzymes that add phosphate groups to their substrates. Phosphorylation of specific target proteins at different stages of the cell cycle is necessary for the cycle to proceed. The kinases bind to a second group of proteins, **cyclins**. These regulate the kinase activity. In the absence of cyclin the kinases are inactive and are therefore called **cyclin dependent kinases**, (**cdks**). Different members of the cyclin family appear at different points of the cell cycle.

Checkpoints have been identified at three stages of the cell cycle, the boundary between G_1 and S phase, the boundary between G_2 and mitosis and within mitosis itself. These ensure that the various processes involved in cell growth are coordinated. For instance, cells with unrepaired damage to their DNA (see Topic B6) are prevented from entering S phase, where replication of damaged DNA could result in the creation of mutations. Cells with uncompleted DNA replication cannot enter mitosis; mitosis is not completed before the spindle is organized and all chromosomes are attached to it.

Mutations of genes involved in the signal transduction pathway can be cancer inducing; these are known as **oncogenes**. Genes involved in the checkpoints are also often mutant in tumors; they are known as **tumor suppressor genes**. This is dealt with in detail in Topic F2.

B3 PROKARYOTIC GENOMES

Key Notes

Organization of prokaryotic DNA

Prokaryotes are comprised of eubacteria (including *E. coli*) and archaebacteria. In prokaryotes, the genome consists of a supercoiled circular DNA molecule called the bacterial chromosome. Prokaryotic cells have a dense central area called the nucleoid composed of a protein core from which loops of supercoiled DNA radiate. Some nucleoid proteins may help to package the DNA. Separation of replicated bacterial chromosomes for cell division may be achieved by separate attachment points on the cell membrane.

Prokaryotic genes

Almost all bacterial genes occur on the chromosome. A few exist on plasmids The *E. coli* chromosome has been sequenced and the position of many of the genes located. The genes are arranged as operons or as single copies and account for about 75% of the DNA sequence. The remainder is noncoding, intergenic DNA which includes important sequences such as the origin of replication. Bacterial chromosomes vary in size reflecting differences in the number of genes and probably the size of the intergenic DNA.

Plasmids

Plasmids carry genes that may confer useful properties to bacteria. They replicate independently. Some integrate into the bacterial chromosome. Plasmids may be stringent (low copy number) or relaxed (high copy number). Many different plasmids exist. These include: resistance (R), fertility (F), col, and virulence plasmids. Bacteria may contain several types of plasmid. Plasmid incompatibility restricts the types of plasmid that can coexist in a bacterium.

Bacterial transposons

These are DNA sequence elements that use recombination to move about the genome. They encode a transposase enzyme that catalyzes their own movement. Insertion sequences are *E. coli* transposons that have a short inverted repeat sequence at either end. When an insertion sequence transposes, a host DNA sequence at the site of insertion is duplicated such that the insertion sequence is flanked by a direct repeat. Transposons cause insertional mutation of genes. When they move to another site a duplication of the original target sequence is left behind and the gene remains mutated.

Archaebacteria

These are organisms found in extreme environments; they differ from eubacteria. They are thought to represent a group of organisms that is distinct from both prokaryotes and eukaryotes.

Related topics

Genes (A2)
Regulation of gene expression in prokaryotes (A10)
Recombination (B7)

Organization of prokaryotic DNA

Living organisms are divided into primitive forms such as bacteria known as **prokaryotes** and higher organisms called **eukaryotes**. The structure and organization of cells in prokaryotes and eukaryotes are different. Eukaryotic cells have a complex internal structure with membrane bound organelles and a separate nucleus. Prokaryotic cells lack this organization and have no distinct nuclear compartment. The prokaryotes are divided into **eubacteria** or true bacteria including *E. coli*, and an unusual group distinct from eubacteria called **archaebacteria**. Most of the information available refers to *E. coli* and other eubacteria. Prokaryotes have a single circular DNA molecule referred to as the **bacterial chromosome** that contains almost all of the genes. The DNA molecule is extremely long relative to the dimensions of the cell. To allow it to fit inside the cell, it is compacted by a process called **supercoiling**. Enzymes called **topoisomerases** introduce additional turns into the double helix that cause the DNA strand to wind up on itself and adopt a more compact form (*Fig. 1*). The topoisomerases act by breaking the DNA polynucleotide and rotating the two ends relative to each other. The enzyme then rejoins the ends and the polynucleotide reacts by winding up on itself. This is called positive supercoiling. Topoisomerases can also remove coiling in a process called negative supercoiling by creating a turn in the opposite direction. Supercoiling also occurs in eukaryotic cells where it is involved in packaging of DNA in eukaryotic chromosomes.

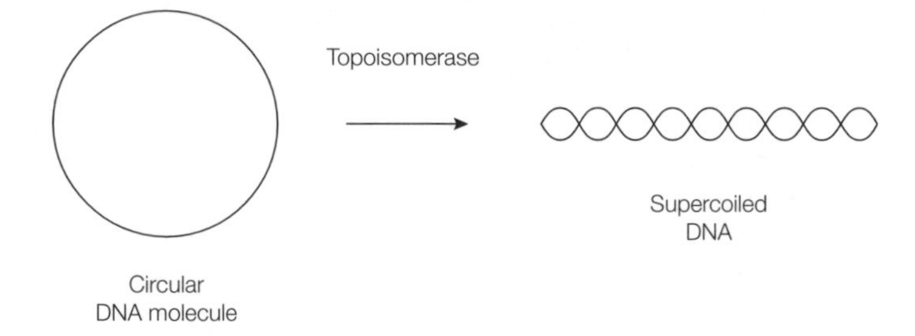

Fig. 1. Supercoiling of circular DNA molecule.

Although prokaryotes lack a distinct nucleus, analysis of bacterial cells by electron microscopy shows a darker central area containing DNA and protein known as the **nucleoid** and an outer area called the cytoplasm. The exact structure of the nucleoid is uncertain but it is known to have a central protein core from which supercoiled loops of DNA radiate (*Fig. 2*). Some of the proteins isolated from the nucleoid resemble the histone proteins found in eukaryotic chromosomes and it is thought that the proteins of the nucleoid may help to organize the folding of the DNA into its compact structure. The length of prokaryotic chromosomes relative to the dimensions of the cell means that replication and partitioning of DNA molecules during cell division is a potentially difficult task. This may be achieved by the DNA molecules having separate attachment points on the cell membrane which move away from each other as the cell divides.

Prokaryotic genes

Almost all the genes present in bacteria occur on the bacterial chromosome. A few other genes exist on small circular DNA molecules called **plasmids** that are present in bacteria in addition to the chromosomal DNA. The organization

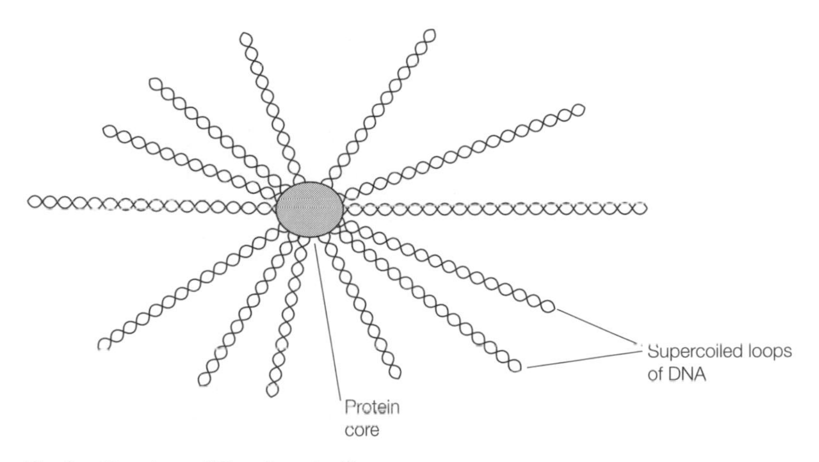

Fig. 2. Structure of E. coli *nucleoid.*

of the genes on the bacterial chromosome is best characterized for *E. coli* which has about 2800 genes carried by a chromosomal DNA molecule of 4.6 million base pairs. The entire sequence of the *E. coli* chromosome and the relative positions on the chromosome of many of the genes have now been determined. Some genes are arranged as families called **operons** that encode proteins with related functions and are regulated in a coordinated way. Other genes occur at random positions along the chromosome with no apparent organization. Most of the genes are present as single copies. The main exception is the ribosomal RNA genes that occur as a cluster of seven copies. A similar pattern of organization is seen in other eubacteria and gene maps are most similar in related species. About 75% of the DNA of the bacterial chromosome is accounted for by the genes. The remaining 25% is **intergenic DNA** which separates individual genes. Some parts of the intergenic DNA have important functions such as the location of the **origin of replication** of the bacterial chromosome. Other intergenic regions may be involved in interactions with DNA packaging proteins. The size of prokaryotic genomes varies between species of bacteria. The size variations reflect differences in the number of genes and possibly larger intergenic regions.

Plasmids

These are circular DNA molecules present in addition to the bacterial chromosome in almost all bacteria. Plasmids carry genes not found on the bacterial chromosome which often confer useful properties, such as resistance to antibiotics, to the bacterium. Plasmids have their own origin of replication and so can replicate independently of the chromosomal DNA. Some small plasmids use the cell's enzymes to replicate. Larger plasmids may carry genes that encode their own replicative enzymes. Other plasmids integrate into the host genome and are copied at the same time as the bacterial chromosome. This integrated form of plasmid is called an **episome** and may pass through many cell divisions before excising itself to exist as a separate plasmid again.

Many different plasmids are found in bacteria and individual species may contain several types. Plasmids can be classified according to the genes they carry and the characteristics they confer on the host cells. Five types have been identified:

- **Resistance (R) plasmids**. These carry genes that make bacteria resistant to antibiotics such as ampicillin and chloramphenicol. The way in which resistance is conferred varies. An example is the RP4 plasmid found in *Pseudomonas* and other bacteria. R plasmids have important consequences for the treatment of bacterial infections as they represent a way in which antibiotic resistance can spread between species of bacteria.
- **Fertility (F) plasmids**. These plasmids allow genes to be transferred between bacterial cells in a process called **conjugation**. The F plasmid contains genes that direct the transfer of the F plasmid from one bacterial cell to another by means of a tube-like structure called a **sex pilus**. The F plasmid may carry additional genes which it acquires from the chromosome and these are transferred to the recipient cell during conjugation.
- **Col plasmids**. These plasmids carry genes that encode proteins called **colicins** that can kill other bacteria. An example is ColE1 of *E. coli*.
- **Degradative plasmids**. These encode proteins that allow the host bacterium to metabolize unusual molecules such as toluene or salicylic acid.
- **Virulence plasmids**. These plasmids confer the ability to the bacterium to cause disease. An example is the **Ti plasmid** that is found in the bacterium *Agrobacterium tumefaciens* that causes Crown Gall disease in plants.

Plasmids vary in size with the smallest around 1 kb in length and the largest up to 250 kb. Individual plasmids vary according to the host cells in which they occur. Some are present in many different species of bacteria and others in just a few species. The number of plasmid molecules in a bacterial cell also varies. Plasmids that are present as just one or two copies are said to have a low copy number and are called **stringent** plasmids. Other plasmids have a high copy number with 10 or more plasmid molecules present and are known as **relaxed** plasmids. There are also restrictions on the types of plasmid that can coexist in bacterial cells. This feature is called **plasmid incompatibility**. Plasmids that occur in the same species of bacteria must belong to different incompatibility groups.

Bacterial transposons

These are DNA sequence elements that are capable of moving around in the genome. They occur in both eukaryotes and prokaryotes where they are present on bacterial chromosomes and on plasmids. The movement process is called **transposition** and depends on recombination between DNA sequences. Transposons are autonomous units and each encodes an enzyme called **transposase** that catalyzes its own transposition. Many different transposons are known. The first to be identified were the **Insertion Sequences** which occur in *E. coli* (*Fig. 3a*). Several types of insertion sequence have been identified and as many as 10 copies of each may be present in bacterial genomes. Transposition is a relatively infrequent process occurring only every 10^3–10^4 cell divisions. Insertion sequences can transfer between bacteria during conjugation and can also transfer between related species. A characteristic feature of insertion sequences is that they have a **short inverted repeat** at either end. These are duplicated sequences in which the two copies point in opposite directions. This means that the same sequence is encountered moving from the flanking sequence at either end towards the insertion sequence. In addition, when an insertion sequence transposes a host DNA sequence at the site of insertion is

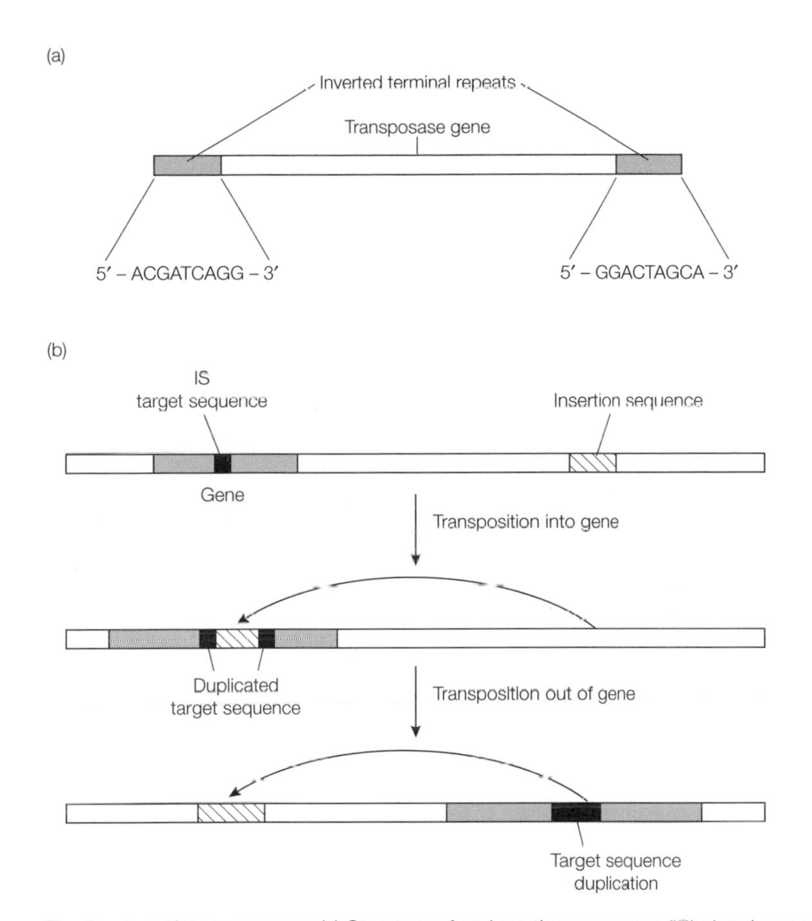

Fig. 3. Insertion sequences. (a) Structure of an insertion sequence (IS) showing a possible inverted repeat sequence. (b) Gene mutation by transposition of insertion sequences.

duplicated such that the insertion sequence is always flanked by a short duplication of the target sequence known as a **direct repeat** (direct means that the two copies are in the same orientation). The ability of transposons to move around in the genome means that they can mutate a gene when they transpose into it. When they move to another site the short duplication of the original target sequence is left behind so means that the gene remains mutated even after the transposon has left (*Fig. 3b*).

Archaebacteria

This is a group of organisms that is distinct from eubacteria. It is now recognized to be as different from other prokaryotes as it is from the eukaryotes. The archaebacteria comprise three groups of related organisms: **methanogens, extreme thermophiles** and **extreme halophiles**. Most archaebacteria live in extreme environments making them difficult to culture and so to study. Biochemically they differ from eubacteria with respect to the structure and composition of their cell wall. Genetically, several distinctions have been identified that are related to the ribosomal RNA genes including the presence of introns and a different structure and organization of the genes.

B4 THE HUMAN GENOME

Key Notes

The human genome

The human genome is 3 billion base pairs long and contains 50–100 000 genes arranged on 23 chromosomes. The genes account for 25% of the DNA. The rest is extragenic DNA.

Genes

The coding information in genes occurs as a series of exons separated by noncoding introns. Genes vary greatly in size and also with respect to the number and sizes of the introns. Leader and trailer sequences occur at the 5′ and 3′ ends of genes; these are transcribed but not translated. Upstream promoter sequences regulate gene transcription.

Gene families

Some genes occur as families containing multiple copies of genes with identical or related sequences. The genes in a family may be present at single or multiple loci. Gene families may also occur as individual clusters at multiple loci.

Pseudogenes

These are diverged members of gene families that have acquired one or more inactivating mutations. Processed pseudogenes are nontranscribed DNA copies of mRNAs probably derived by a mechanism involving reverse transcription. Gene fragments are inactive genes that lack part of the parent gene. They are thought to have arisen by deletion or recombination of the original gene sequence.

Extragenic DNA

This is composed of sequences that are not genes, gene-related sequences or pseudogenes and accounts for about 75% of the genome. Most extragenic sequences (70–80%) are unique or exist as a small number of copies. The rest (20–30%) are moderately or highly repeated sequences present as tandem arrays or dispersed throughout the genome. Extragenic DNA has no known function.

Dispersed repetitive sequences

These consist of SINEs and LINEs (short and long interspersed nuclear elements, respectively). SINEs include Alu sequences. These are a family of sequences about 250 bp long present as about 700 000 highly dispersed copies. They are thought to be derived from processed pseudogenes that acquired the ability to move about the genome. LINEs are longer than SINEs. The L1 LINE is 6500 bp and exists as 60 000 copies. LINEs are retroelements and have the ability to copy themselves using reverse transcriptase and to move about the genome.

Clustered repetitive sequences

The human genome has extensive regions containing long tandem arrays of repetitive sequences. These are classified according to the length of the cluster of repetitive sequence as satellite DNA, minisatellite DNA and microsatellite DNA. Minisatellite and microsatellite clusters have overlapping sizes but microsatellite DNA characteristically has short repeat sequences. CA repeats and mononucleotide repeats account for 0.8% of the entire genome.

Variable number tandem repeats (VNTRs)

VNTRs are repetitive sequences that vary according to the number of times the repeated sequence is present. Variation occurs at a given locus between individuals and between pairs of chromosomes carried by one individual. Polymerase chain reaction (PCR) can be used to detect the variations. VNTRs are used in forensic science to identify individuals at the scene of a crime and in medical genetics to identify carriers of genetic diseases.

Related topics

Genes (A2)
Genetic diseases (F1)
The human genome project (F5)
Genetics in forensic science (F6)

The human genome

This term is used to describe the different types of sequence that together make up the DNA in a human cell. The DNA in the human genome is about **3 billion base pairs** (bp) long and is estimated to contain **50 000–100 000** genes. The DNA is arranged as a set of 23 chromosomes each of which is a single, double-stranded DNA molecule 55 – 250 million bp long. The genes and gene-related sequences account for about 25% of the DNA (*Fig. 1*). The remainder is called extragenic DNA and has no known function.

Genes

The coding information in a gene is present as a series of segments of DNA sequence called **exons** separated from each other by intervening noncoding sequences called **introns**. Genes vary greatly in size and also with respect to the number and sizes of the introns. Some genes such as the histone H4 gene are just a few hundred base pairs long. Others, such as the Factor VIII gene, are several hundred kilobase pairs (kbp) in length and contain many large introns such that the actual coding sequence accounts for just a few percent of the total gene sequence. Additional sequences are present which are associated with genes. **Leader** and **trailer sequences** occur at the 5′ and 3′ ends of the gene which are transcribed but not translated. **Promoter sequences** occur upstream of the point where transcription begins and regulate synthesis of mRNA from the gene. The promoter may extend up to about 1 kbp upstream but other regulatory sequences that influence transcription may occur at sites much further away.

Gene families

Some genes exist as a number of copies with identical or related sequences that can be grouped into families. Gene families may be organized in a number of ways: (i) all of the genes in the family occur at the same chromosomal locus. An example of this is the growth hormone gene family whose five members are clustered on chromosome 17; (ii) the genes belonging to the family may occur at different loci. For example, the five members of the aldolase gene family are on different chromosomes; (iii) the genes of a family may exist as a series of clusters on different chromosomes. An example are the homeobox genes which occur as four clusters on separate chromosomes each containing about 10 individual genes. In some multigene families all the genes are identical and may encode a protein required in large amounts by the cell such as histones. In other families, the genes are not identical but show some sequence divergence. In some cases the divergence is so great that the genes encode proteins that are related but have distinctive properties. An example of this is the α and β globin gene families whose members are expressed at different stages of embryonic development and in adults.

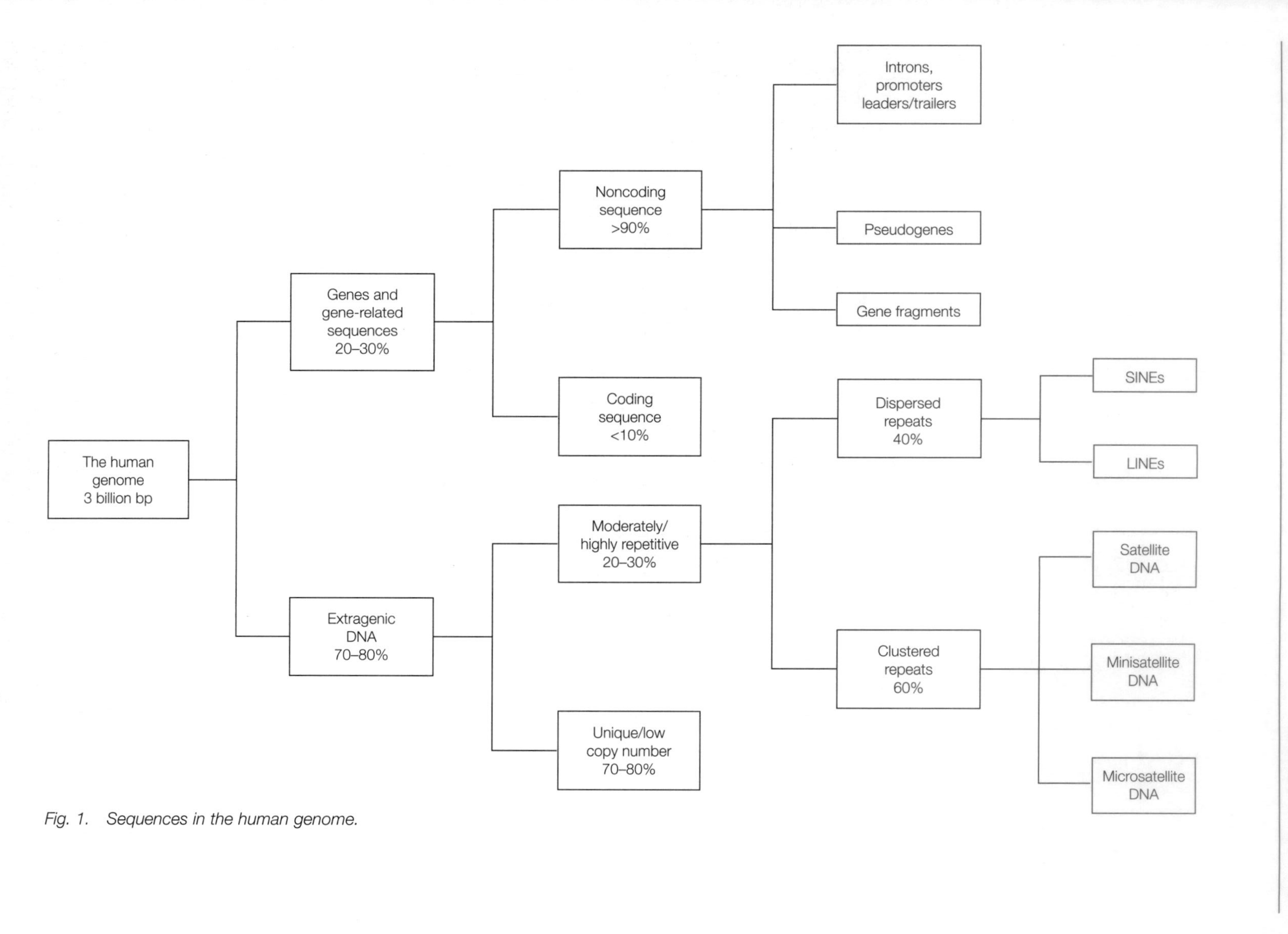

Fig. 1. Sequences in the human genome.

Pseudogenes

These are diverged members of gene families that have acquired one or more inactivating mutations so that they are no longer able to function and do not produce biologically active protein. A pseudogene is simply a mutated version of a parent gene. Often the mutations present are nonsense mutations which generate stop codons and result in premature termination of translation. Related to pseudogenes are **processed pseudogenes** which are DNA copies of mRNA. Processed pseudogenes lack introns and because they do not have a promoter they are not transcribed and do not result in protein production. It is not clear how processed pseudogenes arose in the human genome but they may have been derived by reverse transcription of mRNA into double-stranded DNA which was then inserted into the genome. Another group of inactive genes are **gene fragments** which lack the 5′ or 3′ region of the parent gene. These are thought to have arisen by a deletion event or by recombination that split the parent gene.

Extragenic DNA

This part of the human genome is composed of sequences that exist in addition to the genes and gene-related sequences described above (*Fig. 1*). Extragenic DNA is composed of sequences that are not part of a gene (exons and introns), not associated with a gene (leader and trailer sequences, promoters and distant regulatory elements) and not a pseudogene or a gene fragment. Although extragenic sequences account for most of the DNA in the human genome (70–80%), they have no known function. Most of the extragenic DNA sequences (70–80%) are unique or exist as a small number of copies. The remainder (20–30%) are moderately or highly repeated sequences that may be dispersed throughout the genome or lined up end on end as long tandem arrays.

Dispersed repetitive sequences

Two types of dispersed repetitive sequence exist known as short and long interspersed nuclear elements, abbreviated to **SINEs** and **LINEs**. The best known example of SINEs are **Alu elements**. These sequences are not identical but they are similar enough to be classed as a family. They have an average length of 250 bp and about 700 000 copies exist which are very widely dispersed throughout the genome occurring in most places including the introns of some genes. The origin of Alu elements is uncertain but they are believed to have arisen as one or more processed pseudogenes. It is suggested that at one stage these sequences acquired transposable activity which allowed them to replicate and move to different sites in the genome.

LINEs are similar to SINEs but have longer sequences. A well known example is the **L1 LINE** which is 6500 bp long and is present as 60 000 copies. LINEs are a type of **retroelement**. These are sequences that are capable of moving through the genome by a process called **transposition** which allows them to copy themselves by reverse transcription and to insert the copy into the genome at a distant site (*Fig. 2*). Most L1 elements are truncated, but the full length version has all the components needed for transposition including a gene that could code for reverse transcriptase.

Clustered repetitive sequences

The human genome contains extensive regions in which repetitive sequences are arranged end to end as long tandem arrays. This is called **satellite DNA** and occurs in three forms depending on the length of the cluster of repetitive sequences (*Fig. 1*). Satellite DNA was the first type to be identified in the human genome and consists of clusters of repetitive sequences of between 100 and 5000 kbp in length. **Minisatellite DNA** contains clusters of repetitive sequences

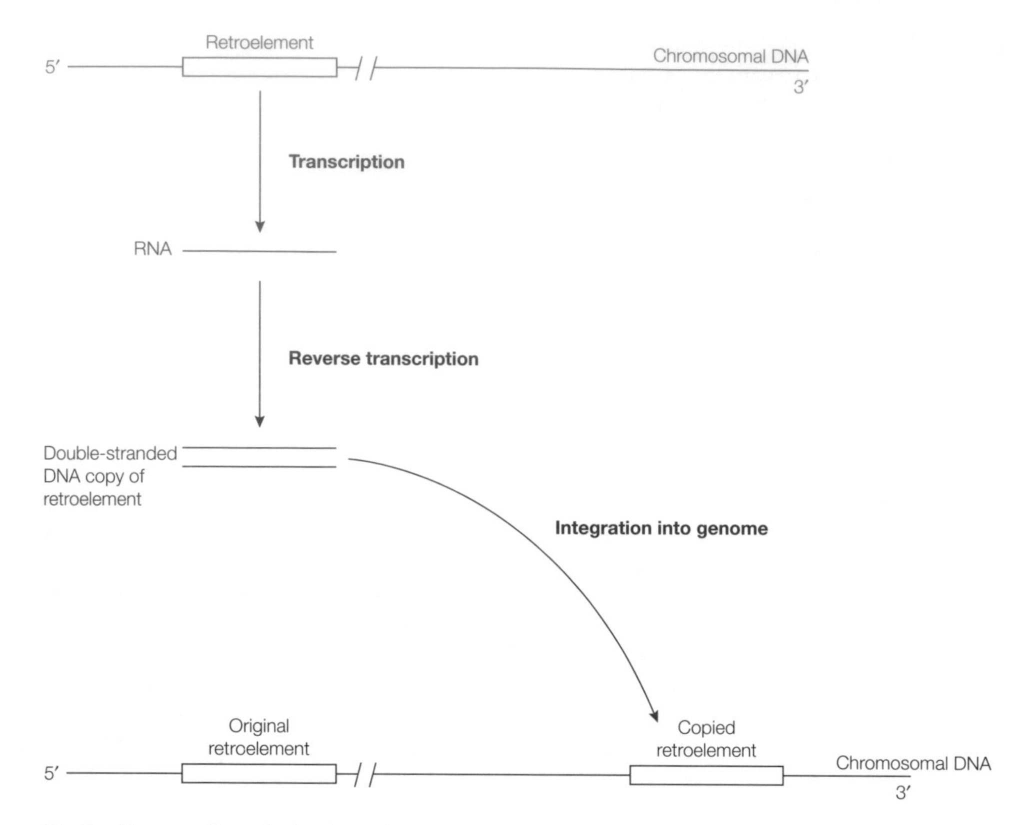

Fig. 2. Transposition of retroelements.

that are shorter, between 100 bp and 20 kbp in length. In the case of the third form, **microsatellite DNA**, the length of the clusters overlaps with minisatellite DNA but microsatellite DNA is distinguished by the short length of the repeated sequence which is rarely more than 4 bp. Dinucleotide repeats such as the **CA repeat** are very common and account for 0.5% of the entire genome. Mononucleotide repeats consisting of a single repeated base account for another 0.3%.

Variable number tandem repeats (VNTRs)

Clustered repetitive sequences such as CA repeats frequently show variations in the number of repeats at a given locus. Variations occur between individuals and between pairs of homologous chromosomes carried by one individual. As such, these sequences are referred to as **variable number tandem repeats (VNTRs)**. Repeat sequences amplified by polymerase chain reaction (PCR) using primers specific for unique sequences flanking the repeat produce amplified products of different lengths, depending on the number of times the repeated sequence occurs (*Fig. 3*). The length of the amplified DNA is used to identify individuals and to distinguish between chromosome pairs carried by one individual. VNTRs have a number of applications. For example, they are used in forensic science to identify individuals using biological material recovered from the scene of a crime (see Topic F6) and they can also be used in medical genetics to identify individuals who are carriers of genetic diseases (see Topic F1).

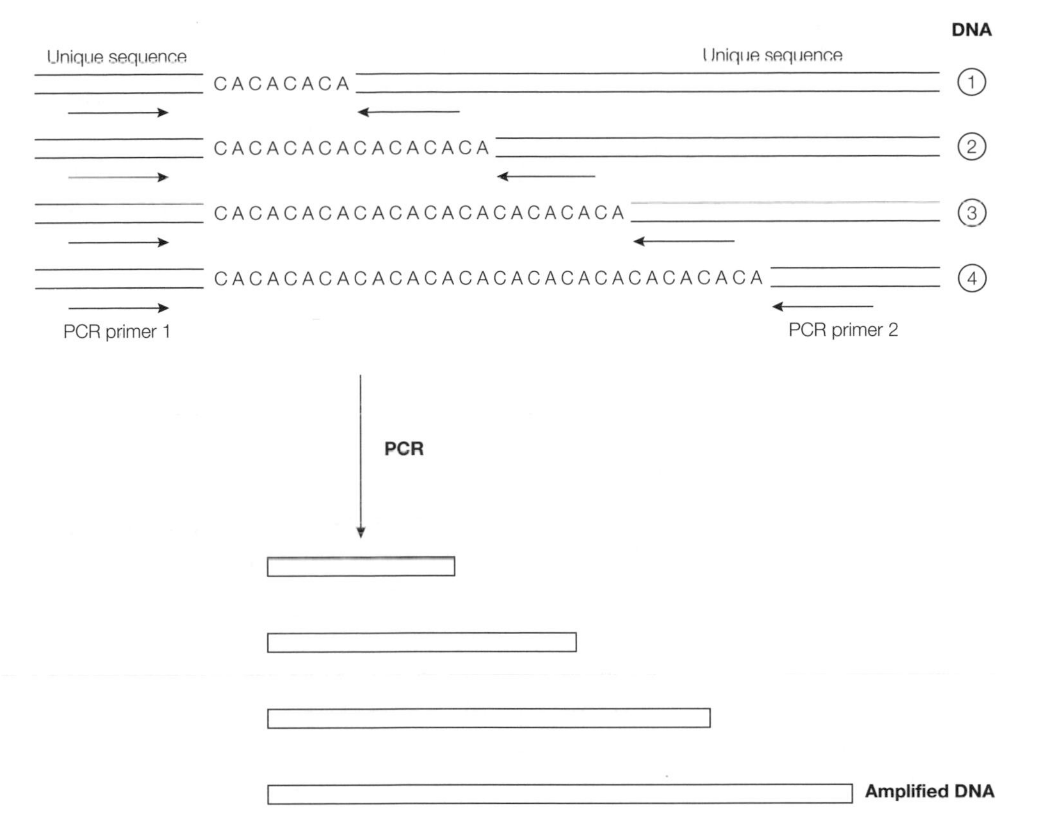

Fig. 3. VNTRs. CA is repeated 4–16 times. PCR using primers specific for flanking unique sequence produces amplified DNA of varying lengths.

B5 DNA MUTATION

Key Notes

Mutations

Mutations are alterations to the usual DNA sequence of an organism that result from the action of chemical and physical agents or errors of DNA replication. Mutations are perpetuated by cell division. The nature of a mutation and its effect on an organism are described by the genotype and phenotype. Organisms may have wild-type or mutant phenotypes. Point and gross mutations exist. Only mutations in the coding regions of genes are likely to affect protein function.

Point mutations

These involve the alteration of a single base. Missense mutations change a single encoded amino acid. Nonsense mutations create stop codons and produce shortened polypeptides. Frameshift mutations involve insertion or deletion of a base producing an altered reading frame. Silent mutations occur at the third base of a codon and do not change the encoded amino acid. Silent mutations accumulate in the DNA as polymorphisms. Point mutations have varying consequences for the biological activity of the encoded protein.

Gross mutations

These involve alteration of longer DNA sequences. Deletion mutations involve the removal of bases. The deleted sequence may vary from a single base to an entire gene sequence. Insertion mutations involve the addition of extra bases. Rearrangements involve segments of DNA sequence within or outside a gene exchanging position with each other. Gross mutations usually result in a complete loss of the biological activity of the encoded protein.

Mutation and disease

DNA mutation is the underlying cause of genetic diseases and of cancer. Genetic diseases are caused by inherited mutations. Usually a single gene is involved. The mutation arises in a germ cell which results in an individual who carries the mutation in all of their cells and will pass it on to subsequent generations. Mutations associated with the development of cancer occur in somatic cells and often in genes that control cell division.

Mutations at the level of the organism

Mutations can be defined in terms of the phenotype they produce. This usually refers to bacterial mutants. Auxotrophic mutants fail to synthesize essential metabolites. Temperature-sensitive mutants fail to grow when the temperature is raised. Antibiotic resistant mutants survive in the presence of antibiotics. Regulatory mutants loose the ability to regulate the expression of genes or operons.

Related topics

The genetic code (A3)
Genetic diseases (F1)
Genes and cancer (F2)

Mutations

The DNA sequence of a gene determines the amino acid sequence of its encoded protein. It is very important that the DNA sequence is preserved because alterations to the amino acid sequence may affect the ability of the protein to function, which in turn may have a deleterious effect on the organism. Alterations to the DNA sequence do occur as a result of the action of a number of chemical and physical agents on DNA and also due to rare errors in DNA replication. These changes are known as **mutations**. Once introduced, the DNA sequence changes are made permanent by DNA replication and are passed on to daughter cells following cell division.

Two important terms that describe an organism carrying a mutation are **genotype** and **phenotype**. Genotype is used to describe the mutation and the gene it occurs in. Phenotype describes the effect on the organism of the mutation. An organism that displays the usual phenotype for that species is called the **wild-type**. An organism whose usual phenotype has changed as the result of a mutation is called a **mutant**. Mutations occur in two forms: **point mutations** which involve a change in the base present at any position in a gene and **gross mutations** which involve alterations of longer stretches of DNA sequence. The location of the mutation within a gene is important. Only mutations that occur within the coding region are likely to affect the protein. Mutations in noncoding or intergenic regions do not usually have an effect.

Point mutations

Point mutations fall into a number of categories, each with different consequences for the protein encoded by the gene.

Missense mutations

These point mutations involve the alteration of a single base which changes a codon such that the encoded amino acid is altered (*Fig. 1a*). Such mutations usually occur in one of the first two bases of a codon. The redundancy of the genetic code means that mutation of the third base is less likely to cause a change in the amino acid. The effect of a missense mutation on the organism varies. Most proteins will tolerate some change in their amino acid sequence. However, alterations of amino acids in parts of the protein that are important for structure or function are more likely to have a deleterious effect and to produce a mutant phenotype.

Nonsense mutations

These are point mutations that change a codon for an amino acid into a termination codon (*Fig. 1b*). The mutation causes translation of the messenger RNA to end prematurely resulting in a shortened protein which lacks part of its carboxyl-terminal region. Nonsense mutations usually have a serious effect on the activity of the encoded protein and often produce a mutant phenotype.

Frameshift mutations

These result from the insertion of extra bases or the deletion of existing bases from the DNA sequence of a gene. If the number of bases inserted or deleted is not a multiple of three the reading frame will be altered and the ribosome will read a different set of codons downstream of the mutation substantially altering the amino acid sequence of the encoded protein (*Fig. 1c*). Frameshift mutations usually have a serious effect on the encoded protein and are associated with mutant phenotypes.

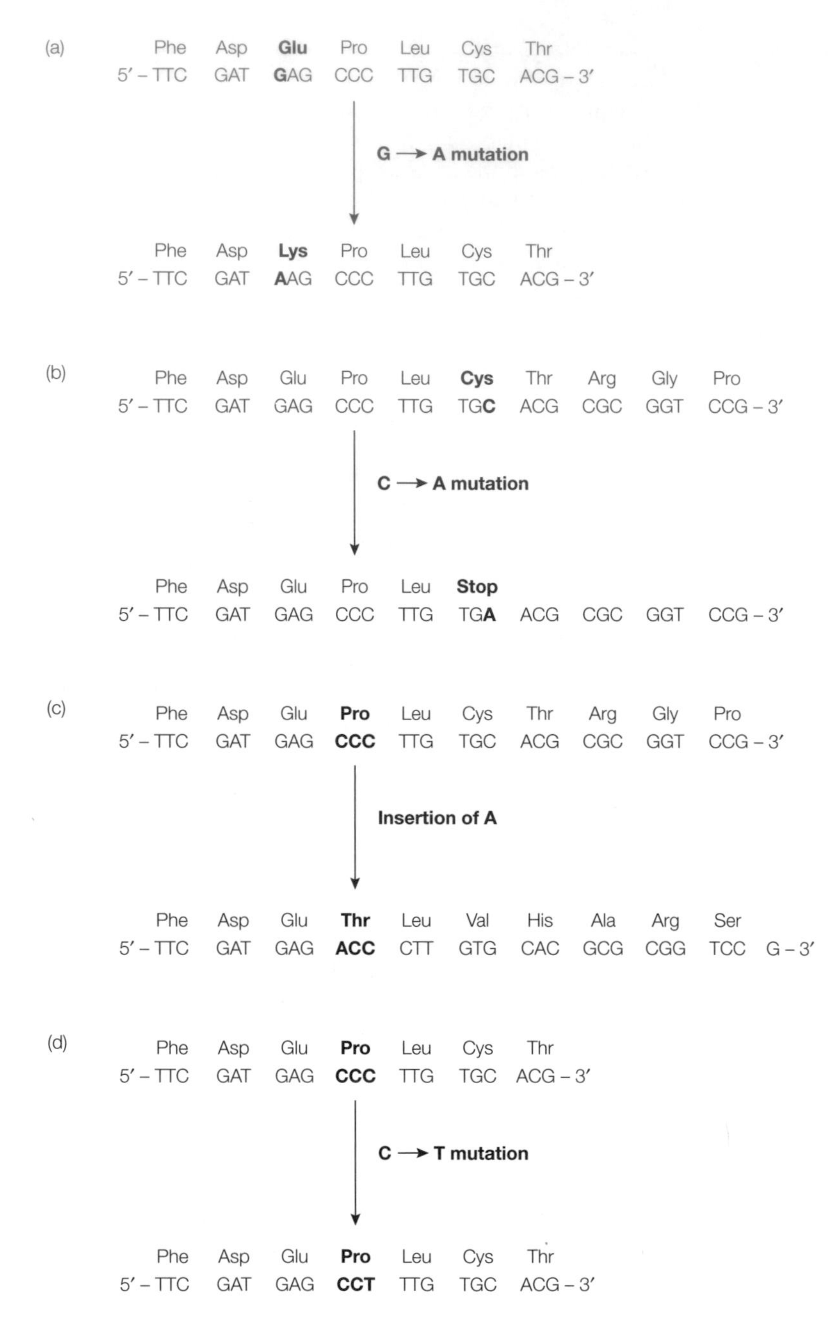

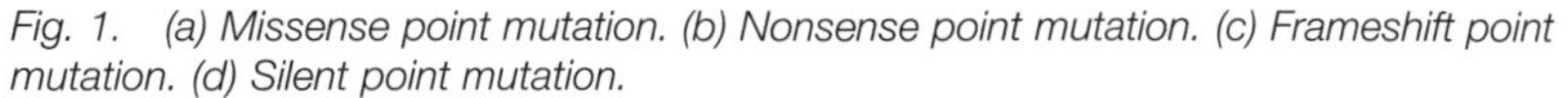

Fig. 1. (a) Missense point mutation. (b) Nonsense point mutation. (c) Frameshift point mutation. (d) Silent point mutation.

Silent mutations

Mutations may occur at the third base of a codon and, due to the degeneracy of the genetic code, the amino acid will not be altered (*Fig. 1d*). Silent mutations have no effect on the encoded protein and do not result in a mutant phenotype. They tend to accumulate in the DNA of organisms where they are known as **polymorphisms**. They contribute to variability in the DNA sequence of individuals of a species.

Point mutations that involve replacement of a purine with a pyrimidine or vice-versa are known as **transversions**. Replacements involving two purines or two pyrimidines are **transitions**.

Gross mutations

Gross mutations cause substantial alterations to the DNA often involving long stretches of sequence. A number of gross mutations can occur.

Deletions

These involve the loss of a portion of the DNA sequence (*Fig. 2a*). The amount lost varies greatly. Deletions can be as small as a single base or much larger – in some cases corresponding to the entire gene sequence.

Insertions

In this case the mutation occurs as a result of insertion of extra bases, usually from another part of a chromosome (*Fig. 2b*). As for deletions, the amount inserted may be one or two bases or may be much larger.

Rearrangements

These mutations involve segments of DNA sequence within or outside a gene exchanging position with each other (*Fig. 2c*). A simple example is inversion mutations in which a portion of the DNA sequence is excised then re-inserted at the same position but in the opposite orientation.

Gross mutations, because they involve major alterations to gene sequences, invariably have a serious effect on the encoded protein and are frequently associated with a mutant phenotype.

Mutation and disease

In humans and other higher organisms DNA mutation plays a significant role in the development of disease. Specifically, mutation is the underlying cause in genetic diseases such as hemophilia and cystic fibrosis and also in the development of cancers.

Genetic diseases are caused by mutations that are inherited and are passed on from parents to offspring. Mutations can occur at random in any cell but normally a mutation in a single cell has no effect on the organism because cells are continually being replaced. However, a mutation in a germ cell (sperm or ovum) can be passed on after conception and will be present in every cell of the resulting offspring which then becomes a carrier of the mutation and can in turn pass the mutation on to subsequent generations. Genetic diseases usually involve mutation of a single gene. The mutated genes vary widely; this accounts for the diverse nature of genetic diseases.

Mutations associated with the development of cancer occur in the normal cells of the body (somatic cells). Cancers often develop in association with mutations in genes that regulate cell division. Such mutations often confer the ability for cells to divide in an uncontrolled way leading to the development of a tumor.

Mutations at the level of the organism

In addition to defining mutations in terms of alterations to the DNA sequence, it is also possible to use definitions that describe the phenotype of the mutant organism. This approach applies to both prokaryotic and eukaryotic cells but is most commonly associated with bacterial mutants. Several mutant phenotypes have been defined.

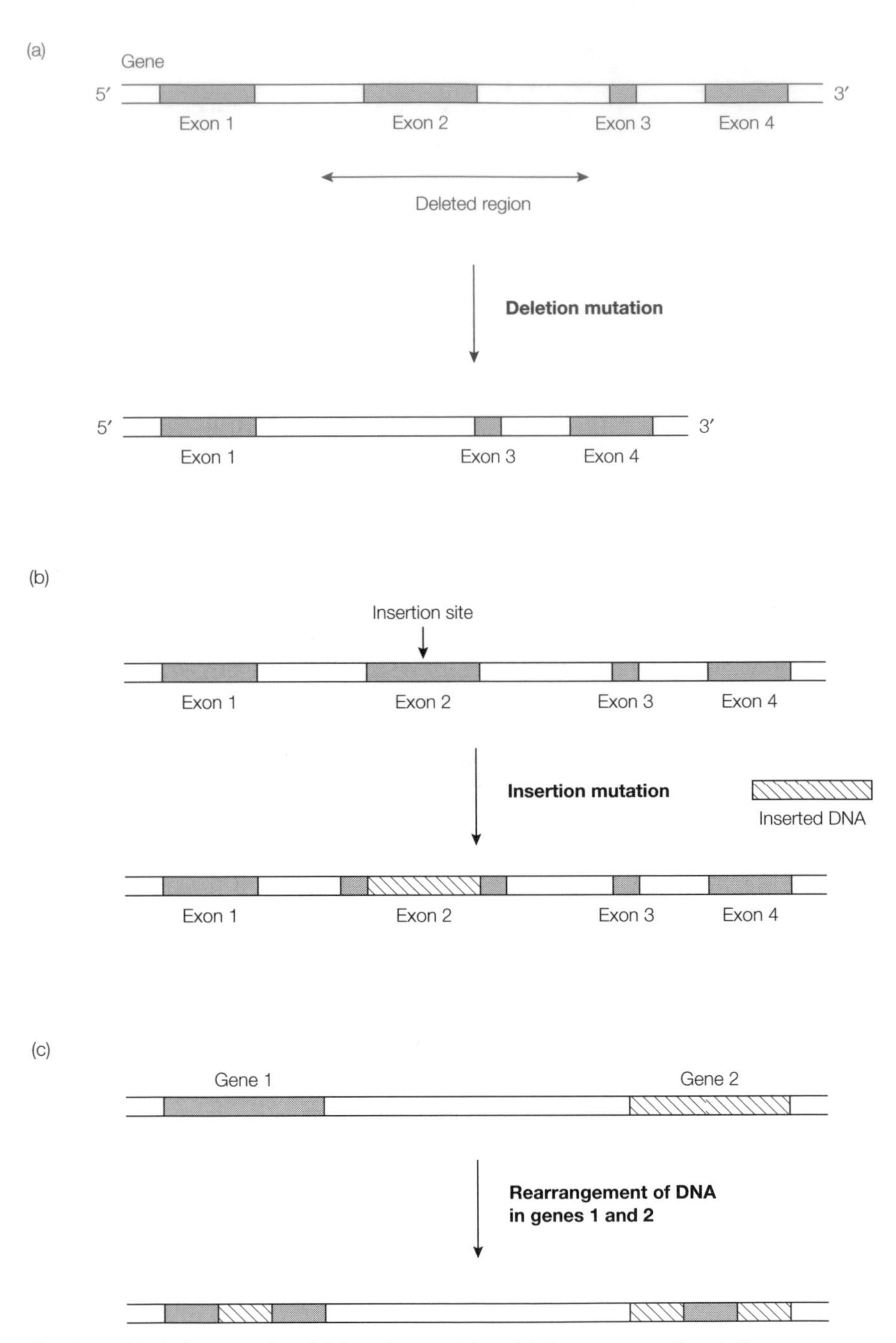

Fig. 2. (a) Deletion mutation. (b) Insertion mutation. (c) Rearrangement mutation.

Auxotrophic mutants

These mutants lack a gene product involved in the synthesis of an essential metabolite, such as an amino acid. They can only be isolated by being grown in medium supplemented with the metabolite. For example, the tryptophan auxotroph of *E. coli* cannot synthesize tryptophan for itself and must obtain it from the growth medium. The opposite of an auxotroph is a **prototroph** which has no special nutritional requirements.

Temperature-sensitive mutants

These mutants are limited by the conditions used to grow the cells and can only survive when cultured at a given temperature. If the temperature is raised, the cells are unable to grow and they die. The mutation carried by a temperature-sensitive mutant usually affects an amino acid that is important for maintaining the structure of the protein. When the temperature is raised the mutant protein becomes denatured and looses its activity resulting in the mutant phenotype.

Antibiotic resistant mutants

Antibiotics kill wild-type bacteria but have no effect on resistant mutants. Antibiotic resistance can arise in several ways but often involves mutation of a gene encoding a protein that is a target for the antibiotic.

Regulatory mutants

In this case mutants loose the ability to regulate expression of a gene or operon. For example, regulatory mutants of *E. coli* exist which express genes required for the metabolism of lactose even when lactose is absent from the medium. These are called **constitutive mutants** and they usually arise by mutation of the *lac* repressor protein.

B6 MUTAGENS AND DNA REPAIR

Key Notes

Mutagens

Mutations arise due to rare errors in DNA replication or by the action of chemical and physical agents called mutagens on DNA. Changes in the structure of nucleotides cause altered base pairing which become permanent after DNA replication. Other mutagens produce physical distortions in the DNA that block replication or transcription.

Chemical mutagens

Many different chemicals act as mutagens. Base analogs substitute for normal bases during DNA replication and cause mutation by having altered base – pairing patterns. Intercalating agents slip between the bases in the double helix. They cause the insertion of an extra base during replication producing a frameshift mutation. Many chemical mutagens modify bases, often by the addition of alkyl or aryl groups or by deamination. DNA also undergoes spontaneous mutation by reaction with normal chemical species in cells. Reactive oxygen species (ROS) present in aerobic cells also damage bases.

Physical mutagens

Ionizing radiation in the form of X-rays and γ-rays damage DNA molecules extensively. Ultraviolet radiation is absorbed by bases and leads to the formation of cyclobutyl dimers between adjacent pyrimidine bases. Heat is a significant mutagen.

DNA Repair

The presence of numerous agents that mutate DNA has led organisms to develop extensive DNA repair mechanisms. The repair mechanisms are complex but essentially three main types occur: excision repair, direct repair and mismatch repair.

Excision repair

Single-strand nicks are created in the DNA adjacent to a damaged nucleotide by a repair enzyme. A nuclease removes the damaged base and adjacent bases. The gap is then filled with new DNA by a DNA polymerase and closed by DNA ligase.

Direct repair

This mechanism involves the reversal of structural alterations that occur in nucleotides. Photoreactivation is an important example which involves repair of thymine dimers by enzymes called DNA photolyases which break the links formed on dimerization.

Mismatch repair

This system corrects errors of DNA replication by identifying mismatched nucleotides and replacing the incorrectly inserted base. The system determines which of the mismatched bases is correct by identifying the parental DNA strand which is methylated.

Related topics

DNA structure (A1)
DNA replication (A9)
DNA mutation (B5)

Mutagens

Mutations can arise due to errors introduced during DNA replication. These are normally very rare, occurring in bacteria about once in every 10^{10} bases incorporated and probably occur more frequently in higher organisms. This is called the **spontaneous mutation rate**. The rate of mutation increases when cells are exposed to chemical or physical agents known as **mutagens** that interact directly with the DNA and alter the structure of individual nucleotides. This may lead to an alteration or a failure of base-pairing such that when DNA replication occurs an incorrect base is inserted opposite the modified base changing the DNA sequence. Subsequent rounds of replication make the change permanent leaving the DNA sequence mutated (*Fig. 1*). Some mutagens work in a different way and cause serious physical distortions in the DNA which block DNA replication or transcription.

A wide variety of natural and synthetic, organic and inorganic chemicals can react with DNA altering its structure and causing mutation. Most chemical mutagens are **carcinogenic** and cause cancer. In addition, a range of physical agents can cause mutation. These include ionizing radiation in the form of X-rays and γ-rays, nonionizing radiation, particularly ultraviolet light, and also heat.

Chemical mutagens

A wide variety of chemicals can cause mutation of DNA. Some chemical mutagens are **base analogs**. These are structurally similar to the normal bases found in DNA and can be incorporated by DNA polymerases into the DNA during replication. They cause mutation by producing altered base-pairing. An example is **5-bromouracil (5BU)**, a base analog derived from thymine. 5BU normally base-pairs with adenine; however, it can undergo a slight change in its structure called a **tautomeric shift** which causes it to base-pair with guanine. After DNA replication, the original TA base-pair is replaced by GC on one of the daughter strands leading to a point mutation (*Fig. 2*).

Other chemical mutagens are **intercalating agents**. These are flat molecules that disrupt DNA replication by slipping between adjacent base pairs of the double helix. An example is **ethidium bromide,** a molecule containing four rings that has dimensions similar to a purine–pyrimidine base pair. Ethidium bromide is said to intercalate into the double helix causing adjacent base pairs to move apart slightly (*Fig. 3*). It is not clear exactly how this disrupts replication but the overall effect is to cause the insertion of a single nucleotide at the intercalation position causing a frameshift mutation in the gene.

Many mutagens act by chemically modifying bases. Some cause the addition of alkyl or aryl groups. Examples include **methylmethane sulfonate** and **ethylnitrosourea** which add methyl groups to bases at a variety of positions (*Fig. 4*). Other base modifications include **deamination**. For example, **nitrous acid** deaminates cytosine to produce uracil (*Fig. 5*) which base pairs with adenine resulting in an alteration of the base pair from GC to AT after replication. Deamination of adenine to a guanine analog called hypoxanthine causes an AT base pair to be replaced by GC.

Mutations can also occur spontaneously due to the chemical reactivity of DNA with some of the normal chemical species in cells. For example, cytosine undergoes spontaneous deamination to uracil. In addition, cytosine, which is sometimes present as 5-methylcytosine, may be deaminated to produce thymine. A very common modification which occurs spontaneously is **depurination** which results from breakage of the link between a deoxyribose sugar and its purine base (*Fig. 6*). Damage to bases can also occur due to the presence of

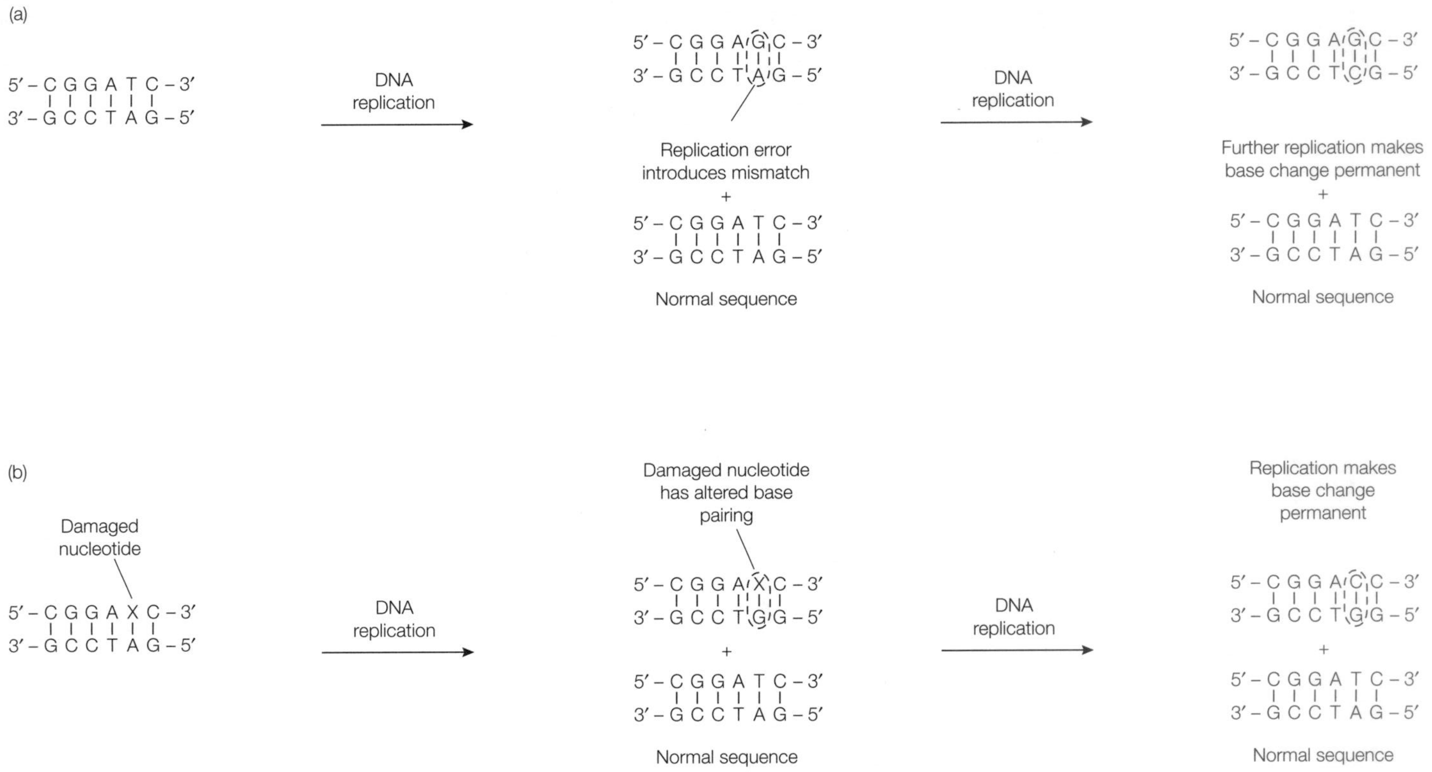

(a)

5′ – C G G A T C – 3′
3′ – G C C T A G – 5′

DNA replication →

5′ – C G G A G C – 3′
3′ – G C C T A G – 5′

Replication error introduces mismatch

+

5′ – C G G A T C – 3′
3′ – G C C T A G – 5′

Normal sequence

DNA replication →

5′ – C G G A G C – 3′
3′ – G C C T C G – 5′

Further replication makes base change permanent

+

5′ – C G G A T C – 3′
3′ – G C C T A G – 5′

Normal sequence

(b)

Damaged nucleotide

5′ – C G G A X C – 3′
3′ – G C C T A G – 5′

DNA replication →

Damaged nucleotide has altered base pairing

5′ – C G G A X C – 3′
3′ – G C C T G G – 5′

+

5′ – C G G A T C – 3′
3′ – G C C T A G – 5′

Normal sequence

DNA replication →

Replication makes base change permanent

5′ – C G G A C C – 3′
3′ – G C C T G G – 5′

+

5′ – C G G A T C – 3′
3′ – G C C T A G – 5′

Normal sequence

Fig. 1. Mechanisms of mutation. (a) Mutations arise due to errors introduced during DNA replication. (b) Mutations arise due to structural changes in nucleotides leading to altered base pairing.

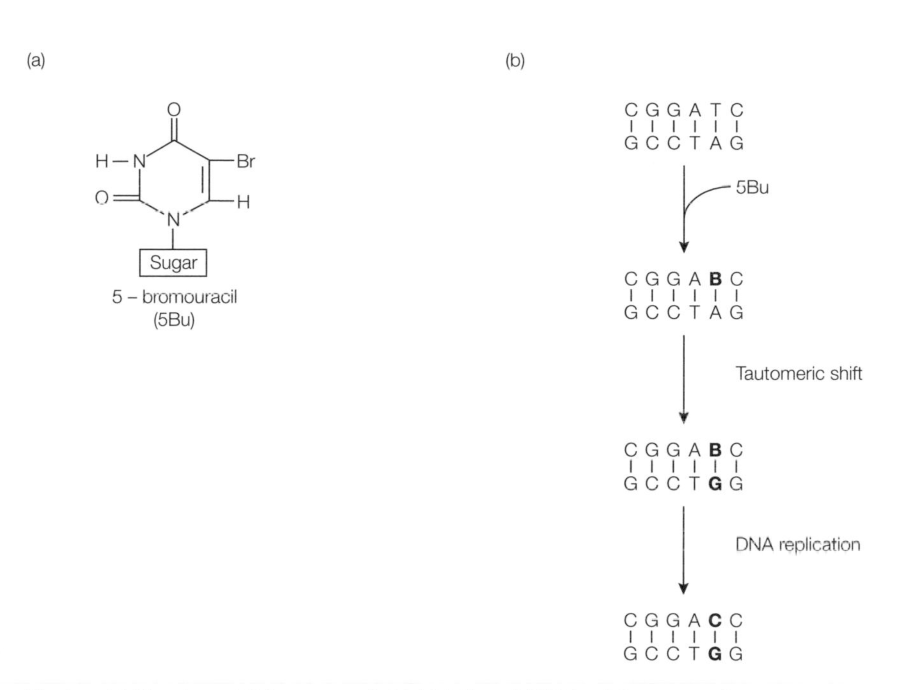

Fig. 2. *(a) Structure of 5-bromouracil. (b) Mutation of DNA by 5-bromouracil.*

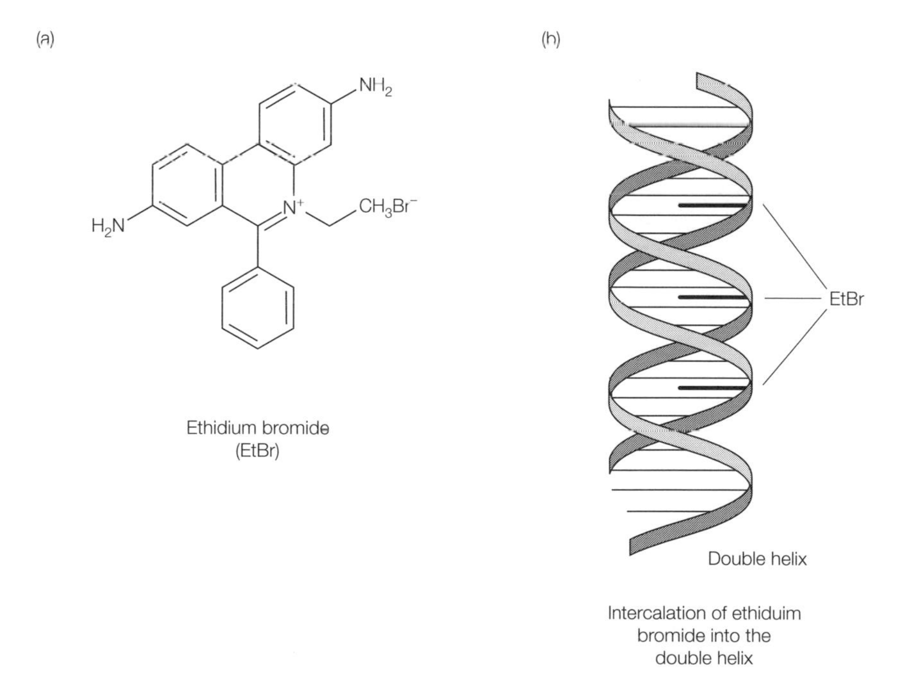

Fig. 3. *(a) Structure of ethidium bromide. (b) Mutagenic effect of ethidium bromide.*

reactive oxygen species (ROS) which occur in all aerobic cells. These include the superoxide, hydrogen peroxide and hydroxyl radicals.

Physical mutagens

High energy ionizing radiation, such as X-rays and γ-rays cause extensive damage to DNA molecules producing strand breaks and the destruction of sugars and bases. Nonionizing radiation in the form of ultraviolet (UV) light is absorbed by bases and can induce structural changes. In particular, UV light can cause the formation of structures called **cyclobutyl dimers** between adjacent pyrimidines, especially thymines (*Fig. 7*). Dimerization causes the bases to stack closer together and can result in deletion mutations following DNA replication. Heat is also a significant environmental mutagen. Its effect on DNA molecules is to produce apurinic sites in the DNA polynucleotide which can cause point or deletion mutations when the DNA is replicated.

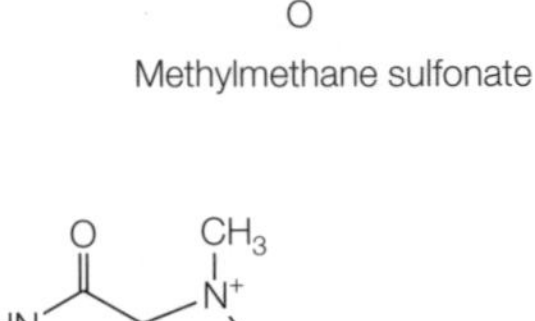

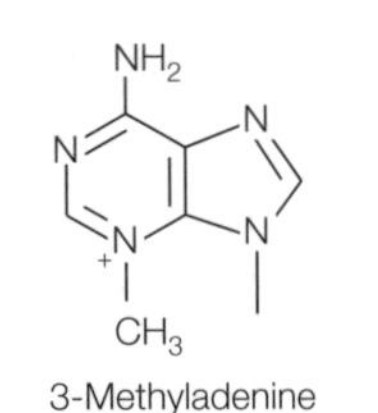

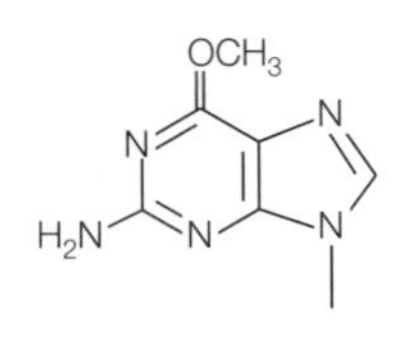

Fig. 4. Examples of (a) alkylating agents and (b) alkylated bases.

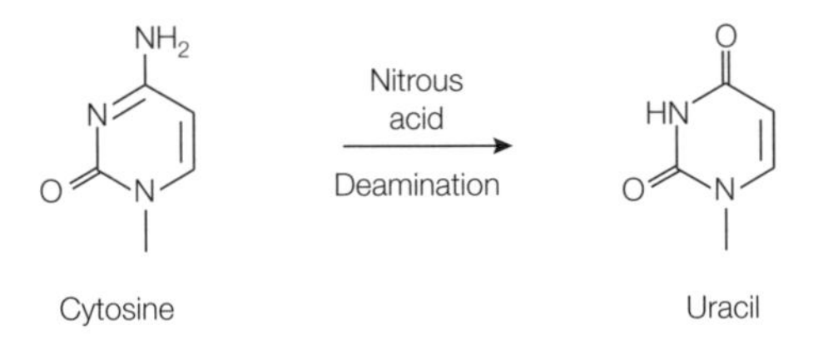

Fig. 5. Deamination of cytosine to uracil by nitrous acid.

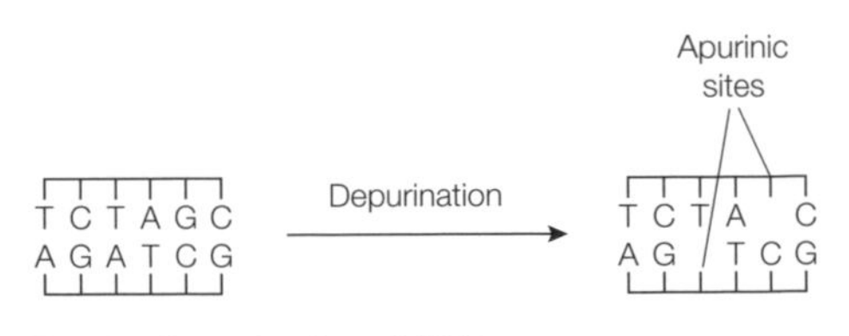

Fig. 6. Depurination of DNA.

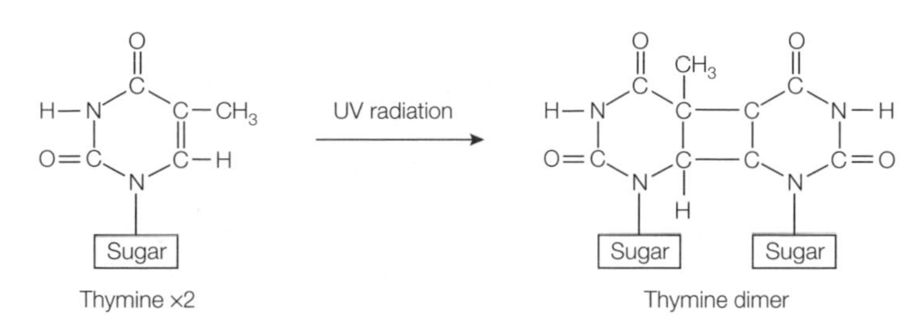

Fig. 7. Formation of thymine dimers by UV radiation.

DNA repair

The requirement that the DNA sequence of a gene is preserved despite the presence of numerous agents that can mutate DNA has caused organisms to develop methods which prevent mutation by repairing damage to DNA. The repair mechanisms are complex but essentially three main types occur.

Excision repair

This is a complex system which is probably the most common form of DNA repair. Many different types of damage are repaired including pyrimidine

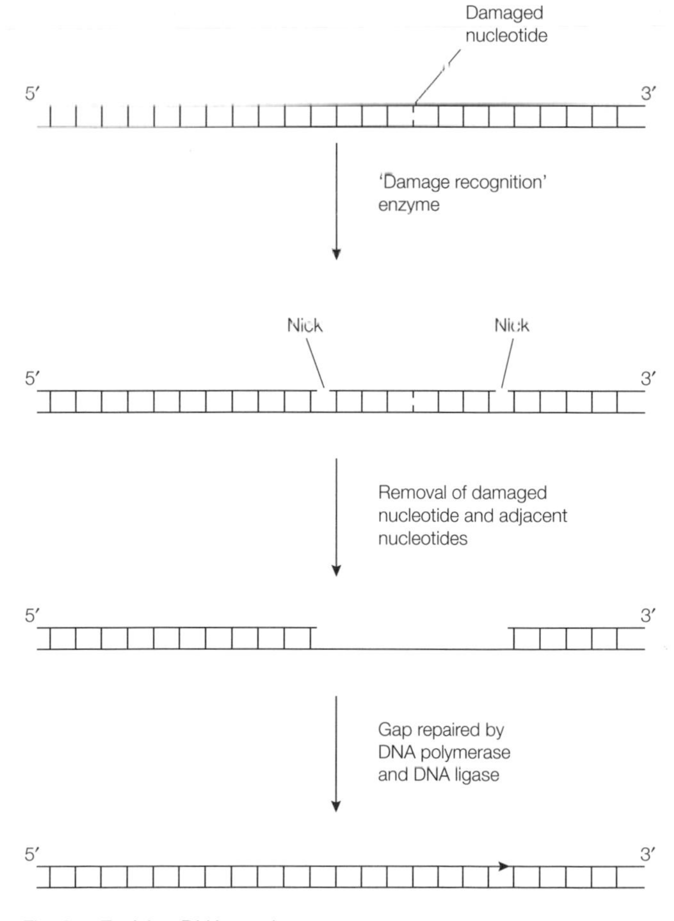

Fig. 8. Excision DNA repair.

dimers. Initially, one of a number of enzymes recognizes nucleotides that are damaged and marks them for repair. The mark can take the form of a nick in one of the strands of the double helix adjacent to the area of damage or a damaged base may be removed leaving a gap. In the next stage, a nuclease removes the marked nucleotide as well as a number of its neighbors. A DNA polymerase (DNA polymerase I in *E. coli*) then synthesizes new DNA to replace the missing bases and DNA ligase joins the new DNA to the existing molecule restoring the DNA to its original structure (*Fig. 8*).

Direct repair

This is a much less common form of DNA repair which involves reversal of structural alterations that occur in nucleotides. An important example of direct repair is photoreactivation which repairs pyrimidine dimers produced by UV radiation. Enzymes called DNA photolyases are induced by visible light and repair pyrimidine dimers by breaking the links that form on dimerization. DNA photolyases occur in bacteria, microbial eukaryotes and plants.

Mismatch repair

This system corrects errors introduced during DNA replication by identifying mismatched nucleotides. A number of enzymes mark the mismatch or can repair it directly. It is important that the system can determine which of the mismatched bases is correct and it does this by distinguishing between the parental DNA strand which has the correct sequence and the daughter strand containing the mutated sequence. In *E. coli* the parental strand is easily recognized because it is tagged with methyl groups attached to adenine bases.

B7 Recombination

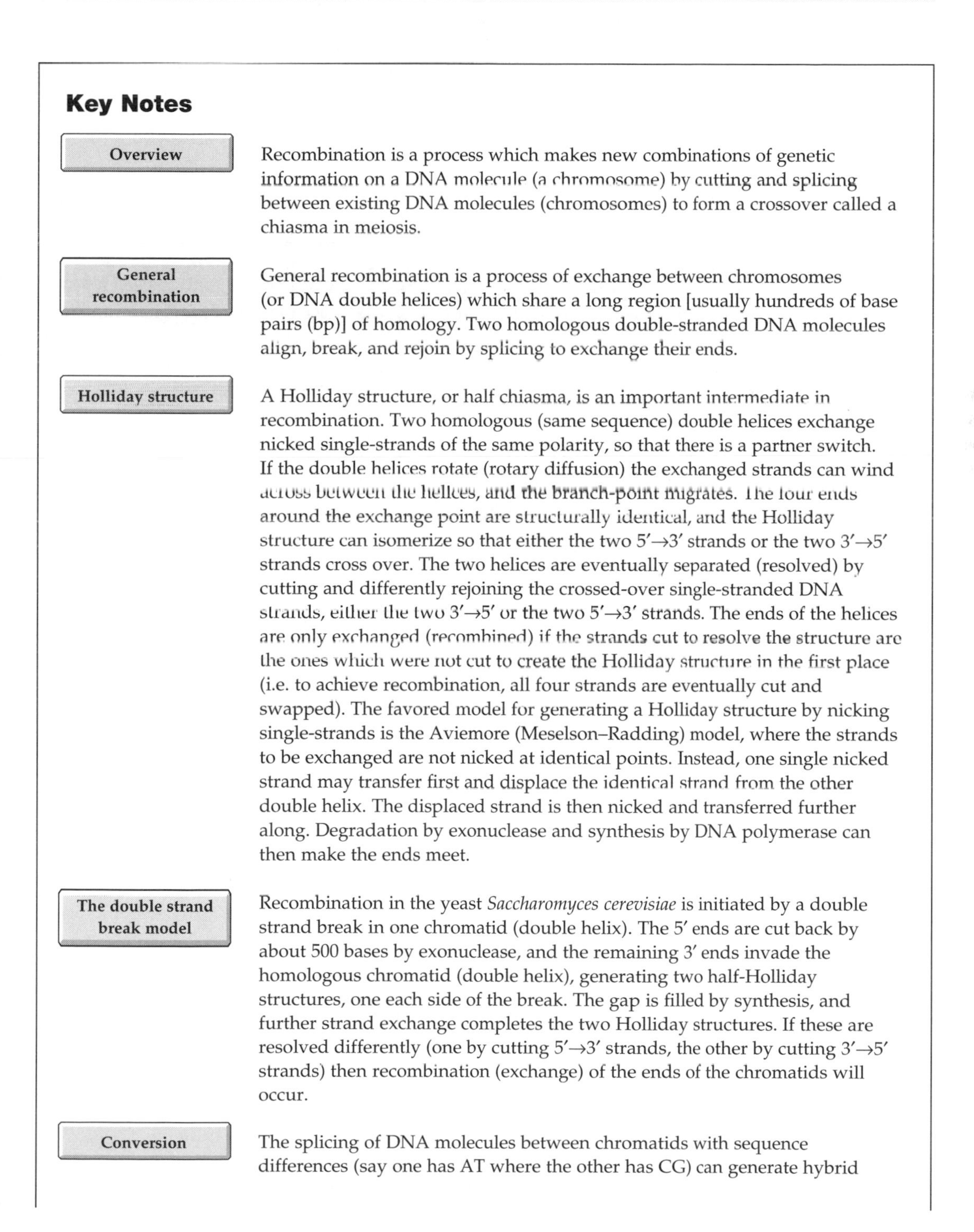

Key Notes

Overview

Recombination is a process which makes new combinations of genetic information on a DNA molecule (a chromosome) by cutting and splicing between existing DNA molecules (chromosomes) to form a crossover called a chiasma in meiosis.

General recombination

General recombination is a process of exchange between chromosomes (or DNA double helices) which share a long region [usually hundreds of base pairs (bp)] of homology. Two homologous double-stranded DNA molecules align, break, and rejoin by splicing to exchange their ends.

Holliday structure

A Holliday structure, or half chiasma, is an important intermediate in recombination. Two homologous (same sequence) double helices exchange nicked single-strands of the same polarity, so that there is a partner switch. If the double helices rotate (rotary diffusion) the exchanged strands can wind across between the helices, and the branch-point migrates. The four ends around the exchange point are structurally identical, and the Holliday structure can isomerize so that either the two 5′→3′ strands or the two 3′→5′ strands cross over. The two helices are eventually separated (resolved) by cutting and differently rejoining the crossed-over single-stranded DNA strands, either the two 3′→5′ or the two 5′→3′ strands. The ends of the helices are only exchanged (recombined) if the strands cut to resolve the structure are the ones which were not cut to create the Holliday structure in the first place (i.e. to achieve recombination, all four strands are eventually cut and swapped). The favored model for generating a Holliday structure by nicking single-strands is the Aviemore (Meselson–Radding) model, where the strands to be exchanged are not nicked at identical points. Instead, one single nicked strand may transfer first and displace the identical strand from the other double helix. The displaced strand is then nicked and transferred further along. Degradation by exonuclease and synthesis by DNA polymerase can then make the ends meet.

The double strand break model

Recombination in the yeast *Saccharomyces cerevisiae* is initiated by a double strand break in one chromatid (double helix). The 5′ ends are cut back by about 500 bases by exonuclease, and the remaining 3′ ends invade the homologous chromatid (double helix), generating two half-Holliday structures, one each side of the break. The gap is filled by synthesis, and further strand exchange completes the two Holliday structures. If these are resolved differently (one by cutting 5′→3′ strands, the other by cutting 3′→5′ strands) then recombination (exchange) of the ends of the chromatids will occur.

Conversion

The splicing of DNA molecules between chromatids with sequence differences (say one has AT where the other has CG) can generate hybrid (heteroduplex) DNA with mismatches (AG, CT). These may be corrected by mismatch repair pathways, converting the sequence on one chromosome to be like the sequence on the other. This gives a nonMendelian ratio, three chromatids can have one sequence, one chromatid still has the other sequence. If the mismatch is not repaired before the next DNA replication phase, the mismatched bases will act as templates, generating different sequences in the two daughter cells after mitosis. This is called post meiotic segregation (pms). Gap-filling synthesis in the double strand break repair model also converts the sequence in the gap.

Recombination enzymes

Most of the enzymes involved in the repair of DNA damage, both excision repair (ultraviolet light induced damage) and double strand break repair (ionizing radiation, X-ray and γ-ray induced damage) also function in recombination in meiosis. There are several enzymes specific to meiosis as well. Deficiencies in some recombination-related enzymes are associated with human disease (e.g. breast cancer 1 gene product and ataxia telangiectasia).

Site specific recombination

Bacteriophage lambda integrates into the host (*E. coli*) genome at a specific 15 bp sequence (*att* or 'O') found on both the bacteriophage and host chromosome. The bacteriophage gene *int* coding for integrase is necessary for this. Excision is also by recombination catalyzed by excisionase.

Related topics

DNA replication (A9)
Mutagens and repair of DNA (B6)
Bacteriophage (B8)
Meiosis and gametogenesis (C3)
Linkage (C4)

Overview

Recombination means making new combinations of genetic information on a DNA molecule (a chromosome). (This is different from reassortment of whole chromosomes, which are separate DNA molecules, into gametes and zygotes.) Recombination mechanisms involve an interaction between two DNA double helices (= two double-stranded (ds) DNAs). The effect is that both dsDNAs break, and different ends rejoin, but the mechanisms are more like splicing events. The event is also described as a **crossover**, visible as a **chiasma** (plural **chiasmata**) at meiosis. A chromatid is the same as one double helix from a genetic viewpoint. Meiotic crossing over (see Topics B3 and B4) involves one double helix (= chromatid) from each of two homologous chromosomes.

General recombination

General recombination is the type of mechanism which eukaryotes use to recombine chromatids in meiosis. Bacteria also use it to incorporate homologous regions of foreign DNA into their chromosome after the DNA enters the cell by conjugation (sex), by transduction (inside a bacteriophage) or by transformation (simply DNA brought into the cell from its surroundings). It is a reciprocal process of exchange between two chromosomes, and in eukaryotes is normally conservative (nothing is gained or lost except for conversions, see below). The effect is that two homologous dsDNA molecules align, break at identical sites, and each broken end rejoins with the matching end from the other molecule, to exchange (recombine) the entire ends of the dsDNA. The molecular event is a splice, not a blunt cut and butt-end join, and sequence information is conserved. In bacteria the DNA which does not end up in a stable circular chromosome will be lost. General recombination requires two dsDNA molecules with a large region of homology (nearly identical sequence), usually hundreds of base pairs long. The minimum length of homology probably varies between different species.

Holliday structure

The most important intermediate in recombination is a **Holliday structure** (named after Robin Holliday, and also called **Holliday junction** or **half chiasma**). This can be formed (theoretically) by making single-strand nicks in two strands with the same polarity, one in each double helix, and exchanging one single-strand from each double helix, again with the same polarity, so that they pair with the intact strand in the other double helix (*Fig. 1*). 3′ Ends are shown being exchanged because these could prime DNA polymerase, but there is no evidence as to which ends cross first. This exchange is effectively a partner switch. In the original model, Holliday proposed that the nicks occurred at identical sites in both double helices (chromatids) as in *Fig. 1*, but genetic evidence suggests that some asymmetrical exchange occurs.

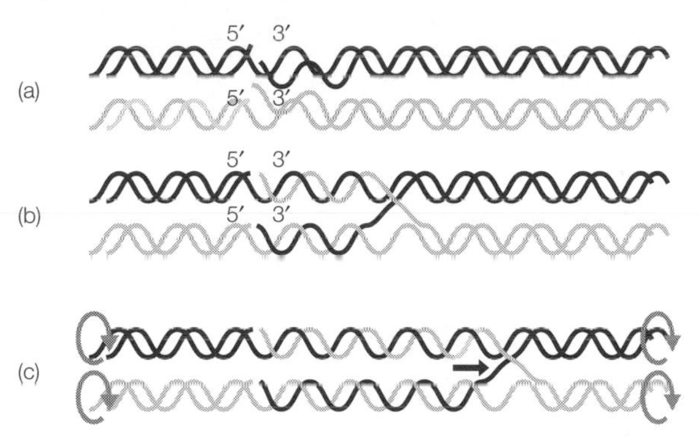

Fig. 1. (a) Formation of a Holliday structure by nicking of single strands and exchange of ends of the same polarity. (b, c) Branch migration by rotary diffusion moving exchange point.

To take account of this asymmetry the model was revised and called the **Aviemore model** after the location of the workshop where it was discussed, or the **Meselson–Radding model** after the publishers of the revision. The major difference from Holliday's model is that one single strand is nicked first and invades the other duplex (double helix). The displaced strand is partially degraded, then transfers into the original donor duplex to complete the Holliday structure (*Fig. 2a*). The single strand gap on the donor is filled by synthesis of DNA by DNA polymerase (dashed line *Fig. 2a*). In this way the transfer is asymmetrical near the original nick. This model is very versatile because any ends that are too long can be digested back by exonuclease, any gaps can be filled by polymerase and free ends can be joined by ligase.

Holliday structures have interesting properties. If the double helices rotate (**rotary diffusion**) in the same direction then both the exchanged single-strands can unwind from one helix and wind onto the other. In this way, the exact point where the strands cross over (the branch point) can move along the chromatids. This is called **branch migration** (*Fig. 1c*). Another feature is that the four double helices going away from the site of the exchanged single-stranded (ss)DNA are structurally identical (*Fig. 2*). The arrangement can undergo **strand isomerization** so that either the two 3′→5′ strands are exchanged or the two 5′→3′ strands (the strands going the other way) are exchanged. Please refer to *Fig. 2* while

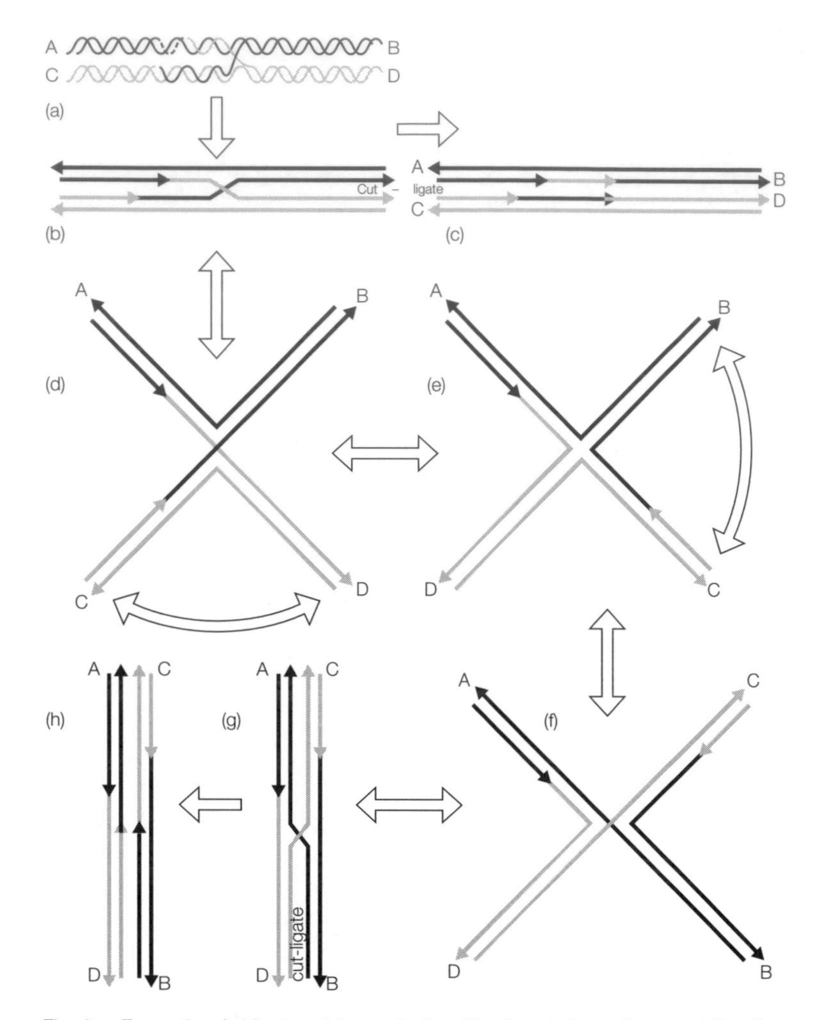

Fig. 2. Formation (a,b), strand isomerization (d,e,f) and alternative resolution (b,c and g,h) of a Holliday junction.

reading this. At the start (*Fig. 2a*) the single strand ends are transferred. This is more easily drawn as straight lines (*Fig. 2b*). In the original arrangement the ends A and B are on one chromatid, ends C and D are on another. The structure can open (Fig 2d), then rotate to interchange ends C and D, uncrossing the cut strands B–C and D–A (*Fig. 2e*). The structure is now open with all four ends equivalent. Interchanging ends B and C now makes the continuous (uncut) strands (A–B and D–C) cross over (*Fig. 2f*). Cutting these strands and ligating their ends gives recombination (a crossover) (*Fig. 2g* and *h*). Now the ends A and D are on one chromatid and the ends C and B on another.

Thus the structure can be **resolved** by cutting two DNA strands of the same polarity and rejoining them differently to separate the two dsDNA helices. If the single strands which were exchanged first are cut and the ends swapped and ligated, then the original dsDNA molecules will be restored without any recombination and the same ends will still be joined together (*Fig. 2c*). If the other two strands are cut and swapped, both single strands in both helices will have been cut and swapped. This is a staggered double strand break–rejoin which will cause **recombination** of the genes on either side, because the ends of the DNA molecules will have been swapped over. That is what recombination is: a re-combination. This can only be understood by following the diagrams in *Fig. 2*, or by making a three-dimensional model from rope or tubing.

The double strand break model

The double strand break model of recombination was developed to explain the observed features of recombination in *S. cerevisiae*, bread yeast. It is quite possible that this model operates in all eukaryotes, and that the Aviemore model never actually occurs in nature. However, there is little evidence relating to the mechanism actually operating in other eukaryotes with larger genomes. Some scientists believe that the greater complexity and size of the genomes of most eukaryotes require a more cautious mechanism of recombination than double strand breaks. Recombination in *S. cerevisiae* is initiated by the generation of a double strand break in one double helix (*Fig. 3a*). The protein which does this, SPO11, is related to a type of topoisomerase II, and forms covalent DNA–protein links across the break. The 5′ ends are digested back from the break to leave

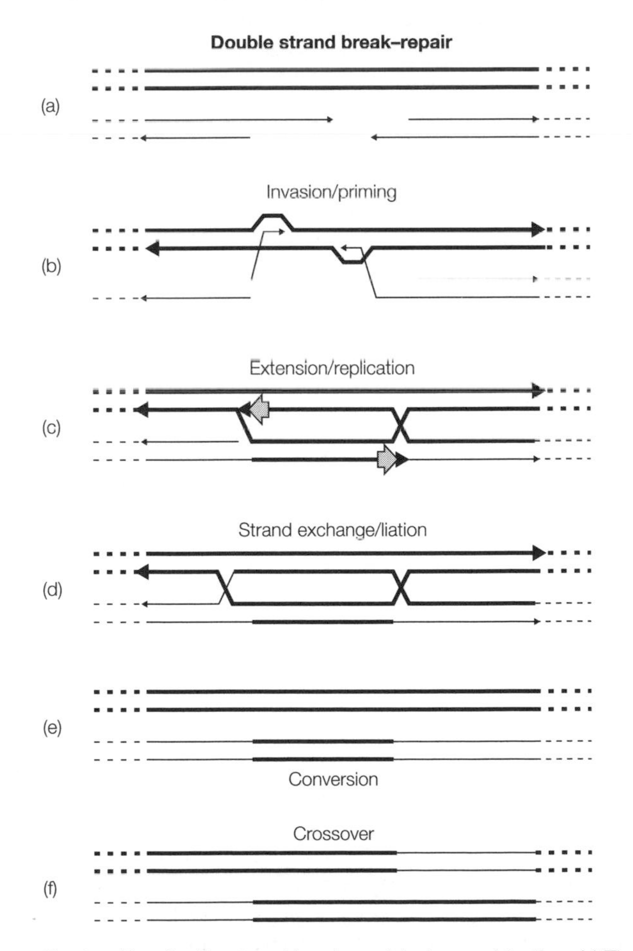

Fig. 3. The double strand break model of recombination. (a) There is a double strand break, then (b) the free ends invade an intact homologue. They are extended across the gap (c, open arrows), and one Holliday structure forms at each side (d). These can resolve to give conversion (e) or recombination (f).

about 500 bp of 3′ ends, detectable biochemically. The model proposes that these 3′ ends invade the intact homologous DNA double helix (*Fig. 3b*), and can prime extension synthesis across the gap by DNA polymerase using the complementary strand of the intact dsDNA as a template (*Fig. 3c*). When the gap is bridged, further strand exchange generates two Holliday structures, one on each side of the original break (*Fig. 3d*). If these are both resolved in the same way (the same single strands are cut/rejoined in each) then there will be no recombination (*Fig. 3e*) but the sequence in the gap in the broken chromatid will have been **converted** to be the same sequence as the other chromosome. If the two Holliday structures are resolved differently, there will be an exchange (crossover) which may be either side of the gap, and the ends of the two double helices will be recombined (*Fig. 3f*).

Conversion

Conversion is a phenomenon which results in nonMendelian ratios in the four haploid cells produced by one meiotic cell. If all genetic information is conserved, then two of the four haploid cells will have copies of one allele and the other two cells will have copies of the other allele, in accordance with Mendel's laws. Sometimes this rule is broken and a **3:1 ratio** occurs. In some fungi there is a second mitotic division which duplicates the haploid products of meiosis, forming eight haploid spores, changing the 3:1 ratio into 6:2. These can also give a **5:3 ratio, aberrant 4:4 ratio** or rarely a **7:1 ratio**. In each case the sister cells from one mitosis are genetically different, an event called **post meiotic segregation** or **pms** (described below). The spliced DNA double helix formed during recombination must have contained a mismatched base pair (*Fig. 4*). (Note that in the asymmetric exchange region only one chromatid of the four can have a mismatch.)

Suppose that one parent chromosome of a pair has an AT base pair, whereas the other (its homolog) has CG. If single strands are exchanged in this region then the **heteroduplexes** (made by one ssDNA strand from each chromosome) will have an AG base pair and a CT base pair. If these **mismatches** are recognized by repair enzymes they may be changed to an AT or a CG base pair, and either mismatch may be repaired either way (or not repaired at all), so three of the four chromatids may contain the same sequence. If a mismatch is not repaired before the next DNA synthesis then the two strands will be used as templates for replication into new duplexes. These will separate at the next mitosis giving daughter cells with different sequences, one AT and the other CG. Because this happens at the mitosis after meiosis it is called **post meiotic segregation**. There are three possibilities on each chromatid, two directions of repair, or no repair, so there are nine possible combinations, one of which restores the parental arrangement. Heteroduplex which could contain mismatches is also shown in *Fig. 2* and *Fig. 3*. Symmetrical exchange without mismatch repair gives the aberrant 4:4 ratios after pms.

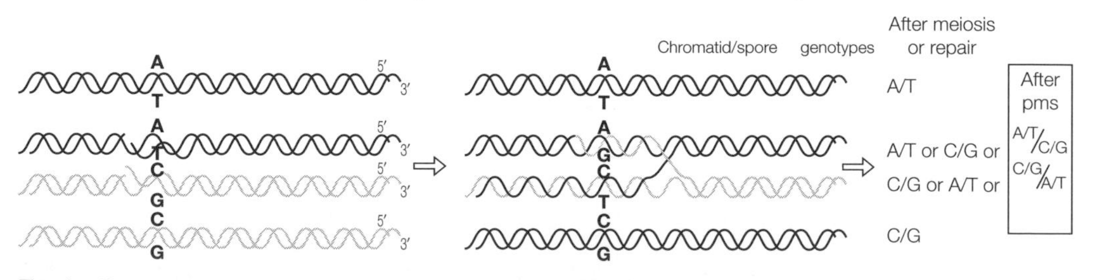

Fig. 4. Gene conversion by mismatch correction in heteroduplex DNA .

Conversion, particularly a 3:1 ratio after meiosis, is easily explained by the double strand break model because the base sequence in the gap in the broken chromatid is always replaced by sequence information from the other homologous chromosome.

Recombination enzymes

Recombination involves many proteins. Several are similar to *E.coli* RecA protein; they bind ssDNA and insert it into homologous dsDNA. The proteins in bacteria, yeast and humans are very similar. Most enzymes involved in DNA repair also function in meiotic recombination. Repair of double strand breaks caused by ionizing radiation is very efficient in yeasts and involves high levels of recombination as the double strand break–repair model predicts. Excision repair enzymes are also widely active in meiotic recombination. Mismatch recognition complexes, which locate errors such as thymine dimers after ultraviolet irradiation, also seem to function in meiosis. Some mismatch repair proteins are present at the stages of pairing and synapsis of the entire homologous chromosomes, and may help to restrict recombination to homologous DNA. Rejection of mismatches may prevent recombination between chromosomes in hybrids between related species, giving post mating isolation of those species (see Topic D7). Others repair proteins localize at recombination events. The genes required for recombination are often essential for normal repair of damage to DNA. When they malfunction, the DNA damage causes mutations which may cause cancer. The normal product of the human breast cancer gene, *Brca1*, is a component of the synaptonemal complex, which has a role in pairing and recombination of chromosomes in meiosis (see Topic C3). The genes which are defective in ataxia telangiectasia are also important in meiotic recombination.

Site specific recombination

Bacteriophage lambda can integrate its DNA into the chromosome of its host *E. coli* staying dormant and being replicated with the host DNA (a state called lysogeny). It does this by recombination between a 15 bp sequence called *att* (for attachment) in the host chromosome and an identical sequence in the circular bacteriophage chromosome (*att* may also be called 'O'). This reaction requires an integrase enzyme (called Int) encoded by the bacteriophage. Excision also occurs by recombination, this time catalyzed by excisionase (coded by the bacteriophage *xis* gene).

B8 BACTERIOPHAGES

Key Notes

Bacteriophages

Bacteriophages (phages) are viruses that infect bacteria. They are composed of a nucleic acid genome of single- or double-stranded DNA or RNA surrounded by a protective protein coat. Three shapes occur: icosahedral (MS2); filamentous or helical (M13); head and tail (T4 and λ). Bacteriophage genomes vary in size and complexity with the largest genomes associated with the more complex phage structures such as head and tail. The phage genes encode capsid proteins and proteins involved in DNA replication. All phages have some requirement for host cell enzymes.

Phage life cycles

Phage infection follows lytic or lysogenic pathways. Lytic infection results in cell lysis. In lysogenic infection, the phage remains quiescent for many generations before inducing cell lysis.

T4 lytic pathway

Phage T4 undergoes lytic infection of *E. coli*. The phage attaches to the cell surface and injects its DNA into the cell. Phage DNA is replicated and phage genes are expressed. Expression of host cell genes and replication of host DNA is arrested. Phage DNA is packaged into capsids and new phage particles are released following cell lysis.

Lambda lysogenic life cycle

Phage λ undergoes lytic and lysogenic infection of *E. coli* but lysogenic infection is more common. The phage attaches to the cell surface and injects its DNA into the cell. The DNA integrates into the host cell chromosome where it remains quiescent. Eventually, a switch occurs to lytic infection triggered by chemical or physical stimuli associated with DNA damage. Phage DNA is excised from the chromosome and is replicated. Phage genes are expressed and new phage particles are produced which are released following cell lysis.

M13 phage

M13 infection of *E. coli* is intermediate between the lytic and lysogenic pathways. The M13 genome is a single-stranded circular DNA molecule which replicates via a double-stranded replicative form. New phage particles are released without cell lysis or are passed on to daughter cells following division of infected cells. M13 is used as a cloning vector to produce single-stranded DNA for sequencing.

Gene expression in lytic infection

Phages regulate expression of their genes to different degrees. Usually genome replication precedes the synthesis of capsid proteins. Lysozyme is produced at the end of the lytic cycle. Simple phages such as ΦX174 express all their genes at the same time. Other phages have early and late phases of expression. Early gene expression is associated with replication of the phage genome and late gene expression with the synthesis of structural proteins. Expression of early genes activates the expression of later genes by modifying the specificity of the host RNA polymerase.

Gene expression in lysogenic infection

During lysogenic infection of *E. coli* by phage λ, gene expression is inhibited by binding of the cI repressor protein to P_L and P_R promoters that control expression of early λ genes. Binding of cI also stimulates its own expression. The λ cro protein acts as a repressor of cI gene transcription. The choice between the lytic and lysogenic pathways depends on the relative levels of cI and cro. If cro levels are high, cI is repressed and the block on λ gene expression is removed resulting in lytic infection. If cI levels are high lysogeny is maintained. The switch to lytic infection following lysogeny can be induced by ionizing radiation which induces expression of the RecA protein which cleaves cI.

Related topics

Regulation of gene expression in prokaryotes (A10)
Eukaryotic viruses (B9)
DNA sequencing (E4)

Bacteriophages

Bacteriophages (phages) are viruses that infect bacteria. Many different types of bacteriophage exist which tend to be species specific infecting only certain bacterial hosts. Like other viruses, their structure consists of an inner nucleic acid genome surrounded by an outer protective protein coat. Three types of phage structure are seen (*Fig. 1*): (i) **icosahedral**, in which the individual polypeptide molecules form a geometrical structure that surrounds the nucleic acid. An example is the phage MS2 that infects *E. coli*; (ii) **filamentous or helical**, in which the polpeptide units are arranged as a helix to form a rod like structure. An example is the *E. coli* phage M13; (iii) **head and tail**, the most complex

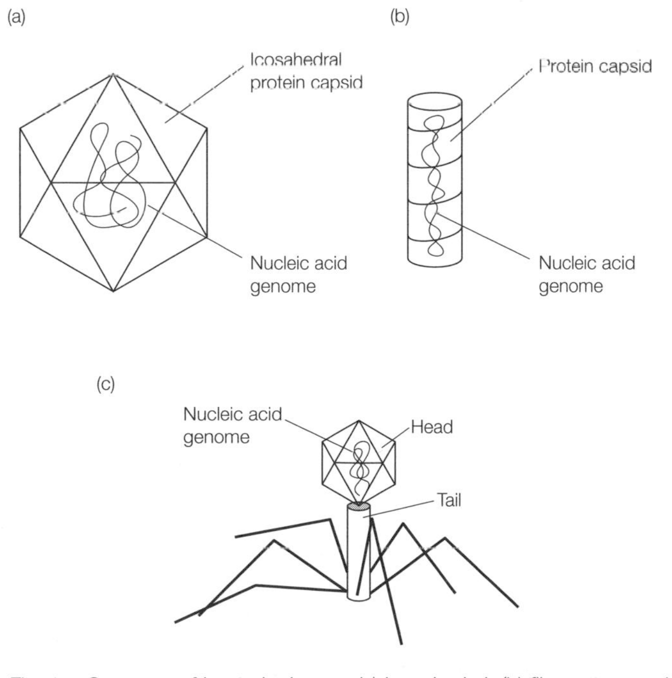

Fig. 1. Structure of bacteriophages. (a) Icosahedral, (b) filamentous or helical, (c) head and tail.

phage structure consisting of an icosahedral head and a filamentous tail that allows the phage nucleic acid to enter the cell. Examples of this type are the *E. coli* phages T4 and λ. The nucleic acid in phages may be DNA or RNA which may be single-stranded or double-stranded. Usually all the genes are present on a single nucleic acid molecule, however, in some RNA phages the genes are present on more than one molecule and these phages are said to be **segmented**. Phage genomes vary greatly in size from just a few kilobase pairs (kbp) to about 150 kbp. The number of genes varies roughly in proportion to genome size. The genome of the phage M13 is about 6 kbp and contains just 10 genes. Larger phages, especially those with complex structures (e.g. head and tail), have many more. For example, phage T4 at 166 kbp has 150 genes. Phage genes encode proteins required for the construction of the capsid and enzymes involved in phage DNA replication. All phages also require at least some host proteins and RNAs. A feature of some phages (e.g. ΦX174), also seen in viruses of eukaryotic cells, is that they have **overlapping genes** translated in different reading frames.

Phage life cycles

The events that follow infection of a cell by a phage vary but essentially two patterns, known as **lytic** and **lysogenic infection**, are seen. In the lytic pathway, the phage causes lysis of the host cell soon after infection. In the lysogenic pathway, the phage causes cell lysis only after an extended period during which it remains quiescent. Different types of phage may undergo either type of infection. Phages that undergo lytic infection are called **virulent**. Those undergoing lysogenic infection are said to be **temperate**.

T4 lytic pathway

Phage T4 undergoes lytic infection of *E. coli* (*Fig. 2*). Phage particles attach to a receptor protein on the surface of the cell called *omp*C and phage DNA is injected into the cell via the T4 tail structure. Inside the cell, transcription of the phage genes begins and synthesis of host cell DNA, RNA and protein is arrested. Host cell DNA is depolymerized and the nucleotides are used to replicate phage DNA. Phage capsid proteins are synthesized and new phage particles are assembled. Finally, the host cell is lysed and 200–300 new phage particles are released.

Lambda lysogenic life cycle

Temperate phages such as λ can undergo lytic infection of *E. coli* but more commonly follow the alternative lysogenic pathway (*Fig. 3*). After the phage injects its DNA into the cell, the DNA circularizes then integrates into the host cell chromosome. Integration occurs by recombination and involves a 15 bp phage sequence homologous to a sequence on the *E. coli* chromosome. Integration always occurs at the same position and the integrated form is known as a **prophage**. It remains undisturbed for many generations and is replicated with the host cell chromosome. Eventually after numerous cell divisions, a switch occurs to the lytic mode of infection. This is induced by a number of physical or chemical stimuli all of which are linked to DNA damage and possibly signal the imminent death of the cell. In response to these stimuli, the phage DNA is excised from the chromosome by a reverse recombination event. Phage genes are expressed and capsid proteins are synthesized. Phage DNA is replicated and is packaged into the capsids. Eventually the cell lyses and new phage particles are released.

M13 phage

Some phage life cycles show variations on the lytic or lysogenic pathways. An example is M13 which infects *E. coli*. Its genome is a single-stranded circular

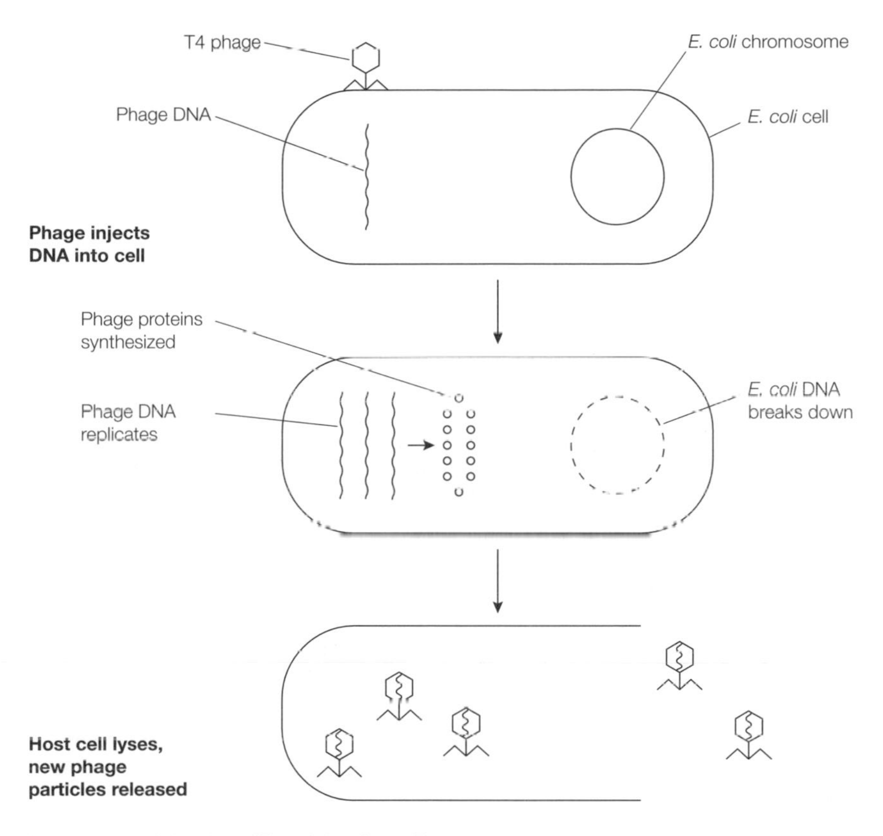

Fig. 2. Lytic infection of E. coli *by phage T4.*

DNA molecule. On infection, a double-stranded **replicative form** (RF) is produced by host cell enzymes. Multiple copies of the RF are synthesized which act as templates for the production of single-stranded forms which are packaged into capsids and released from the cell without cell lysis. Phage particles are also passed to other cells following cell division and cells infected by M13 continue to grow and divide giving rise to infected daughter cells. Several features of the M13 life cycle have made it a useful cloning vector for molecular biology. The double-stranded circular RF can be easily purified like other plasmids and the single-stranded phage DNA is an ideal template for DNA sequencing.

Gene expression in lytic infection

Phages show different degrees of regulation of gene expression. Usually genome replication precedes the synthesis of capsid proteins and lysozyme, the enzyme that causes cell lysis, is produced at the end of the lytic cycle. For simple phages such as ΦX174, regulation of gene expression is minimal. All of its 11 genes are transcribed by the host RNA polymerase upon infection. However lysozyme production, which is required last, is delayed by being translated more slowly than the other transcripts. With most other phages there are two phases of gene expression known as **early** and **late**. Early gene expression is usually associated with replication of the phage genome and late gene expression with the production of structural proteins. The early genes are transcribed first and are responsible for activating expression of the late genes. In T4, the earliest genes are transcribed by the host RNA polymerase using phage promoter sequences

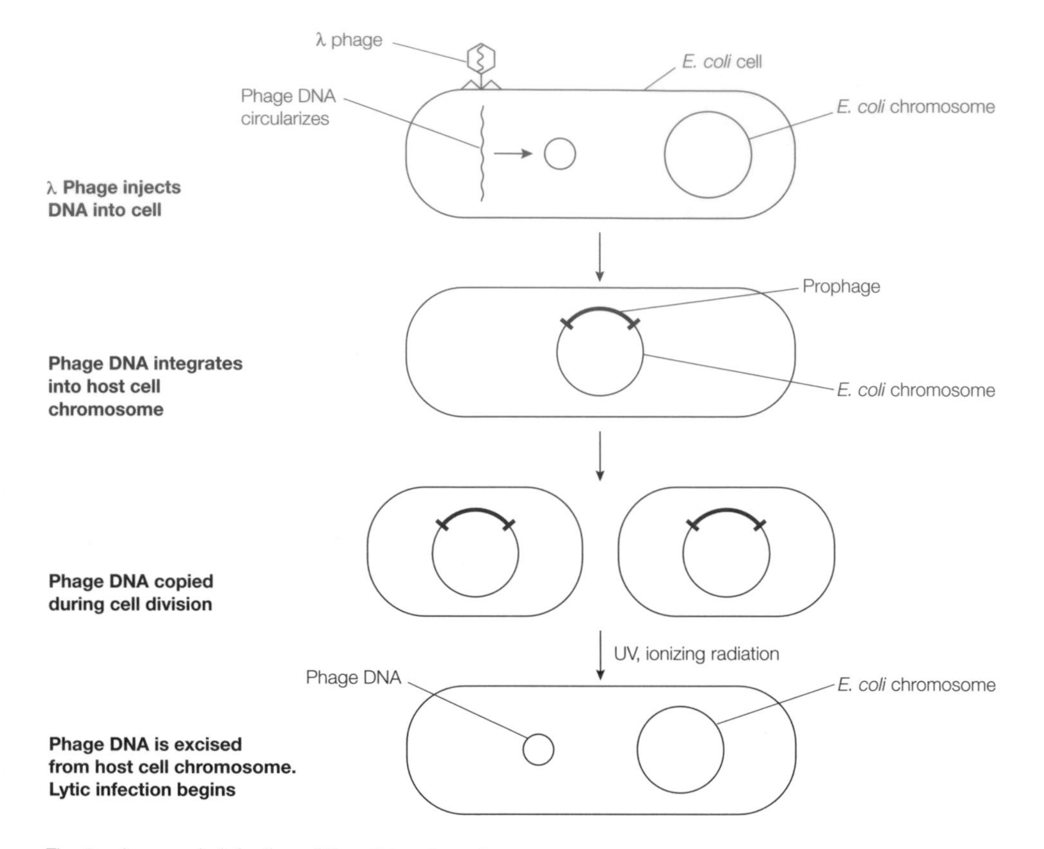

Fig. 3. Lysogenic infection of E. coli *by phage λ.*

that normal *E. coli* genes possess. Some of the genes transcribed encode proteins that modify the specificity of the host RNA polymerase such that it no longer recognizes host promoters. This causes host gene expression to be switched off but also leads to the transcription of a second set of phage genes (see Topic A10). Some of these encode proteins that modify the specificity of the polymerase again so that it transcribes a third set of genes. In this way expression of phage genes is organized into phases with the products of early genes switching on later genes.

Gene expression in lysogenic infection

The bacteriophage λ has been studied extensively as a model of phage gene expression. On infection of *E. coli*, λ may follow the lytic or lysogenic pathways. There are three phases of λ gene expression: immediate early, delayed early and late. Expression of the early genes is regulated from two promoters, P_L and P_R, present on either side of a regulatory gene called **cI** which encodes a repressor protein (*Fig. 4*). During lysogeny transcription from P_L and P_R is blocked by the cI protein which also stimulates its own expression when bound to P_R which overlaps with the cI promoter. As long as cI is present, expression of the early genes and of later genes is repressed and lysogeny is maintained. The choice between the lytic and lysogenic pathways involves a λ protein called **cro** which acts as a repressor of cI gene transcription. After infection, the host RNA polymerase starts transcribing λ genes from a number of λ promoters. cI is expressed and prevents early gene transcription blocking progression of

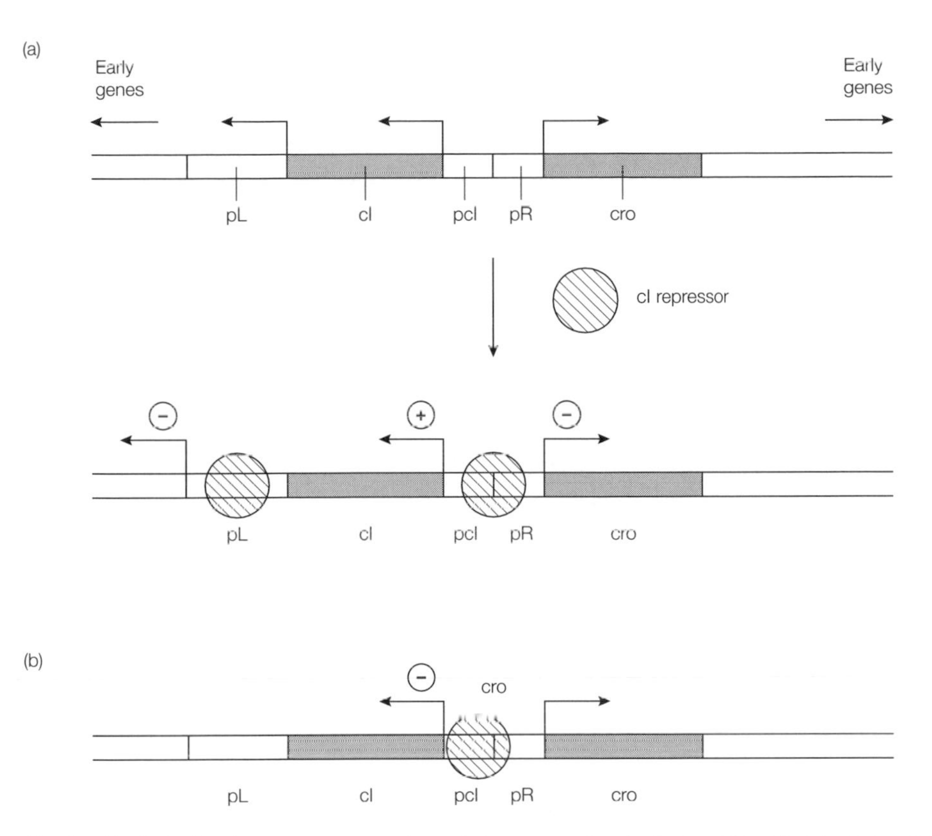

Fig. 4. Control of λ phage gene expression. (a) cI repressor blocks transcription of early genes and stimulates cI gene transcription. (b) cro repressor blocks transcription of the cI gene allowing early genes to be expressed.

the lytic pathway. Cro is also expressed and if sufficient cro protein accumulates cI is repressed and the block to early gene transcription is removed allowing progression to the lytic pathway. The competition between cI and cro and the decision between lysis and lysogeny seems to be a random process depending on how quickly each protein accumulates. As long as enough cI protein is present lysogeny will be maintained. If cI levels fall, however, lysis will be induced spontaneously. The induction of lysis following lysogeny can be triggered by agents such as ionizing radiation that activate a general protective mechanism in *E. coli* called the SOS response. This involves the ***RecA*** gene whose protein inactivates the cI repressor by cleaving it. When cI is inactivated in this way, the early genes are activated, lysogeny ends, and the lytic pathway is induced.

B9 Eukaryotic viruses

Key Notes

Viruses

Viruses are infectious agents composed of a nucleic acid genome surrounded by protein. They reproduce by using the host cell's ability to synthesize nucleic acids and protein. Many viruses are pathogenic.

Viral genomes

These may be single- or double-stranded, DNA or RNA, linear or circular. Genomes may be positive or negative or both senses. Viral genomes may be segmented. Some viral particles have incomplete genomes and only replicate in the presence of wild-type virus or by complementation.

Structure of viruses

The protein coat surrounding the viral nucleic acid is the capsid; this may be icosahedral or filamentous in shape. Some viruses have a host cell-derived lipid envelope containing host cell proteins and viral glycoproteins that play a role in infection. Structural viral proteins are antigenic and are useful in the development of vaccines.

Replication strategies

Viruses use many different strategies for replication. All depend on the host cell completely for translation and to different extents for transcription and replication. DNA viruses are more dependent on host cell enzymes than RNA viruses, and small DNA viruses are more dependent than large DNA viruses. RNA viruses require virus encoded RNA dependent RNA polymerases. Retroviruses are RNA viruses that replicate via a double-stranded DNA intermediate produced by virus encoded reverse transcriptase that copies RNA into DNA. Host cells that support infection are said to be permissive. The capacity of a virus to cause disease is called virulence.

DNA viruses

The genomes of DNA viruses vary greatly in size (5–270 kbp) and may be linear or circular, single- or double-stranded DNA (ssDNA or dsDNA). Large DNA viruses have a complex life cycles with coordinated patterns of gene expression and genome replication. Herpes simplex virus type 1 has a 150 kbp genome with over 70 genes present on both strands with some genes overlapping. The viral genes are expressed in three phases (immediate early, early and late) and expression is coordinated with viral genome replication. Small DNA viruses have fewer genes and are more dependent on the host cell for replication and gene expression. The SV40 virus has a 5 kbp genome containing five overlapping genes present on both strands of the DNA. Viral gene expression makes use of overlapping reading frames and alternative splicing. The genes are expressed in two phases (early and late). Early gene transcription produces the large and small T antigens which stimulate viral and host cell transcription and replication and are responsible for the tumorigenic properties of the virus.

RNA viruses

The genomes of RNA viruses may be single- or double-stranded. Single-stranded genomes may be positive or negative or both senses. The viral *pol* gene encodes an RNA-dependent RNA polymerase required for the transcription of viral genes. RNA viruses are not dependent on host polymerases and so may replicate in the cytoplasm. Due to the lack of proofreading ability of the RNA polymerase, RNA viruses have a high mutation rate allowing them to evolve rapidly causing changes in antigenicity and virulence. Some RNA viruses exist as a collection of mutant forms that can only replicate by complementation. The high mutation rate also imposes an effective limit on the viable size for an RNA viral genome of about 10 000 nucleotides.

Retroviruses

This group includes human immunodeficiency virus (HIV). Retroviruses have a single-stranded, positive sense RNA genome. Two copies are present in each viral particle. On infection, viral reverse transcriptase synthesizes a dsDNA copy of the viral RNA which integrates into the host cell genome and acts as a template for viral replication and gene expression. Some retroviruses cause cancer and carry genes originally derived from the host cell called oncogenes. Retroviruses carry three genes called *gag*, *pol*, and *env*; *gag* encodes capsid proteins; *pol* encodes enzymes involved in viral replication; *env* encodes proteins found in the lipid envelope. The viral genome has repeat sequences at either end called long terminal repeats (LTRs). Host cell RNA polymerase II binds to the 5′ LTR and transcribes viral genes. Viral mRNAs are translated as polyproteins that are cleaved proteolytically to give mature proteins. HIV creates extra proteins by differential splicing of mRNAs. The reverse transcriptase of some retroviruses has a high error rate producing defective copies of the viral genome. Replication depends on complementation between the two genome copies present in each viral particle. The high turnover rate of HIV and its high mutation rate allow the virus to adapt to selective pressures.

Related topics

Bacteriophages (B8)
Genes and cancer (F2)

Viruses

These are submicroscopic infectious agents composed of a nucleic acid genome surrounded by a protein coat. They are parasitic and reproduce themselves by infecting cells and making use of the cell's ability to replicate DNA and synthesize protein. It is difficult to define them as living organisms because they are incapable of existing independently. Many viruses are pathogenic and cause destruction of the host cell leading to disease in humans and other organisms. In addition to eukaryotes, prokaryotic organisms such as bacteria can be infected by viruses called **bacteriophages** (see Topic B8).

Viral genomes

There are many different types of virus that infect eukaryotic cells. These differ with respect to their nucleic acid genomes which may be ssDNA, dsDNA, ssRNA or dsRNA. The nucleic acid may also be linear or circular. In some cases the viral genome is a single molecule of nucleic acid but in others the viral genes exist on more than one molecule and the genome is said to be **segmented**. Single-stranded genomes may be **positive sense** or **negative sense**. The sense refers to the sequence of the genome and the mRNAs transcribed from it. If the genome is positive sense, its sequence will be the same as its transcribed mRNAs. Conversely, if it is negative sense then the genome sequence is complementary to the mRNAs. In some cases the genome will encode mRNAs which are of either sense. Many viral particles do not have a complete or functional genome and so are only capable of replication when they are rescued by coinfection of the cell with a wild-type replication competent **helper virus** or by a process known as **complementation** in which coinfection occurs by a defective virus with a different mutation.

Structure of viruses

The viral genome is surrounded by a protective protein coat known as the **capsid** which is assembled from individual virus-encoded polypeptides. The combination of the genome and the capsid is known as the **nucleocapsid**. Two basic shapes occur (*Fig. 1*): (i) **icosahedral**. In this case the individual polypeptide molecules form a geometrical structure that surrounds the nucleic acid. An example is poliovirus; (ii) **filamentous** or **helical**. In this case the polypeptide units are arranged as a helix to form a rod-like structure surrounding the nucleic acid genome. In many viruses the capsid is surrounded by a **lipid envelope** derived from the host cell membrane as the virus is released from the cell. Virus-encoded glycoproteins and proteins derived from the host cell may be inserted into the membrane. The viral glycoproteins play an important role in facilitating infection by interacting with receptor proteins on the surface of the host cell. **Matrix proteins** also occur which allow interaction between the nucleocapsid and the lipid envelope. The matrix and the capsid proteins also have other roles associated with virus replication and transcription of viral genes. The structural proteins of a virus are often antigenic and are of interest in the development of vaccines.

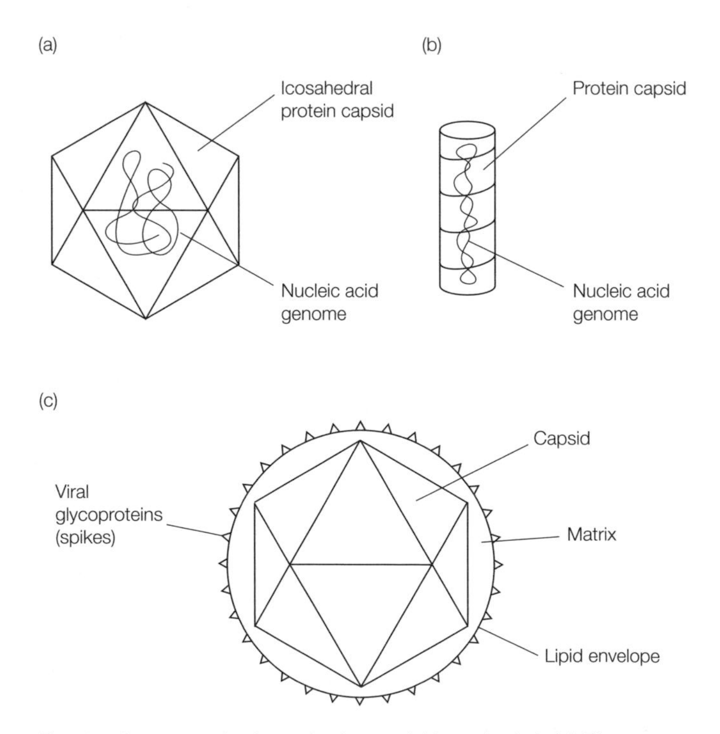

Fig. 1. Structure of eukaryotic viruses. (a) Icosahedral. (b) Filamentous or helical. (c) Enveloped.

Replication strategies

Viruses make copies of themselves by exploiting the ability of cells to replicate DNA and to synthesize proteins. The strategies adopted by viruses for replication of the viral genome and transcription of the viral genes vary greatly. All viruses rely on the host cell for translation and to differing extents for transcription and replication. DNA viruses are more dependent on the host cell enzymes than RNA viruses and small DNA viruses more dependent than large DNA viruses on the host cell's replication and transcription machinery. RNA viruses require RNA dependent RNA polymerases for replication which are not present in host cells but must be encoded by the virus. Other RNA viruses called **retroviruses** replicate via a dsDNA intermediate produced from the viral RNA by a unique virus encoded enzyme called **reverse transcriptase** that copies RNA into DNA.

The susceptibility of a cell to infection by a virus is determined by its ability to support viral replication and the presence of cell surface receptors specific for the virus. Cells capable of supporting replication are called **permissive**.

The consequence of viral infection is not always disease. The primary function of the virus is to replicate itself which may coincidentally cause cell damage. If enough cells are damaged, this may result in disease. The capacity of viruses to cause disease is called **virulence** and may be disadvantageous to the virus because it may reduce the capacity for viral replication.

DNA viruses

The genomes of DNA viruses vary greatly in size from 5 kbp to as large as 270 kbp. The DNA may be single-stranded or double-stranded and may be linear or circular. Almost all DNA viruses replicate in the nucleus and make use of the host cell machinery for genome replication and viral gene expression. Large DNA viruses encode as many as 200 genes and have a complex life cycle with coordinated patterns of gene expression and genome replication. An example of large DNA viruses are the **herpes viruses** which infect a range of vertebrates including humans where they cause diseases such as chicken pox, shingles and glandular fever. **Herpes simplex virus 1** is a well-studied herpes virus that causes cold sores. It has a 150 kbp dsDNA genome which contains over 70 genes present on both DNA strands, some overlapping with each other. The transcription and replication of the viral genome are tightly regulated. The viral genes fall into three groups called immediate early (α), early (β) and late (γ) which are expressed in a defined sequence following infection of a host cell. Expression is also coordinated with replication of the viral genome.

DNA viruses with small genomes have many fewer genes and consequently are more dependent on the host cell for replication and gene expression. Examples include polyoma virus and the tumorigenic monkey virus **SV40** which are members of the papovavirus family. The SV40 virus has a small 5 kbp genome which contains five genes. The genes are accommodated in the small genome by being present on both strands of the DNA and by having overlapping sequences. Viral proteins are produced by a combination of the use of overlapping reading frames and alternative splicing. The genes are expressed in two phases after infection known as **early** and **late**. The early genes produce proteins that activate transcription called the **large T antigen** and the **small T antigen**. These stimulate both viral and host cell transcription and replication and are responsible for the tumorigenic properties of the virus.

RNA viruses

The genomes of RNA viruses may be single-stranded or double-stranded. If single-stranded, the genome may be positive or negative sense. In some cases the genome will encode mRNAs of both senses. Transcription of viral genes from an

RNA genome requires enzymes called **RNA dependent RNA polymerases** which do not occur in host cells but are encoded by a viral gene called *pol*. Because RNA viruses do not require host cell polymerases, transcription and replication need not take place in the nucleus as is the case for DNA viruses. Consequently, many RNA viruses replicate in the cytoplasm. Unlike host cell polymerases, RNA dependent RNA polymerases are not capable of proofreading and replicate their templates with a much higher error rate. The mutation rate is one base in 10^3–10^4 per replication cycle which is much higher than the rate for DNA viruses of one base in 10^8 to 10^{11}. The consequences of the much higher mutation rate for RNA viruses are that they are capable of evolving rapidly and can develop changes in antigenicity and virulence quickly allowing them to adapt to changing environments and attempts to eliminate them by the host's immune system.

Some RNA viruses mutate so rapidly that they exist as a population of genome sequences which replicate by complementation and are known as **quasi-species**. Many of the mutations that occur in RNA viruses will be deleterious to viral replication. This effectively creates an upper limit to the genome size of about 10^4 nucleotides, equivalent to the size that, on average, would produce one mutation per genome.

Retroviruses

These are an important group of RNA viruses with single-stranded positive sense RNA genomes. The group includes human immunodeficiency virus (**HIV**), the virus that causes the acquired immune deficiency syndrome (**AIDS**). Retroviruses contain two copies of the genome in each viral particle. On infection of a host cell, the ssRNA enters the cell, is converted to a dsDNA copy by **reverse transcriptase**, and is integrated into the host cell genome by a viral **integrase** enzyme. The integrated form is known as a **provirus** and acts as the template for replication of the viral genome and expression of the viral genes (*Fig. 2*). The genomes of reteroviruses have structural similarities and some sequence homology with sequence elements present in the human genome called **retrotransposons**. Retroviruses vary in complexity. Some are extremely simple and differ from retrotransposons essentially only by having an *env* gene that encodes envelope glycoproteins required for infectivity. Others, such as HIV, have larger genomes with more complex life cycles, encode proteins that are active at different stages of the replication cycle and are involved in regulating both viral and cellular functions.

Some retroviruses are known to cause cancer in animals although they are linked only rarely to human cancers. These retroviruses are said to be **oncogenic** and frequently carry a gene called an **oncogene** which was originally acquired from the host cell genome by recombination with the viral genome. Oncogenes encode proteins involved in regulating cell growth and have cellular counterparts called **proto-oncogenes**. Expression of the oncogene following viral infection results in uncontrolled cell division associated with the formation of a tumor (see Topic F2).

The genomes of retroviruses all have the same basic structure (*Fig. 3*). Most of the genome is taken up by three genes called ***gag***, ***pol*** and ***env***. *Gag* encodes the capsid proteins, *pol* encodes the enzymes involved in viral replication (reverse transcriptase, RNAse H, integrase and protease) and *env* encodes the proteins present in the lipid envelope. At either end of the genome are duplicate sequences called **LTRs** (long terminal repeats) that are involved in viral replication, integration and gene expression. Transcription of the viral genes from the integrated provirus depends on host cell RNA polymerase II which binds to sequences in the 5′ LTR. The mRNAs are translated into **polyproteins** that are cleaved proteolytically to give individual viral proteins. Some retroviruses such as HIV produce additional

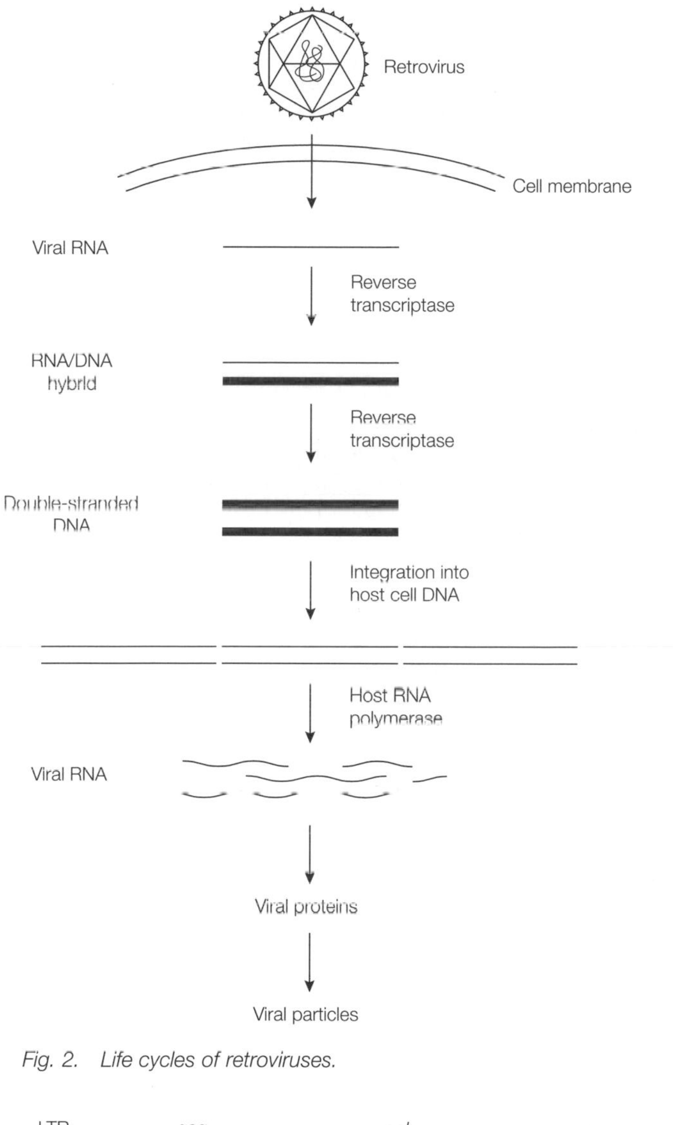

Fig. 2. Life cycles of retroviruses.

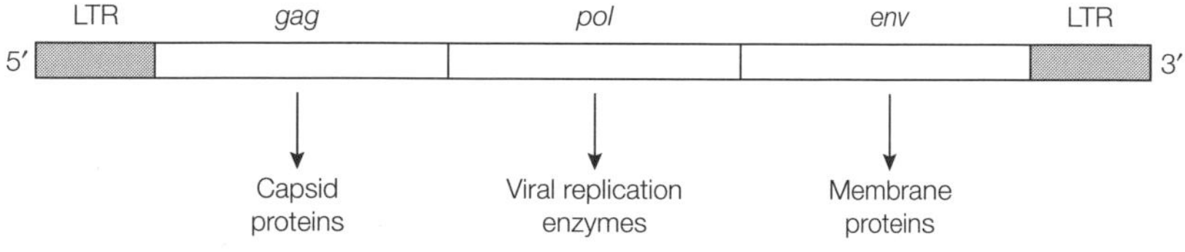

Fig. 3. Retrovirus genome.

mRNAs by differential splicing. The reverse transcriptase of some retroviruses has a high error rate. This means many of the genome copies produced are incapable of replicating themselves. This can be overcome by complementation between the two genome copies present in each viral particle which can also recombine with each other during reverse transcription. These features, combined with a high turnover rate, (10^9–10^{10} new viral particles per day) enable HIV-1 to adapt to new environments such as selective pressure from antibodies and drug treatments.

C1 BASIC MENDELIAN GENETICS

Key Notes

Basic concepts

Any inherited trait such as eye color is referred to as a phenotype. All phenotypes result from the presence of a specific gene or combination of genes, the genotype. In hybrids one phenotype may be dominant to another. Pure-breeding lines are strains of a species which have been bred for many generations and have maintained the same phenotype.

The monohybrid cross

Crosses between pure-breeding (homozygous) lines differing in one inherited character yield progeny that all have the same phenotype. This is termed the F1 generation and is composed of heterozygous individuals. When these are inter-crossed, the next generation, the F2, shows both of the original phenotypes in the ratio 3 : 1 with the dominant phenotype being the majority class. Each individual carries two copies (alleles) of each gene. Homozygous individuals carry two identical alleles, heterozygous individuals carry two different alleles.

Detection of heterozygotes

When an individual that is heterozygous for one gene is crossed to a recessive homozygote only two classes of progeny are observed. The dominant and recessive phenotypes arise at equal frequency. This is a test cross.

Variations of the 3 : 1 ratio

The 3 : 1 ratio depends on complete dominance of one phenotype over the other. If the phenotypes under study show partial or codominance, a 1 : 2 : 1 ratio will be obtained. If either allele has a negative effect on viability this will also distort the ratio. Alleles that can cause lethality when homozygous are called lethal alleles. Semi-lethal alleles have a quantitative effect on viability.

Multiple alleles

Most genes exist in several different forms, multiple alleles. This is caused by mutations of bases at different sites within the same gene, thus effecting different amino acids in the encoded protein. These arise at random within the population.

Related topic

More Mendelian genetics (C2)

Basic concepts

The first clear evidence pointing to what we now call genes came from the work of Gregor Mendel who carried out experiments on inheritance in pea plants in the middle of the nineteenth century. Before we examine his results it is necessary to establish an understanding of some of the basic terms that are used in the study of inheritance in higher organisms.

Phenotype

Any character (trait) which can be shown to be inherited, such as eye color, leaf shape or an inherited disease, such a cystic fibrosis, is referred to as a phenotype. A fly may be described as having a red-eyed phenotype or a child as displaying the cystic fibrosis phenotype. The pattern of genes that are responsible for a particular phenotype in an individual is referred to as the **genotype**.

Pure-breeding lines
This refers to organisms which have been inbred for many generations in which a certain phenotype remains the same. Pedigree breeds of dogs or cats are commonplace examples of pure-breeding lines.

Dominance
Within a species there may be differences in the phenotype for one inherited character. In hybrids between two individuals displaying different phenotypes only one phenotype may be observed. For instance, in crosses of pure-bred fruit flies with short wings with pure-bred long-winged flies the progeny will all have long wings. The phenotype expressed in the hybrids is said to be dominant and the other recessive. In the example above long wings are dominant to short wings.

The key ingredients for success in Mendel's experiments were the use of pure-breeding strains of pea plants and the fact that he subjected his results to simple mathematical analysis.

The monohybrid cross

Mendel studied inheritance of several phenotypes in pea plants, but we will concentrate on only one of these, petal color. He made a cross between two pure-breeding lines of plants, one of which had violet petals and the other white petals. The hybrids produced in this cross were referred to as the **F1 (first filial)** generation. These all had violet flowers. Thus violet was dominant to white. He then allowed these plants to self-fertilize to produce the **F2 (second filial)** generation. Some plants had white flowers and others violet flowers. The ratio of violet to white flowered plants was close to 3 : 1. In crosses between plants differing in seed color, pod shape or other phenotypes, the same pattern was observed. The recessive phenotype always reappeared in the F2 generation and made up approximately one quarter of the plants.

These experimental data led Mendel to suggest that heredity was due to the action of specific factors, which we now call genes. This apparently simple conclusion was, however, in complete opposition to the conventional view that heredity was due to blending of fluids from both parents. Clearly no blending had occurred in Mendel's experiments, neither in the F1 where only one phenotype was expressed nor in the F2 where both were expressed separately.

From our modern knowledge of genes and gene structure it is easy for us to appreciate how Mendel explained what was happening in his experiment. This is set out diagrammatically in *Fig. 1*. He suggested that the pure-breeding violet flowered plants carried two copies of a gene for violet pigment, *V*. The white flowered plants carried two copies of a variant of this gene that codes for white flowers, *v*. We refer to individuals with two identical copies of a gene as being **homozygous**. The F1 hybrids inherited two different copies of the pigment gene, *V v* and are referred to as **heterozygous**. As violet is dominant to white these plants had violet flowers. When F1 plants self-fertilize three different classes of genotype, *V V*, *V v* and *v v* are possible. These arise in the ratio 1 : 2 : 1 (*Table 1*). This is how the 3 : 1 phenotypic ratio is established (i.e. 3 violet : 1 white).

The validity of this hypothesis was strengthened when individual plants from each of the F2 classes were self-fertilized. All white-flowered plants were found

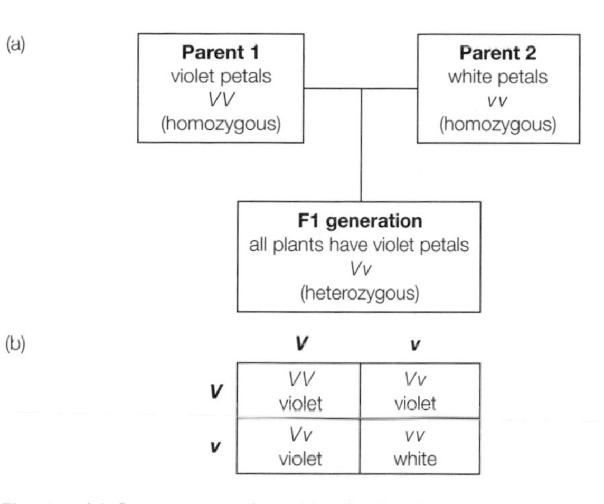

Fig. 1. (a) Gametes produced by the F1 plants carry V or v alleles these fuse at random to give offspring with the genotypes VV, Vv or vv as shown in (b). (b) Each of the four fusions arise with equal frequency. Hence the phenotype ratio in the F2 generation is 3 purple: 1 white and the genotype ratio is 1VV : 2Vv : 1vv.

Table 1. Products of F1 self-fertilization in pea plants

Genotype	Phenotype (petal color)	Ratio
VV	Violet	1
Vv	Violet	2
vv	White	1

Table 2. Examples of inheritance is controlled by a single gene

Species	Character (phenotype)
Mice	Albino/normal coat, pale/normal ears
Red clover	Red/white flowers
Fruit flies	Normal/vestigial wings
Humans	Blue/brown eyes, cystic fibrosis, sickle cell anaemia, phenylketoneuria

to be pure-breeding. One-third of the violet-flowered plants were pure-breeding and two-thirds give purple- and white-flowered plants in the ratio 3 : 1. There have subsequently been many examples of the 3 : 1 ratio for the inheritance of characters controlled by a single gene in many different species. A few of these are listed in *Table 2*.

One new term needs to be defined at this stage. Genes become altered through the process of **mutation**. The different variants of a gene are referred to as **alleles**. Students often are confused between the terms genes and alleles. In the previous example it is better to refer to *V* and *v* as two alleles of a petal color gene. It is conventional to denote dominant alleles with upper case and recessive alleles with lower case letters. The 3 : 1 ratio is referred to as the **monohybrid ratio**, and is the basis for all patterns of inheritance in higher organisms.

Detection of heterozygotes

One simple extension of the 3 : 1 phenotype ratio is a 1 : 1 ratio produced when an F1 individual is crossed to the homozygous recessive parent. As shown in *Fig. 2* the heterozygous F1 can produce only two classes of gamete, carrying either the dominant or the recessive allele. The parent with the recessive phenotype can only produce gametes with recessive alleles, and so the progeny of the cross have the dominant and recessive phenotypes in equal numbers, a 1 : 1 phenotype ratio. This type of cross is termed a **testcross**, and is useful in any situation where it is necessary to determine if an individual is heterozygous. It is also the expected phenotype ratio in families where one parent carries a rare dominant allele, such as Huntington's disease. Because the dominant allele is rare, the affected individual is unlikely to be homozygous.

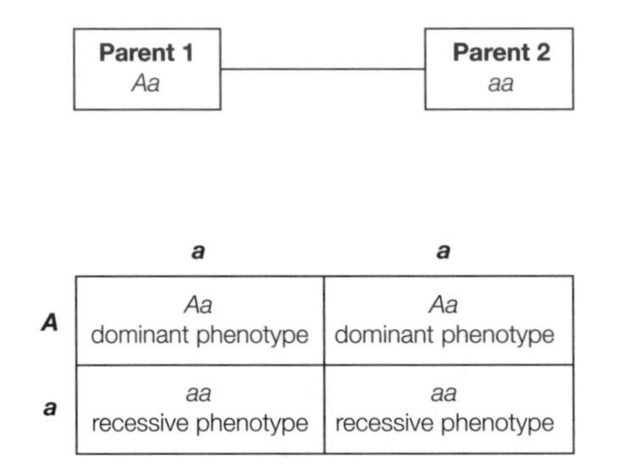

Fig. 2. The genotype and phenotype ratios are both 1 : 1 because the alleles contributed by the doubly recessive parent do not affect the phenotype.

Variations of the 3 : 1 ratio

The simple 3 : 1 monohybrid ratio is not always observed in instances where only one gene is responsible for a particular phenotype. This may be due to a number of factors.

Partial or incomplete dominance

In the preceding section the example used showed complete dominance. In other words the phenotype of the F1 generation was identical to that of one of the parents (the dominant phenotype). This is not always the case. Often the F1 is clearly intermediate between two parents. A simple example of this is the inheritance of petal color in snapdragons. When pure-bred white and pure-bred red-flowered plants are crossed the F1 generation has pink rather than white or red petals. The F2 comprises three classes of plants (*Table 3*). This 1 : 2 : 1 ratio of three phenotypes is clearly different from the 3 : 1 but it is easy to see how the two relate. The cross is set out in *Fig. 3*. The genotype of the

Table 3. Inheritance of petal color in the snapdragon

Genotype	Phenotype	Ratio
rr	White petals	1
Rr	Pink petals	2
RR	Red petals	1

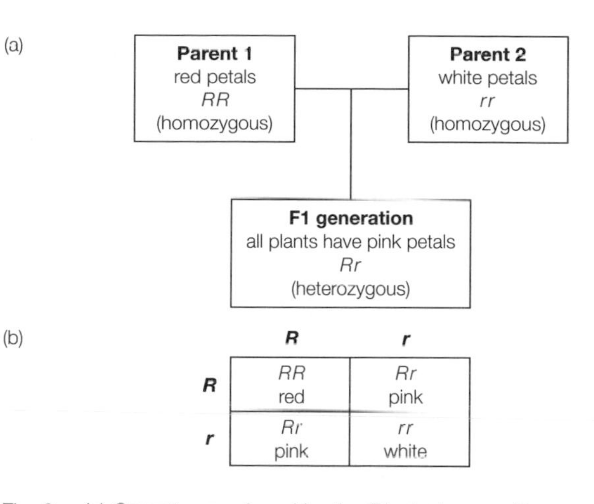

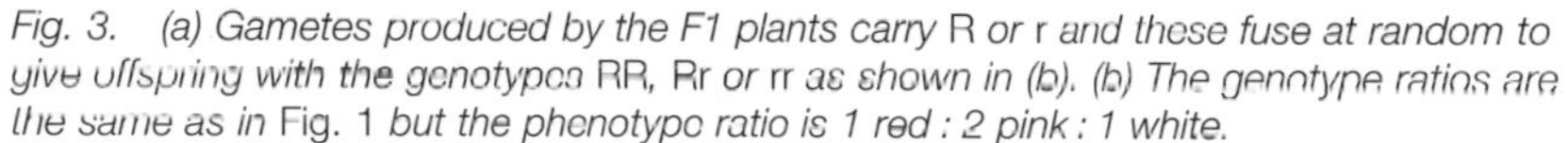

Fig. 3. (a) Gametes produced by the F1 plants carry R *or* r *and these fuse at random to give offspring with the genotypes* RR, Rr *or* rr *as shown in (b). (b) The genotype ratios are the same as in* Fig. 1 *but the phenotype ratio is 1 red : 2 pink : 1 white.*

red-flowered plants is *RR* and that of the white-flowered plants is *rr*. The F1 generation consists of *Rr* heterozygotes. In the F2 generation the ratio of genotypes is the same as that described previously. The difference here is that the *Rr* heterozygotes are pink thus altering the ratio of phenotypes.

The red- and white-flowered F2 classes are homozygous and are therefore pure-breeding, whilst the pink heterozygotes when self-fertilized will always produce a 1 : 2 : 1 ratio of white-, pink- and red-flowered plants.

Co-dominance is similar to incomplete dominance, but here the heterozygote displays both alleles. Examples of this are found frequently in the inheritance of blood groups. In humans the MN blood group is controlled by a single gene. Only two alleles exist, *M* and *N*. Children whose father is an *NN* homozygote with group N blood and whose mother is a *MM* homozygote with group M blood are *MN* heterozygotes and have group MN blood. Both phenotypes are identifiable in the hybrid. Codominance also modifies 3 : 1 ratios to 1 : 2 : 1 ratios. Alleles that are differentiated by molecular methods such as polymerase chain reaction (PCR), or Southern blotting (see Topics F1 and F3) are also codominant.

Lethal alleles

Some alleles affect the viability of individuals that carry them. In most cases the homozygous recessive does not survive but the heterozygotes may have a normal life-span. To be detected the heterozygotes need some observable phenotype. The best known example of this is the inheritance of yellow coat color in mice. Yellow varieties can arise in strains of mice with different coat colors, for instance, black mice. Yellow coat color is dominant to black coat color; *BB* mice are black, BB^y mice are yellow. When two yellow mice are mated the progeny would be expected to be in the proportion shown in *Table 4*. However, BB^y is lethal and any mice with this genotype die *in utero*. Hence liveborn progeny from this cross are in the ratio given in *Table 5*. The 3 : 1 phenotypic ratio has been distorted to a 2 : 1 ratio.

There is a concept here which sometimes causes problems to students. The allele B^y is recessive in relation to its effect on viability, but dominant in relation to coat color. It is important that you recognize this difference. It is quite common

Table 4. Expected inheritance of yellow coat color in a cross of $BB^y \times {}^{BB}y$ *mice*

Genotype	Phenotype	Ratio
BB	Black fur	1
BB^y	Yellow	2
B^yB^y	Yellow	1

Table 5. Actual inheritance of yellow coat color in a cross of $BB^y \times {}^{BB}y$ *mice*

Genotype	Phenotype	Ratio
BB	Black fur	1
BB^y	Yellow fur	2

for a gene to be involved in two different phenotypes and an allele which is dominant for one phenotype may be recessive for another. Other examples where alleles are lethal when homozygous, but which have a dominant effect when heterozygous include tailless Manx cats and short-legged Creeper chickens. Genes that are involved in developmental processes are often found to have lethal alleles. The presence of one mutant allele alters development so as to produce characteristic changes to the animal, but when two are present development is so aberrant as to cause death. This may occur *in utero* as described above or result in shortened life expectancy as found in several examples in humans, such as Tay–Sach's disease, Huntington's syndrome or sickle cell anemia.

In other instances a homozygote may not be absent from a cross, but appear in reduced numbers. An example of this is vestigial wings in fruit flies, a condition that is caused by a recessive allele (vg). Alleles with this effect are referred to as **semi-lethals**.

These examples of alteration to the 3 : 1 ratio may at first appear complicated. It is essential that you realize, that in these cases, the behavior of the genes remains the same as in the 3 : 1 monohybrid ratio but the phenotypic ratio may change. Phenotypes depend on how genes act through protein synthesis and how specific proteins interact in the cells and tissues of an organism. You know a great deal about this from Section A. You will see further examples of interaction between gene products in the following sections. However, if you concentrate on genotypes and the inheritance of alleles it will be much easier to understand how specific genes are influencing the inheritance of phenotypes.

Multiple alleles

All the examples used so far have employed genes with only two alternative alleles. For some genes only two alleles have been identified, but for the majority of genes, a large number of alleles have been found. Examples of this include the human β-globin gene where a specific mutation at codon 6 results in an allele responsible for the hereditary syndrome sickle cell anaemia, whilst mutations at several other sites in the gene cause a different syndrome, β-thalassemia. Although they are alterations of the same gene, the changes are to different codons. The resulting proteins have variant β-globins with discrete differences in amino acid sequence and so behave differently.

In rabbits, multiple alleles of one gene are responsible for a number of different coat color phenotypes. There are four members of this allelic series:

agouti, chinchilla, Himalayan and albino. When homozygous each produces a distinct coat pattern. In heterozygotes there is a clear pattern of dominance. Agouti is dominant over all the other alleles, chinchilla is dominant over Himalayan and albino, while Himalayan is dominant only over albino, which fails to produce any pigment and hence is recessive to all the others.

Another well known example of multiple alleles is the human ABO blood group system. Here a single gene codes for an enzyme that is responsible for the addition of sugar residues to a specific glycoprotein on the membrane of red blood cells. Three different alleles of the gene are known. One form of the enzyme adds a molecule of *N*-acetyl-galactosamine to the glycoprotein resulting in blood group A. A second allele codes for a variant enzyme that adds galactose instead of *N*-acetyl galactosamine resulting in blood group B. A third allele codes for a nonfunctional enzyme that cannot add any sugar to the glycoprotein, resulting in blood group O. All three alleles have arisen by mutation from a single ancestor.

The major histocompatability complex which determines the suitability of donor organs for transplantation is an example of a complex multiple allele system.

Many other examples of multiple alleles are known and in some cases it is possible to differentiate between the alleles directly using molecular technology. Such approaches are discussed in Topics F6 (Forensic genetics) and Topic F5 (The human genome project).

C2 MORE MENDELIAN GENETICS

Key Notes

The dihybrid cross

When two genes are studied in the same cross the phenotype ratio of the F2 generation is 9 : 3 : 3 : 1. This represents the combination of two 3 : 1 ratios. The 9 : 3 : 3 : 1 ratio requires that the two genes are not involved in the same biochemical pathway nor situated close together on the same chromosome. When the F1 generation is crossed with the double recessive parent a ratio of 1 : 1 : 1 : 1 is obtained. The number and frequency of phenotypic classes from crosses involving any number of genes can be calculated by the expression $(3:1)^n$, where n is the number of genes involved.

Epistasis

The F2 phenotypic ratios of the dihybrid cross are altered if the two genes concerned are involved in the same biochemical pathway. Several different alterations of the 9 : 3 : 3 : 1 ratio are observed depending on how the genes interact. In all cases the number of phenotypic classes are reduced, and the new ratios are made up by summing certain of the classes from the original ratio.

Mendel's laws

Two basic rules underpin transmission genetics in eukaryotes. These stem directly from Mendel's experiments. Alleles of the same gene separate to different gametes, and alleles of two different genes segregate (separate) independently of each other in gametogenesis.

Handling problems

When analyzing data obtained from crosses of this nature first look for the phenotypic ratios that are most likely to fit with the information you have. Then check the goodness-of-fit of each of these using the χ^2 statistic.

Related topics

Basic Mendelian genetics (C1) Linkage (C4)

The dihybrid cross

Topic C1 showed how the 3 : 1 monohybrid ratio could be used to explain the inheritance of a phenotype where only a single gene was involved. This is the basic Mendelian ratio and everything that follows depends upon it.

The obvious next step is to look at a situation where the inheritance of two different inherited characters are studied at the same time, **a dihybrid cross**. The simplest experimental system for this is to cross two pure-breeding strains of a species, one of which is homozygous for recessive alleles of two genes and the other homozygous for dominant alleles of the same two genes. Again Mendel was the first to carry out such experiments. In one of these he used pea plants that differed in two properties of the seed.

The choice of seed characteristics aided his work for two important reasons.

- It was easy to analyze large numbers of seeds accurately, thus improving the statistical basis of the work.

- By using seed characters it was possible to reduce the length of time the experiments would take. The F1 generation became the seeds produced by crossing the parental plants. When these seeds were planted out, grown on and allowed to self-fertilize, the F2 generation was the seeds present in the pods. Thus he was saved the time (1 year) and effort of having to plant out these seeds and assess the characters in the plants that they produced.

Seed shape is determined by a single gene that has alleles for round, *R* (smooth), or wrinkled seeds, *r*. The round phenotype is completely dominant over wrinkled. Seed color can be yellow or green. Again this is determined by two alleles of a single gene. Yellow, *Y*, is completely dominant over green, *y*. The experiment is set out in *Fig. 1*. A cross between the two pure-breeding (homozygous) parental lines yielded an F1 generation which consisted only of round yellow seeds. As with the monohybrid cross (see Topic C1) the F2 generation showed considerable diversity. Four different phenotypes could be identified. Of 556 seeds analyzed, he found 315 round yellow seeds, 108 round green seeds, 101 wrinkled yellow seeds and 32 round green seeds. This is close to a ratio of 9 : 3 : 3 : 1, which is referred to as the **dihybrid ratio**. Mendel obtained ratios close to this for several different combinations of pairs of genes, and since that time a great many other examples of this ratio have been demonstrated in crosses in plants, animals and fungi.

The predicted ratios of phenotypes and genotypes can be determined graphically using what is known as a Punnett square. Simple examples of this were given in Topic C1 to show the possible classes of progeny in the monohybrid cross. A Punnett square for the dihybrid cross is set out in *Fig. 1*. The 9 : 3 : 3 : 1 ratio is simply two 3 : 1 ratios combined, and shows that the alleles of the two genes behave (segregate) independently of each other. This is demonstrated more easily by looking at one gene at a time (*Fig. 2*). Take the seed shape gene first: in the F2 generation a ratio of 3 round to 1 wrinkled would be expected. Now look at the seed color gene in those seeds that have the wrinkled phenotype. These should have a ratio of 3 yellow to 1 green. The same is true for wrinkled seeds, these should also have a ratio of 3 yellow to 1 green. If the ratios for the two phenotypes are multiplied across the 9 : 3 : 3 : 1 ratio is obtained.

As noted earlier (see Topic C1) the 3 : 1 phenotypic monohybrid ratio can be distorted by factors such as incomplete dominance or lethal effects of certain alleles. These also affect the 9 : 3 : 3 : 1 ratio, but other factors can also modify this ratio. The 9 : 3 : 3 : 1 ratio depends on two conditions.

- The two different genes must not act on the same character. For instance if the proteins encoded by the two genes are involved in the same biochemical pathway then the ratios of phenotypes resulting from the genotypes in the F2 generation will be altered. This is discussed in detail later in this section, under **epistasis**.
- If the two genes lie close together on the same chromosome the four classes of gamete are not produced at equal frequencies. This is the basis of gene mapping studies and is examined under **linkage** (see Topic C4).

As with the monohybrid cross it is also possible to conduct a testcross with the F1 generation of the dihybrid cross. If, in the round/wrinkled, yellow/green example of the 9 : 3 : 3 : 1 ratio, an F1 plant is crossed with the homozygous recessive parent for both genes the progeny fall into four phenotypic classes,

(a)

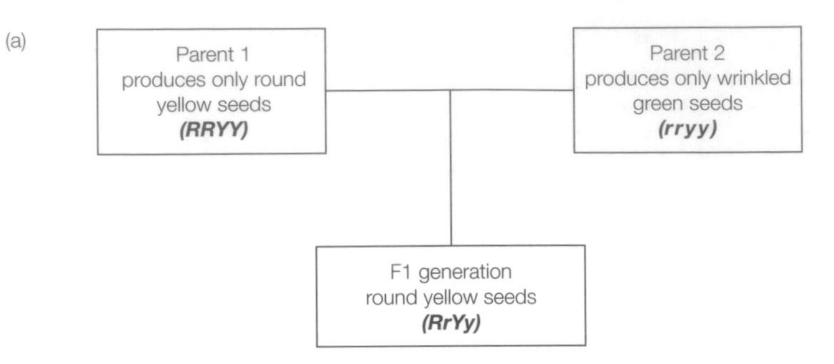

(b)

Gametes	***RY***	***Ry***	***rY***	***ry***
RY	***RRYY*** Round Yellow	***RRYy*** Round Yellow	***RrYY*** Round Yellow	***RrYy*** Round Yellow
Ry	***RRYy*** Round Yellow	***RRyy*** Round Green	***RrYy*** Round Yellow	***Rryy*** Round Green
rY	***RrYY*** Round Yellow	***RrYy*** Round Yellow	***rrYY*** Wrinkled Yellow	***rrYy*** Wrinkled Yellow
ry	***RrYy*** Round Yellow	***Rryy*** Round Green	***rrYy*** Wrinkled Yellow	***rryy*** Wrinkled Green

Phenotype class	*Genotype class*
Round, yellow seeds	*RRYY* (1), *RRYy* (2), *RrYY* (2), *RrYy* (4)
Round green seeds	*RRyy* (1), *Rryy* (2)
Wrinkled yellow seeds	*rrYY* (1), *rrYy* (2)
Wrinkled green seeds	*rryy* (1)

Fig. 1. (a) The production of F1 plants. These can produce four different gametes RY, Ry, rY, ry. The matrix in (b) (Punnett square) shows all the possible genotypes and phenotypes that can arise when these plants are self-fertilized. The figures in brackets indicate the number of times each genotype appears in the Punnett square.

round yellow, round green, wrinkled yellow and wrinkled green. These classes occur with equal frequencies.

It would be a useful exercise for you to work this out for yourself. However this type of cross is most important in studies of gene mapping and linkage and is covered in detail in Topic C4.

At this point it is necessary to make a brief comment about systems where more than two genes are studied simultaneously. In the case of three genes, **a trihybrid cross**, in which complete dominance is observed for alleles of all three genes, the phenotypes observed fall into the ratio 27 : 9 : 9 : 9 : 3 : 3 : 3 : 1 (see Topic C12). The predicted ratios can be determined for crosses involving any number of genes by use of the expression $(3:1)^n$, where n is the number of genes.

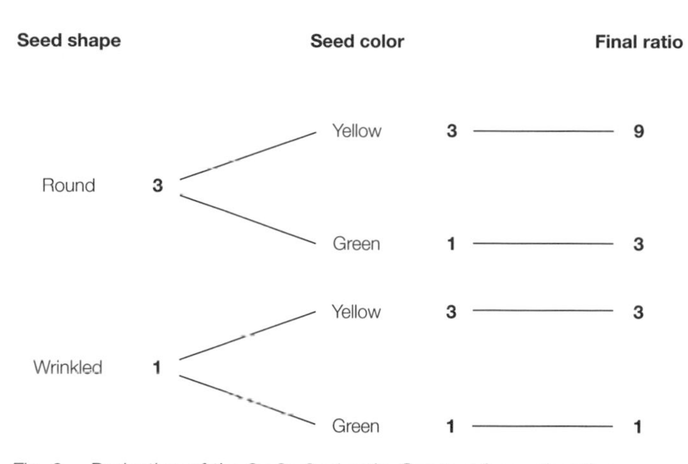

Fig. 2. Derivation of the 9 : 3 : 3 : 1 ratio. See text for explanation.

Epistasis

Epistasis refers to situations where the expected ratio of phenotypes is not observed due to some form of physiological interaction between the genes involved. This is usually seen as a distortion of the 9 : 3 : 3 : 1 ratio with a reduction in the number of different phenotypes observed. Many different ratios can be derived from the original 9 : 3 : 3 : 1 ratio. Three of these are shown in *Figs 3, 4* and *5*. Although each can be explained by specific examples from the literature it is simpler if the different ratios are described here by reference to fictitious biochemical pathways in an unnamed plant species in which two enzymes, that are coded for by separate genes, are both involved in the production of petal pigments. The genes are denoted *A* and *B* with capital letters representing a functional dominant allele and lower case letters representing nonfunctional recessive alleles. All of the examples involve the F2 generation produced from a doubly heterozygous *AaBb* F1 generation. In working through these you may find it useful to construct a Punnett square for the genotypes, as shown in *Fig. 1*, and determine for yourself how the phenotypes are distributed.

In a strict sense 12 : 3 : 1 is the only ratio which was originally referred to as epistasis, because the presence of the *A* allele can completely mask the genotype of the *B* gene, but the term is used now wherever genes interact to alter the expected ratios.

There are several other variations of the 9 : 3 : 3 : 1 ratio caused by interaction between the gene products. These include 9 : 6 : 1, 15 : 1 and 13 : 3. You should attempt to think of biochemical pathways that would yield these ratios. Remember that in every case the ratios are derived by summing together the four phenotype classes 9, 3, 3 or 1 of the basic ratio.

The examples described here are deliberately made simplistic but illustrate the basic principles. You should realize that the term phenotype is capable of different interpretations. A plant breeder may be happy to simply use flower color as we have done to describe phenotype. On the other hand a plant biochemist might wish to interpret the results differently and assay enzymes A and B *in vitro*. The phenotypic ratios determined in this way would be different from those given above. If this is not obvious to you, work them out for yourself. The genotypes would, of course, not change.

Mendel's laws

The monohybrid and dihybrid ratios come directly from the work of Mendel. Remember that this was carried out without any knowledge of chromosomes

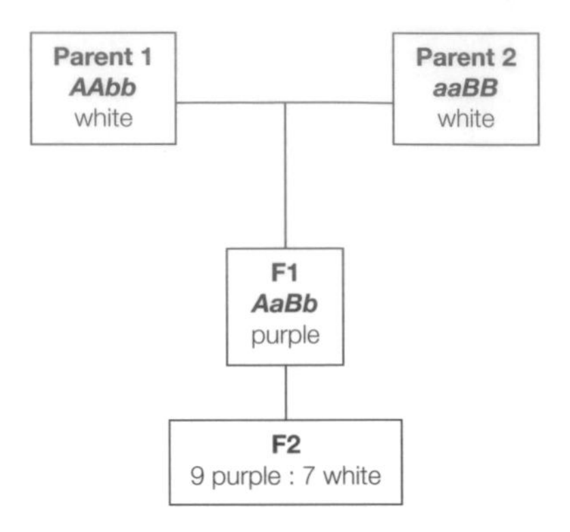

In this example the biochemical pathway would be a simple chain where enzyme A converts its substrate into a white product which is, in turn, the substrate for enzyme B which converts it to a purple product.

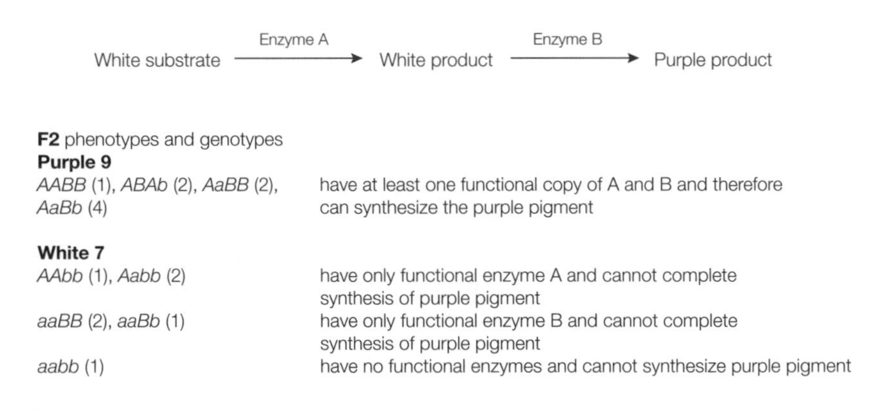

This is known as **complementation**, because the genes in the two white strains each complement the deficiency in the other strain, producing the purple product in the F1.

Fig. 3. 9 : 7 ratio.

or DNA. Mendel's work is often expressed as the two laws or principles which he inferred to explain the inheritance of phenotypes.

The first of these states that in gamete formation the two alleles of the same gene **segregate** (separate) so that each gamete receives only one allele. This is clearly demonstrated in the monohybrid 3 : 1 ratio.

The second law states that alleles at any one gene segregate independently of alleles at any other gene. This derives from the dihybrid cross data where, in the example described in *Fig. 1.* the four alleles *R*, *r*, *Y* and *y* must act independently of each other so that the four different classes of gametes *RY*, *Ry*, *rY* and *ry* arise in equal numbers. These two principles form the basis of our knowledge of transmission genetics.

Handling problems

As either a student in an examination, or a geneticist carrying out research you may be faced with data obtained from F1 and F2 generations of crosses. You would need to be able to recognize ratios in order to decide how many genes are involved and whether or not epistasis is taking place. You may for instance be faced with the following example.

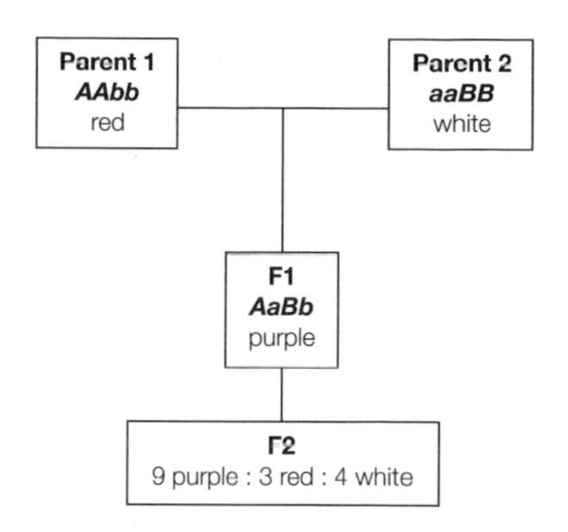

In this example the biochemical pathway would again be a simple chain, but in this case the product of enzyme A would be red in color

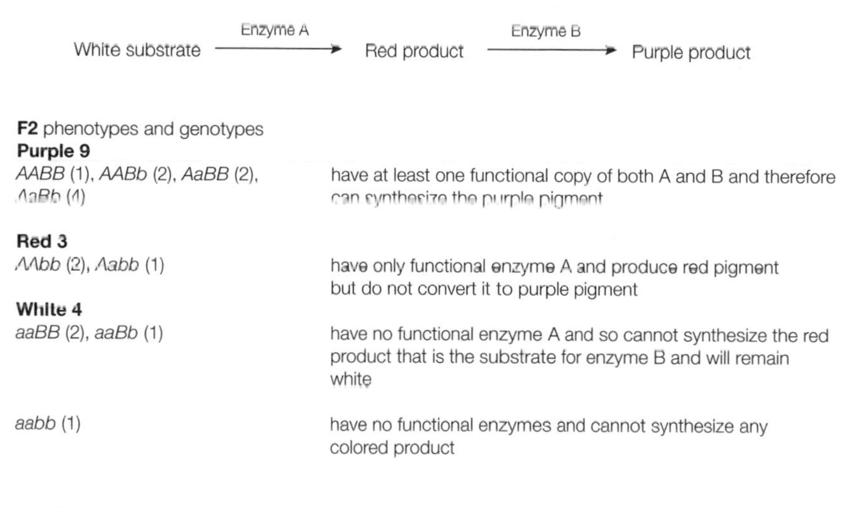

Fig. 4. 9 : 3 : 4 ratio.

A cross between pure-breeding white-fruited and purple-fruited tomato plants produced an F1 generation in which all plants had purple fruit. In the subsequent F2 generation 160 plants were obtained, of these 99 had purple fruit, 25 had red fruit and 36 had white fruit. How would you approach such a problem?

As you know nothing about the genes controlling fruit color in tomato you must first ask yourself the question, 'does the data fit any of the known Mendelian ratios?' The simplest way to proceed is to exclude those ratios which obviously do not apply. Clearly because there are three different phenotypes in the F2 generation any ratio with only two classes such as 9 : 7 or 3 : 1 are excluded. On examination of ratios with three phenotypic classes 9 : 3 : 4 looks a possible candidate, but a 1 : 2 : 1 ratio may also apply. How do you decide which ratio is the best fit to your data?

This is done using the χ^2 statistic, which is well suited to determining the goodness-of-fit to ratios. Detailed explanation of this test is given (see Topic C13), but this example will be worked through here. The observed data is

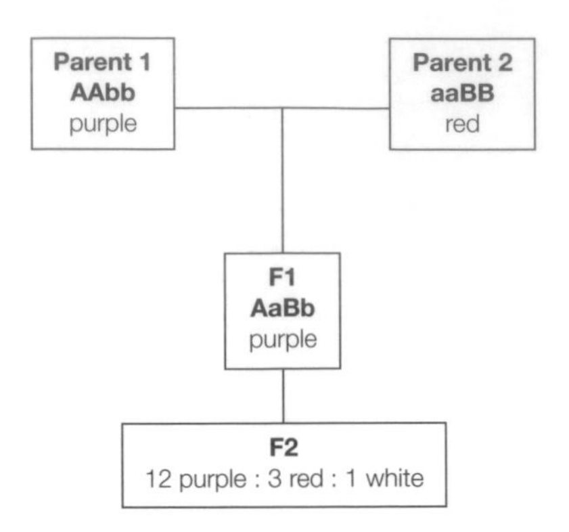

In this case we need a more complex biochemical pathway. Here two enzymes compete for the same substrate. Enzyme A which converts the substrate to a purple product has much higher affinity for the substrate than enzyme B which converts the substrate to a red product. The difference in affinity for the substrate is so marked that enzyme B can only work effectively if no enzyme A is present.

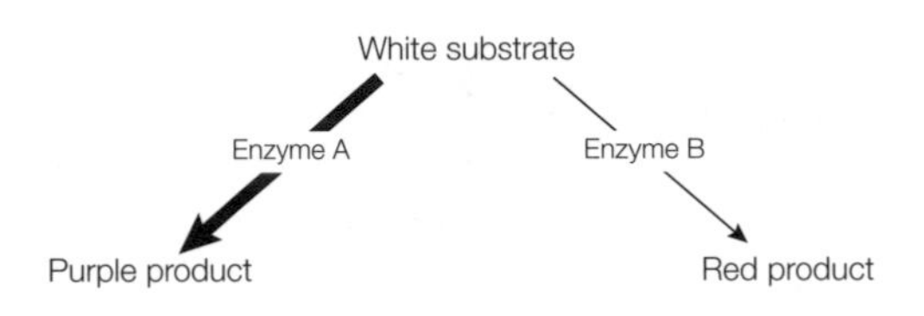

F2 phenotypes and genotypes

Purple 12
AABB (1), *ABAb* (2), *AaBB* (2), *AaBb* (4), *AAbb* (1), *Aabb* (2) — these have at least one functional allele A and convert all the substrate to purple product

Red 3
aaBB (2), *aaBb* (1) — lack any functional enzyme A, but have functional enzyme B which converts the substrate to a red product

White 1
aabb (1) — have no functional enzymes and cannot synthesize any colored pigment

Fig. 5. 12 : 3 : 1 ratio.

compared with that which would have been predicted by the ratio. This is set out for both possible ratios.

Observed result:	99 purple	25 red	36 white	160 total

Result predicted by a 9 : 3 : 4 ratio

$\frac{9}{16} \times 160$	$\frac{3}{16} \times 160$	$\frac{4}{16} \times 160$
90	30	40

$$\chi^2 = \frac{(99-90)^2}{90} + \frac{(30-25)^2}{30} + \frac{(40-36)^2}{40}$$

$= 2.13$ with two degrees of freedom

This does not differ from the 9 : 3 : 4 ratio at the 5% level of probability

Result predicted by a 1 : 2 : 1 ratio

¼ × 160	²⁄₄ × 160	¼ × 160
40	80	40

$$\chi^2 = \frac{(40 - 25)^2}{40} + \frac{(99 - 80)^2}{80} + \frac{(40 - 39)^2}{40}$$

$= 10.53$ with two degrees of freedom

This is different from the 1 : 2 : 1 ratio at the 5% level of probability.

On this basis the 1 : 2 : 1 ratio is rejected in favor of the 9 : 3 : 4 ratio.

C3 MEIOSIS AND GAMETOGENESIS

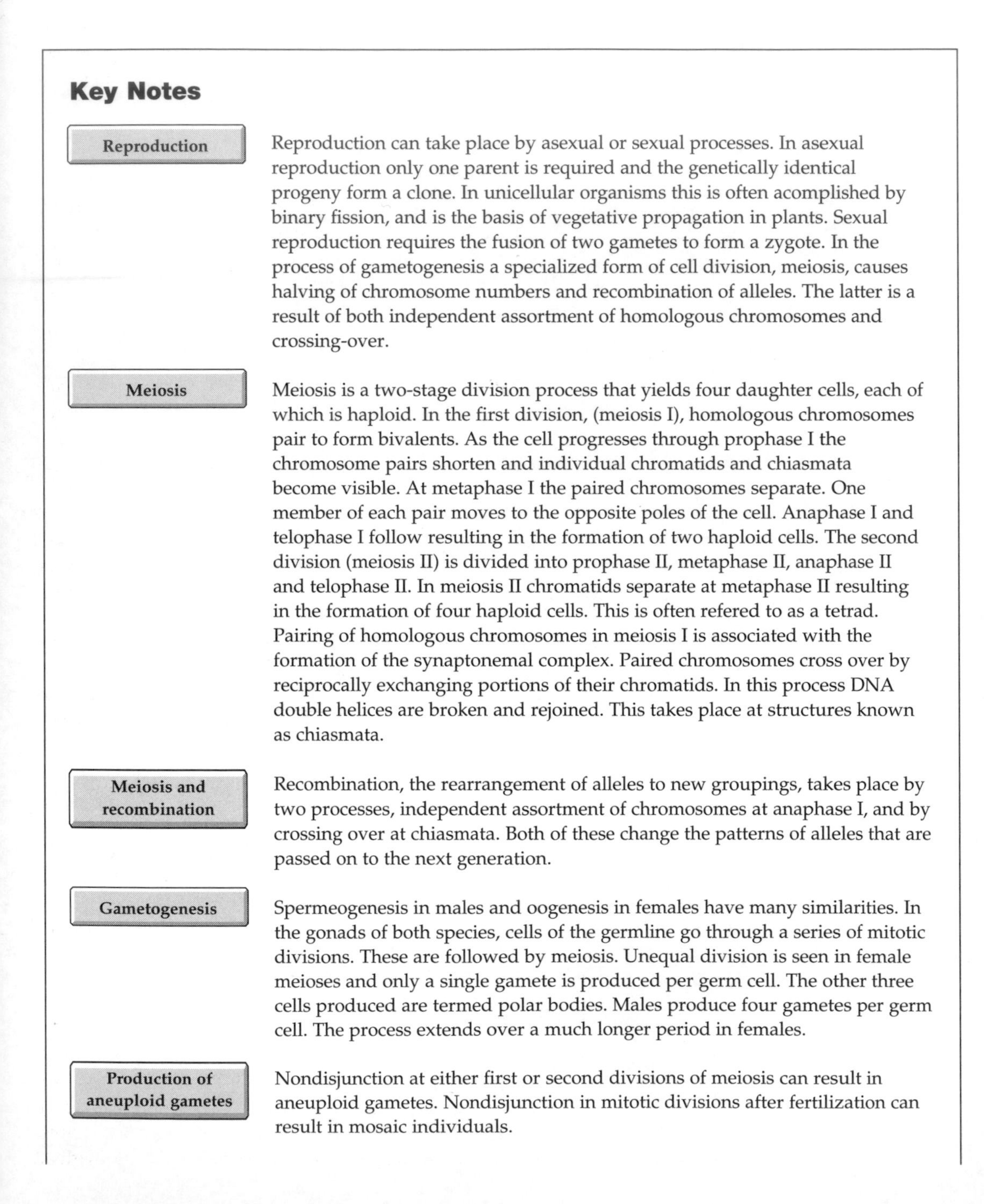

Key Notes

Reproduction

Reproduction can take place by asexual or sexual processes. In asexual reproduction only one parent is required and the genetically identical progeny form a clone. In unicellular organisms this is often acomplished by binary fission, and is the basis of vegetative propagation in plants. Sexual reproduction requires the fusion of two gametes to form a zygote. In the process of gametogenesis a specialized form of cell division, meiosis, causes halving of chromosome numbers and recombination of alleles. The latter is a result of both independent assortment of homologous chromosomes and crossing-over.

Meiosis

Meiosis is a two-stage division process that yields four daughter cells, each of which is haploid. In the first division, (meiosis I), homologous chromosomes pair to form bivalents. As the cell progresses through prophase I the chromosome pairs shorten and individual chromatids and chiasmata become visible. At metaphase I the paired chromosomes separate. One member of each pair moves to the opposite poles of the cell. Anaphase I and telophase I follow resulting in the formation of two haploid cells. The second division (meiosis II) is divided into prophase II, metaphase II, anaphase II and telophase II. In meiosis II chromatids separate at metaphase II resulting in the formation of four haploid cells. This is often refered to as a tetrad. Pairing of homologous chromosomes in meiosis I is associated with the formation of the synaptonemal complex. Paired chromosomes cross over by reciprocally exchanging portions of their chromatids. In this process DNA double helices are broken and rejoined. This takes place at structures known as chiasmata.

Meiosis and recombination

Recombination, the rearrangement of alleles to new groupings, takes place by two processes, independent assortment of chromosomes at anaphase I, and by crossing over at chiasmata. Both of these change the patterns of alleles that are passed on to the next generation.

Gametogenesis

Spermeogenesis in males and oogenesis in females have many similarities. In the gonads of both species, cells of the germline go through a series of mitotic divisions. These are followed by meiosis. Unequal division is seen in female meioses and only a single gamete is produced per germ cell. The other three cells produced are termed polar bodies. Males produce four gametes per germ cell. The process extends over a much longer period in females.

Production of aneuploid gametes

Nondisjunction at either first or second divisions of meiosis can result in aneuploid gametes. Nondisjunction in mitotic divisions after fertilization can result in mosaic individuals.

Related topics	Chromosomes (B1)	Recombination (B7)
	Cell division (B2)	Linkage (C4)

Reproduction

Reproduction takes place by one of two methods, asexual or sexual. Asexual reproduction involves the production of a new individual(s) from cells or tissues of a pre-existing organism. This process is common in plants and in many microorganisms. It can involve simple binary fission (splitting into two) in unicellular microbes or the production of specialized asexual spores. These processes may be exploited for commercial purposes as in the vegetative propagation of plants. More recently it has been possible to artificially regenerate whole organisms from a single cell. This was first shown in carrots and frogs, but it has now been reported in sheep, and, by implication, is possible in all mammals including humans. Asexually reproduced organisms are genetically identical to the individual from which they were derived. A group of such genetically identical organisms is known as a **clone**.

Sexual reproduction differs, in that it involves fusion of cells (**gametes**) one derived from each parent, to form a **zygote**. The genetic processes involved in the production of gametes allow for some genetic changes in offspring. Sexual reproduction is limited to species that are diploid or have a period of their life cycle in the diploid state.

The production of gametes is referred to as **gametogenesis**. This may be a complex process involving sexual differentiation and the production of highly differentiated male and female gametes, or in lower eukaryotes identical cells may fuse (isogamy). Whatever the biology of the process, one fact is obvious: gametogenesis must involve a halving of the chromosome number otherwise each succeeding generation would have double the chromosome number of its parents. Halving of chromosome numbers is achieved in a specialized form of cell division, **meiosis**, that is only observed in gametogenesis. During the process of reducing the number of chromosomes by half the combinations of alleles are rearranged to give **recombinant gametes**. Two distinct processes are involved. These are **independent assortment** of chromosomes and **crossing-over**. These are described below.

Meiosis

The pattern of the meiotic cell cycle is very different from the typical mitotic cycle (see Topic B2). Meiosis involves two successive divisions resulting in the production of four cells each of which has half the number of chromosomes of the mother cell. There is no replication of DNA between the two divisions, which are known as **meiosis I** and **meiosis II**.

Meiosis I is divided into **prophase**, **metaphase**, **anaphase** and **telophase**. These are the same phases that are used to describe mitosis (see Topic B2), but beware, behavior of the chromosomes in the first division of meiosis is very different from that in mitosis.

In prophase each chromosome pairs with its **homolog** (copy of the same chromosome inherited from the other parent). The paired chromosomes are called **bivalents** (the term tetrad is sometimes used to describe these structures but should be reserved for the four cells produced in at the end of meiosis). Each pair is held together by **chiasmata** (singular chiasma). These are structures where homologous chromosomes exchange reciprocal portions of chromatids.

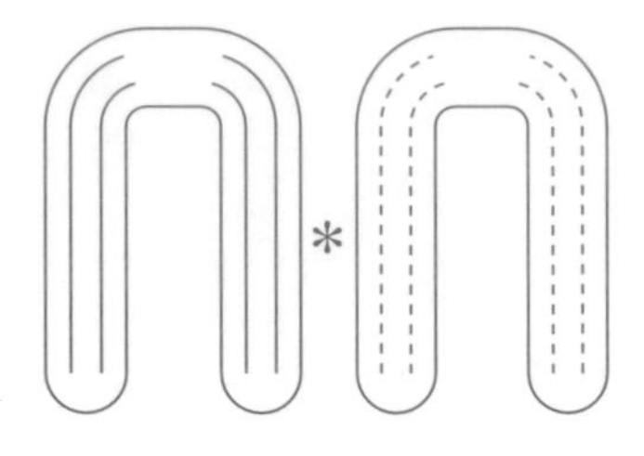

The chromatids break at the asterisk, and reciprocally exchange resulting in transfer of DNA and alleles of the genes that it encodes

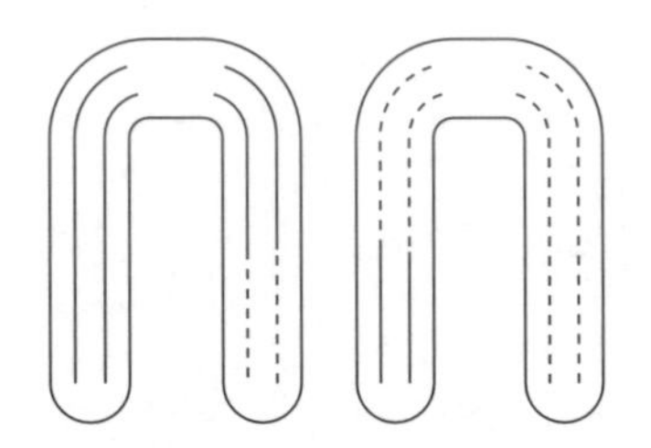

Fig. 1. The molecular basis of crossing-over.

The process involves breaking of the DNA double helix in each chromatid at corresponding sites and then joining the DNA of one chromatid to the DNA of the other. This exchange of genetic material is known as **crossing-over**. The molecular basis of this process is outlined in *Fig. 1.*

The cytological progression of meiosis is set out schematically in *Fig. 2.* Prophase of meiosis I is subdivided into five stages. In the first of these, **leptotene**, the chromatin is seen to condense into very long thin strands, that appear tangled in the nucleus. As prophase proceds the chromosomes become shorter and thicker. At **zygotene**, homologous chromosomes are seen as partially paired structures. The chromosomes are still very elongated at this stage and chromosome pairs may overlap or intertwine. By the third stage, **pachytene**, pairing is complete. It is still not possible to identify clearly the individual chromatids in each bivalent.

The pairing or **synapsis** of homologous chromosomes is a very precise process. The molecular mechanisms of this are still not completely resolved, but examination of paired chromosomes under the electron microscope reveals a structure known as the **synaptonemal complex**. This is a proteinaceous structure which is found only between homologs where they are synapsed (paired). Under the electron microscope it has a zipper-like appearance and is thought to be a major factor in the pairing process. The transition from pachytene to **diplotene** occurs as the homologous chromosomes begin to separate. This process begins at the centromeres and the bivalents are seen to be held together by chiasmata. The nuclear membrane and the nucleolus break down. As the chromosomes continue to condense, the cell moves into **diakinesis**, the final subdivision of prophase I. In diplotene and diakinesis the four chromatids (daughter chromosomes) in each bivalent are identifiable and individual chiasma can clearly be identified. At **metaphase I** the bivalents lie across the equator of the cell with their centromeres attached to the spindle. This is similar to the spindle in mitosis (see Topic B2). The dynamic action of the spindle causes one member of each homologous pair to move to opposite poles of the

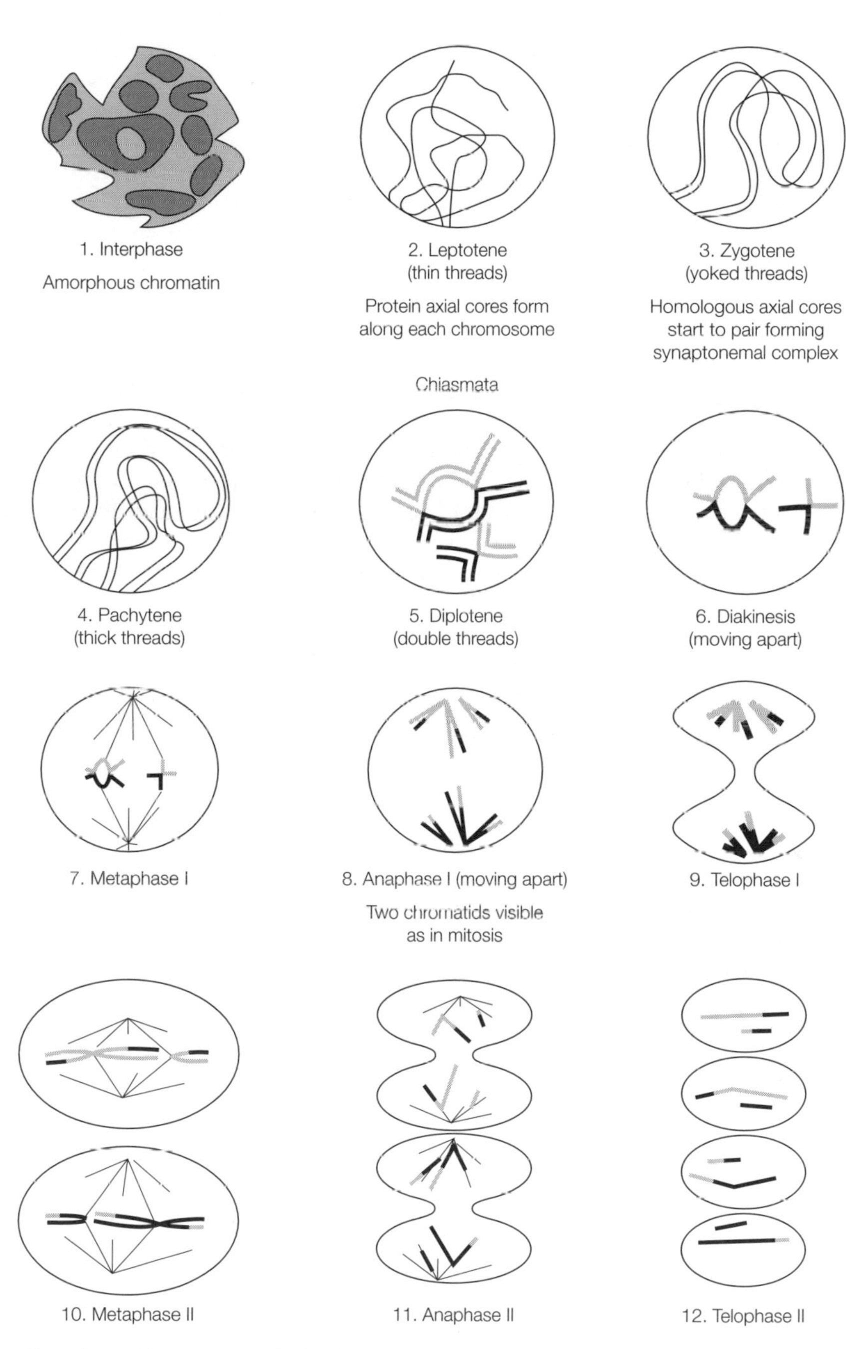

Fig. 2. Diagrams 1–6 outline the changes in a nucleus undergoing meiosis I. Only two pairs of homologous chromosomes are shown. One of each pair is darker than the other. The terms in brackets under each of the prophase stages are literal translations of the meaning of the stage names. Steps 7–9 represent the remaining stages of meiosis I. After this the cells go into a transient interphase, and prophase of meiosis. Steps 10–12 outline the major stages of meiosis II. The result is the formation of four haploid nuclei each of which carries recombined chromosomes.

cell. At **anaphase I**, the two sets of chromosomes are seen with their centromeres moving to the poles and the telomeres trailing behind. This is followed by **telophase I**, where two nuclei form around the segregating chromosomes and a degree of chromosome decondensation is observed. The two nuclei produced at meiosis I contain half the number of chromosomes found in the original cell. The diploid has been reduced to a haploid. The second meiotic division proceeds shortly after telophase I. Note that there is no replication of DNA between the two divisions, and that each chromosome still contains two chromatids.

The second meiotic division closely resembles mitosis. In **prophase II** the chromosomes are seen to recondense within the two nuclei. The nuclear membrane breaks down and the chromosomes are arrayed on the cell equator at **metaphase II**. At this point the centromere of each chromosome splits and the spindle pulls one chromatid of each chromosome to opposite poles.

Note that in metaphase I the spindle pulled complete chromosomes to each pole, while in metaphase II it is chromatids that are moved to opposite poles. This is followed by **anaphase II** and **telophase II**. At this stage the initial diploid cell has divided to give four haploid cells, each with a different genotype. In many instances the group of four haploid cells may remain together and is known as a **tetrad**.

Meiosis and recombination

From the stand-point of inheritance the crucial events in meiosis are those which are responsible for **recombination**, which means that the combinations of alleles passed by individuals to their offspring differ from the combinations they received from each of their parents. This helps to maintain the level of genetic variation within a population (see Topic D4). As noted above, recombination in meiosis arises by two distinct processes.

Each pair of homologous chromosomes consist of one inherited from the father and one from the mother. When paired homologous chromosomes separate (segregate) at anaphase I one member of each pair moves to the opposite poles of the cell. The process does not differentiate between maternally and paternally inherited chromosomes and so both haploid nuclei are likely to contain a random combination of maternally and paternally inherited chromosomes. Hence we have recombination due to **independent assortment** of chromosomes. The likelihood that the two daughter cells contain new combinations of maternally and paternally inherited chromosomes depends on the chromosome number of the species, as set out in *Table 1*.

In the dihybrid ratios obtained by Mendel (see Topic C2) the pairs of genes showed independent assortment. Hence all four classes of gamete were produced in equal numbers, resulting in the 9 : 3 : 3 : 1 ratios he observed. The frequency of recombination in gametes is further increased due to crossing-over. This only affects recombination between genes located on the same chromosome (synteny). The exchange of regions of chromatids inherited from

Table 1. The likelihood of recombination according to chromosome number

Haploid chromosome number	Likelihood of recombination
1	0
2	0.5
3	0.75
n	$(2^n - 2)/2^n$

different parents (observed at meiosis as chiasmata) will result in a new combination of maternally and paternally inherited alleles being inherited on the same chromosome by the next generation. This is discussed in great detail in Topic C4. The breakage and rejoining of DNA double helices involved in this process is generally referred to as the mechanism of recombination, and is fundamental to many genetic phenomena in both eukaryotes and prokaryotes. For greater detail see Topic A7.

Gametogenesis

Meiosis in animals is found only in ovaries and testes, and even in these tissues is restricted to cells that are destined to form gametes (the germline). Although the mechanisms of gametogenesis differs somewhat between organisms, the steps involved in gametogenesis in mammals outlined below are generally representative. These are summarized in *Fig. 3*. In male gametogenesis (**spermeogenesis**), precursors of germ cells go through many rounds of mitotic divisions in order to maintain a pool of **spermatagonia**, which subsequently differentiate into **primary spermatocytes**. It is in these cells that meiosis takes place. After the first meiotic division the cells are referred to as **secondary spermatocytes**. These are haploid. The products of the second meiotic division are **spermatids** which differentiate into motile, tailed **spermatozoa**. The final activation of spermatozoa takes place after copulation.

In female mammals the pattern of **oogenesis** is superficially similar. Here **oogonia** go through mitotic divisions before differentiating into **primary**

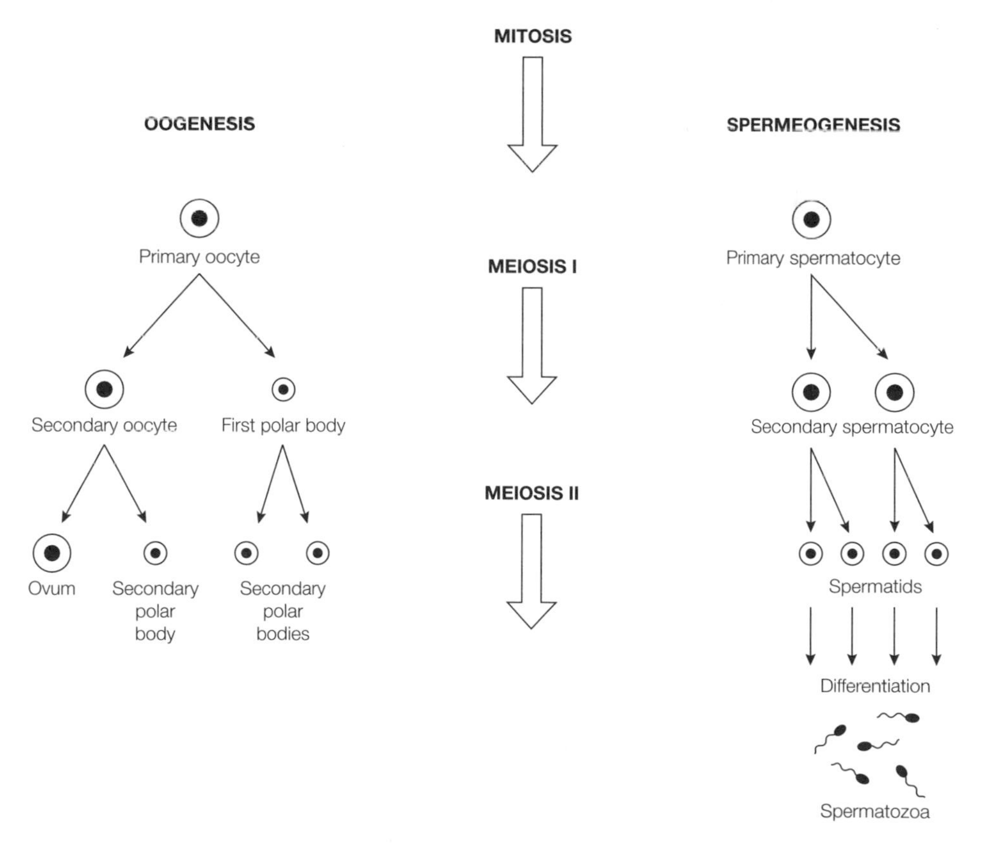

Fig. 3. Pattern of divisions in gametogenesis. For details see text.

oocytes. These then undergo meiosis. Both daughter cells of the first meiotic division are haploid, but differ greatly in size. The larger daughter cell is the **secondary oocyte**, the smaller the **first polar body**. The two cells remain attached. Both undergo a second meiotic division. The secondary oocyte undergoes an unequal division producing a large **ovum** that contains almost all the cytoplasm, and a small **secondary polar body**. The primary polar body also divides to form two further secondary polar bodies. Only the ovum will transmit genes into the next generation. Thus oogenesis differs from spermeogenesis in that only one of the four meiotic products functions as a gamete. Another difference involves the timescale during which the processes take place. Spermeogenesis is a relatively continuous process beginning at puberty. In oogenesis, production of primary oocytes is completed in the fetus. Cells are held in prophase of meiosis I until puberty. The first meiotic division is only completed at ovulation, and the second division occurs after fertilization. In human females, primary oocytes can be held in meiotic arrest for up to 45 years. This may be important in the increased frequency of aneuploid births observed in older mothers (see Topic B1).

Production of aneuploid gametes

Aneuploidy was described in Topic B1. The major cause of aneuploidy is aberrant chromosome behavior at meiosis. Errors can arise at either of the two meiotic divisions. Failure of chromosomes to segregate properly is known as **nondisjunction**.

First division nondisjunction occurs when two homologous chromosomes move to the same pole at anaphase I. The end result of this is that of the four cells arising from meiosis, two will be disomic (contain two copies of the chromosome) and two nullisomic (contain no copy of the chromosome) for the chromosome in question. Where two chromatids remain together at anaphase II the tetrad contains two normal monosomic cells (contain one copy of the chromosome), one nullisomic cell and one disomic cell. Aneuploidy can also arise due to nondisjunction at an early mitosis in the embryo. This results in two populations of cytogenetically different cells in the individual, which is known as a **mosaic**. This is common in Turners syndrome (see Topic B1).

C4 LINKAGE

Key Notes

Definition

Linkage is the tendency for alleles of different genes to be passed together from one generation to the next. Only genes situated on the same chromosome can show linkage. Genes on nonhomologous chromosomes are, by definition, unlinked and always show 50% recombination. Parental gametes carry the same set of alleles as were inherited together from one parent. Recombinant gametes carry alleles derived from both parents. The degree of linkage between two genes depends on the frequency of cross-overs that occur between them during meiosis. The closer they are together the less likely a cross-over will occur between them. Groups of genes that are linked are called linkage groups.

Simple measurement of linkage

Two-factor crosses involve crossing a double heterozygote to a double recessive. The frequencies of different phenotypes in the progeny equal the frequencies of different gametes in the double heterozygote. The proportion of recombinant gametes is the recombination frequency. This is used to give a measure of the distance between two genes. The percentage recombination between two genes is taken as the distance they are apart. One percent recombination equals one map unit or centimorgan (cM). In linkage analysis in humans it is more common to utilize molecular techniques to analyze genotypes than to study phenotypes.

Three-factor crosses

These are more accurate than two-factor crosses, in that they identify and utilize many of the double cross-overs that are missed by the two-factor cross. They allow ordering of genes and generate additive map distances.

Interference

It is difficult for chiasmata to form close to one another. For this reason the number of double cross-overs observed in three-factor crosses may differ from that predicted from the frequency of single cross-overs. If fewer cross-overs are observed this is termed positive chromatid interference. This is measured as the coefficient of coincidence.

Linkage analysis in fungi

In ascomycete fungi, that have ordered asci, first and second division segregation patterns can be used to determine distance between a gene and the centromere. A similar approach can be used to estimate distances between genes.

Recombination frequency and physical distance

The distance between genes as measured by recombination frequencies is not a precise measure of the physical distance because the frequency of crossing-over varies in different parts of the genome. It is useful for deciding on gene order. Physical maps provide distance between genes in absolute terms.

Related topics

Meiosis and gametogenesis (C3) The human genome project (F5)

Definition

Linkage is the tendency for alleles of two or more genes to pass from one generation to the next in the same combination. This usually means that the closer together any two genes lie on the same chromosome the more likely they are to show linkage and the stronger that linkage will be.

Genes located on different (nonhomologous) chromosomes cannot, by definition, show any linkage. As discussed in Topic C3 genes on different chromosomes assort independently at the first division of meiosis.

Consider a double heterozygote *Aa, Bb* where the two genes involved are on different chromosomes. Assume that this individual has inherited the dominant alleles *A* and *B* from one parent, and the recessive alleles *a* and *b* from the other. In meiosis four genetically different gametes, *AB, Ab, aB* and *ab* will be produced in equal proportions, Those gametes carrying *AB* or *ab* are referred to as **parental** and those carrying *aB* or *Ab* as **recombinant**. The recombination frequency is obviously 50%. This is the maximum recombination frequency that can be obtained between any two genes, and genes that show this recombination frequency are said to be **unlinked**.

What would have happened had the two genes been located on the same chromosome? Now the parental allele combinations *AB*, or *ab* can only be rearranged (recombined) if a cross-over occurs at a chiasma between the two genes (see Topic C3). The closer the two genes are together the less likely it is that a cross-over will take place between them. The frequency of recombinants observed will therefore give an indirect measure of how close the two genes lie to each other.

In practice the frequency of cross-overs on most chromosomes is high, and this means that genes that lie far apart of the same chromosome show 50% recombination. If two genes show linkage and a third shows linkage to only one of the original two, by definition they must all be on the same chromosome. They are said to constitute a **linkage group**. Thus linkage maps can be built up even though some genes in the group may not show linkage to all other members of the group.

Simple measurement of linkage

Except in certain circumstances where linkage can be measured to the centromere it is necessary to have heterozygosity for at least two genes in any study of linkage. Such a system is called a two-factor cross. An example is set out below.

An F1 hybrid was made by crossing two pure-breeding strains of tomato, one of which was homozygous for purple fruit and hairy stems, *PP HH*, and the other was homozygous for red fruit and smooth stems, *pp hh*. The hybrid had purple fruit and hairy stems. Thus purple fruit is dominant to red fruit, and hairy stems is dominant to smooth stems. To determine if the genes for fruit color and hairiness of stem are linked, the doubly heterozygous hybrid was crossed to the double recessive (red-fruited and smooth-stemmed) parent. This is a test cross (see Topics C1 and C2). Four classes of progeny were obtained (*Table 1*): classes 1 and 2 are the same phenotypes as the original parental strains and are therefore **parental**: classes 3 and 4 represent new combinations of the alleles at the two genes and are therefore **recombinant**.

If the genes were unlinked a ratio of 1 : 1 : 1 : 1 would be expected between the four possible phenotypes (see Topic C2) Clearly this is not the case. The total number of progeny is 500 of which only 70 are recombinant. The recombination frequency is 70/500, 14%. This is strong evidence that the two genes are linked. They have a map distance of 14 map units between them

Table 1. Phenotypes produced in a test cross between F1 generation purple-fruited, hairy-stemmed tomatoes (Pp, Hh)*, and the double recessive red-fruited, smooth-stemmed parent* (pp, hh)

Phenotypic class	Frequency
(1) Purple-fruited, hairy-stemmed	220
(2) Red-fruited, smooth-stemmed	210
(3) Purple-fruited, smooth-stemmed	32
(4) Red-fruited, hairy-stemmed	38
Total	500

(1% recombination is equivalent to 1 map unit). Map units are referred to as centimorgans (cM) in memory of Thomas Hunt Morgan who was the first geneticist to explain linkage.

Is it necessary to use a test cross to determine linkage? It is not essential, and other crosses can be used, for instance, crossing the double heterozygotes together as in a dihybrid cross (see Topic C2) will give data on linkage, as the numbers in recombinant classes of the 9 : 3 : 3 : 1 ratio will be reduced. However, because the double recessive parent contributes nothing to the phenotypes of the progeny in the test cross, the ratios obtained represent exactly the ratios of gametes and so this cross gives maximum information on linkage. In current studies of linkage in humans (see Topic F5) recombination frequencies are usually determined by the use of molecular methods to identify differences in DNA sequence (the genotype) rather than the phenotype as described in the examples given here. Working with genotypes allows more information to be obtained from crosses. The two-factor cross has limited use in determination of linkage and gene mapping studies. To map the order of genes along a chromosome and to give more accurate estimates of the distances between these genes at least three genes should be studied in the same cross.

Three-factor crosses

A major advantage of a three-factor cross is that genes may be placed in order and that the map distances, at least over relatively short distances, are additive. The worked example given below again involves a test cross. For simplicity, in this example alleles denoted by capital letters are completely dominant over the lower case alleles, and the three genes are simply named after their dominant alleles.

An individual heterozygous at three genes *Aa, Nn* and *Rr* is crossed with the homozygous recessive parent *aa, nn,* and *rr.* The frequency of progeny with different phenotypes is given in *Table 2.*

If there were no linkage between these genes each of the eight classes of progeny should have arisen with equal frequency. This is clearly not the case: classes (1) and (2) represent progeny from gametes where no recombination had taken place between the three genes. These are parental gametes because they retain the parental combination of alleles of the three genes; classes (3) and (4) represent progeny from gametes where recombination has occurred between the genes *N* and *R*; classes (5) and (6) represent recombination between the genes *A* and *N*; classes (7) and (8) are the most important because they represent gametes where recombination has taken place both in the interval between *A* and *N* and also between *N* and *R*. These are referred to as double cross-overs; they are less frequent than any of the other classes because they

Table 2. Phenotypes of progeny produced from the cross between an individual heterozygous for three genes, and one homozygous recessive for the same three genes

Phenotypes of progeny	Frequency	Class
Parental		
A N R	347	(1)
a n r	357	(2)
Recombinant		
A N r	52	(3)
a n R	49	(4)
A n r	90	(5)
a N R	92	(6)
A n R	6	(7)
a N r	7	(8)
Total	1000	

require two independent events to take place. The two least frequent classes of progeny in a test cross of this nature can be used to identify which gene lies between the other two. In this case the central gene is *N*.

To determine the map distance between the genes it is necessary to quantify all the recombination events that have occurred.

(i) To determine the map distance between *A* and *N*:

add progeny in classes (5) and (6) (recombination between *A* and *N*)
and progeny in classes (7) and (8) (double cross-over, one of which is between *A* and *N*)

Express the total as a percentage of all progeny.

$$\frac{(90 + 92 + 6 + 7) \times 100}{1000} = 19.5\% \text{ recombination, or } 19.5 \text{ cM}$$

(ii) To determine the map distance between *N* and *R*:
By the same logic as used for the map distance between *A* and *N* the distance between genes *N* and *R* is:

$$\frac{(52 + 49 + 6 + 7) \times 100}{1000} = 11.4\% \text{ recombination, or } 11.4 \text{ cM}$$

(iii) To determine the map distance between *A* and *R*:
This requires a summation of all recombinants.

$$\frac{(52 + 49 + 90 + 92 + 6 + 7 + 6 + 7) \times 100}{1000} = 30.9\% \text{ recombination, or } 30.9 \text{ cM}$$

Note that the distances are additive.

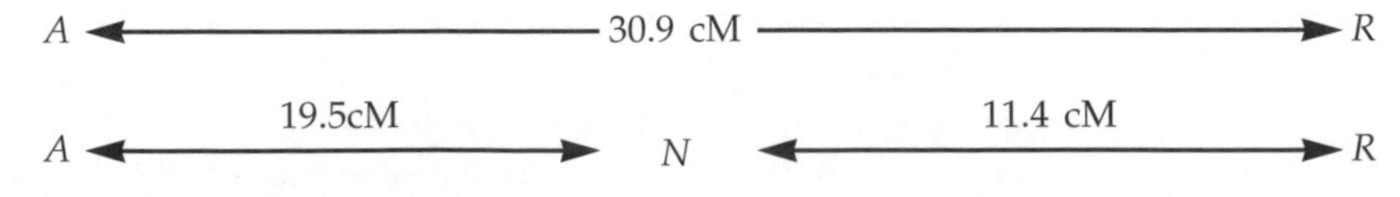

Three-factor crosses are more accurate than two-factor crosses because they detect, and utilize double cross-overs that would go undetected in two-factor crosses. Because more recombination events are detected the calculated recombination frequencies are greater. This is not a problem as recombination frequencies merely reflect the frequency of cross-overs detected rather than act as a physical measurement of actual distance between genes. This point is referred to again in Topic F5.

Interference

Crossing-over takes place at chiasmata. These are physical structures involving two chromatids. Not surprisingly the presence of one chiasmata in a particular chromosome region can reduce the frequency of others forming close to it. This can result in a reduction in the number of double cross-overs observed. In the previous example the observed frequency of double cross-overs is 13 / 1000 (0.013). By using the data for the observed single cross-over between *A* and *N* and between *N* and *R* we can predict the expected number of double cross-overs.

Single cross-overs between A and $N = 195$

Single cross-overs between N and $R = 114$

Predicted number of double cross-overs

$$(195/1000) \times (114/1000) \times 1000 = 22.3$$

This is greater than the observed number, 13, suggesting that **positive chromatid interference** is observed in this region. The extent of interference is calculated as the **coefficient of coincidence (S)**, the observed number of double cross-overs divided by the expected number of double cross-overs. In this case it would be:

$$S = \frac{13}{22.3} = 0.58$$

Linkage analysis in fungi

Several ascomycte fungi produce asci which hold the haploid ascospores produced after meiosis in a specific linear order, an **ordered tetrad**. This order reflects the organization of the bivalents at meiosis. The distal and proximal pair of ascospores each contain the products of one bivalent after the completion of both meiotic divisions. In some species the ascus contains eight spores. This is simply due to each spore having duplicated by mitosis (see Topic B2) so that the original pattern is displayed by pairs rather than single spores. The example shown in *Fig. 1.* represents the products of meiosis in a fungus heterozygous for pale and dark spores. The asci contain eight spores.

Where the four asci of the same color are found together there have been no cross-overs between the spore color gene and the centromere. This pattern is known as **first division segregation**, because the two phenotypes are physically separated at the first meiotic division. In order to show other patterns a cross-over must have taken place between the centromere and the gene for spore color. The color phenotypes are now separated after the second meiotic division, **second division segregation**. Depending on which chromatids are involved in the cross-over different patterns of light and dark spores can be observed in asci showing second division segregation (recombinant asci). These are shown in *Fig 2*. The percentage recombination is determined as:

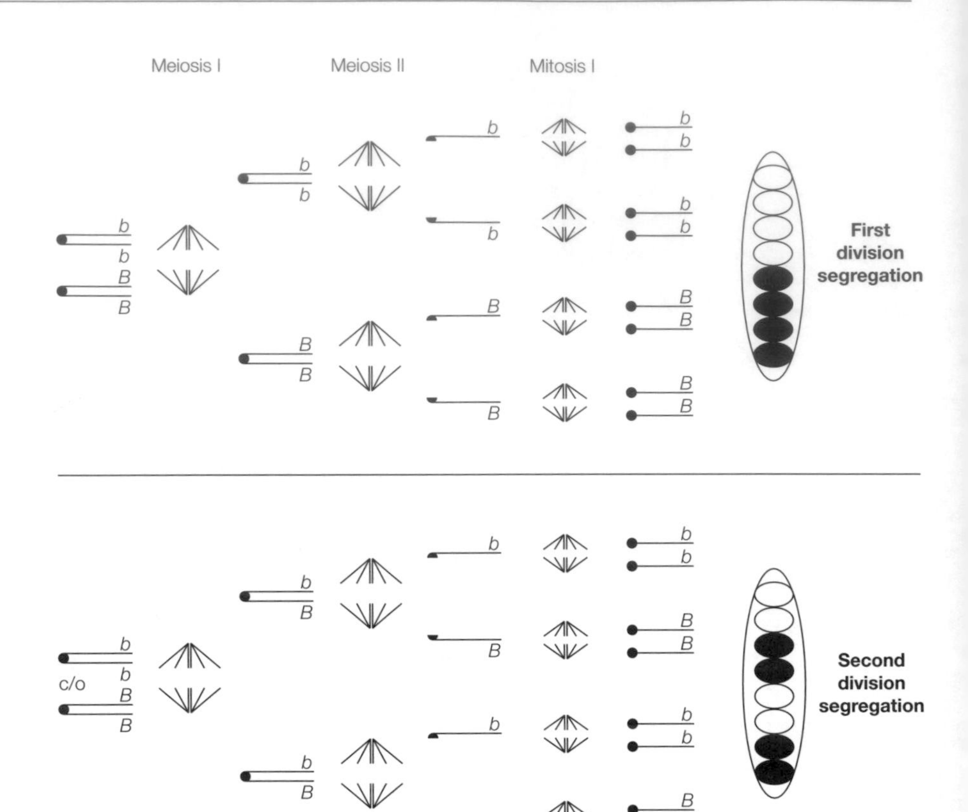

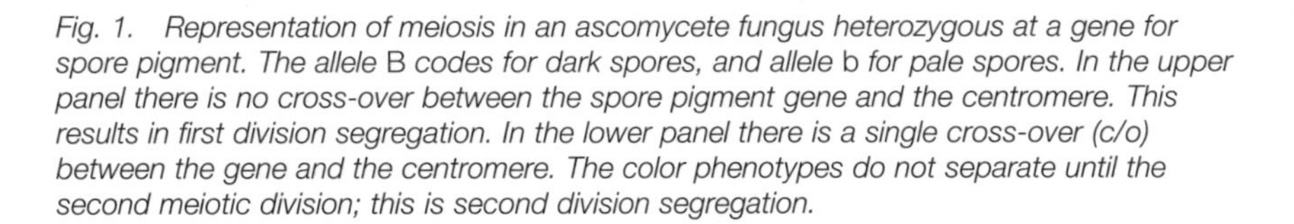

Fig. 1. Representation of meiosis in an ascomycete fungus heterozygous at a gene for spore pigment. The allele B *codes for dark spores, and allele* b *for pale spores. In the upper panel there is no cross-over between the spore pigment gene and the centromere. This results in first division segregation. In the lower panel there is a single cross-over (c/o) between the gene and the centromere. The color phenotypes do not separate until the second meiotic division; this is second division segregation.*

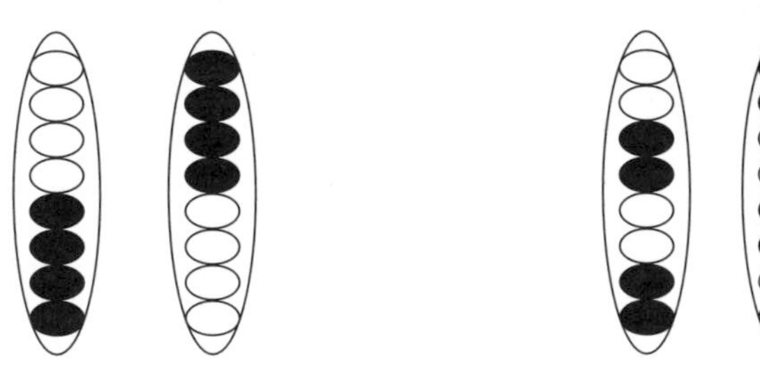

First division segregation

Second division segregation

Fig. 2. Different arrangements of asci in ascospores showing first or second division segregation. As shown each pattern is present twice depending on which member of the bivalent is distal at the time of meiosis. The two different patterns of second division segregation asci depend on which pairs of chromatids were involved in the cross-over.

$$\frac{1/2\ (\text{number of second division segregating asci}) \times 100}{(\text{total number of asci})}$$

The number of second division segregating asci is halved because in each of these asci only half of the chromatids have had a cross-over between the spore color gene and the centromere. As well as the distance to the centromere the distance between two genes can also be mapped in ordered tetrads using a similar procedure. Map distances can also be determined by counting phenotypes of individual ascospores in fungal species which do not produce ordered tetrads, or when the geneticist is too lazy to micro-dissect each ascospore individually from the ascus. However only analysis of ordered tetrads allows mapping of the distance between a gene and the centromere.

Recombination frequency and physical distance

We have assumed that recombination frequency is a measure of the actual distance between genes. This is only true in a semi-quantitative sense. It is now known that the frequency of cross-overs varies in different regions of the genome. It tends to be lower near centromeres and higher near telomeres. It can also differ between males and females. However gene maps based on linkage data give a reasonably accurate indication of distance between genes and they are most useful in determining the order of genes along the chromosome.

C5 Transfer of genes between bacteria

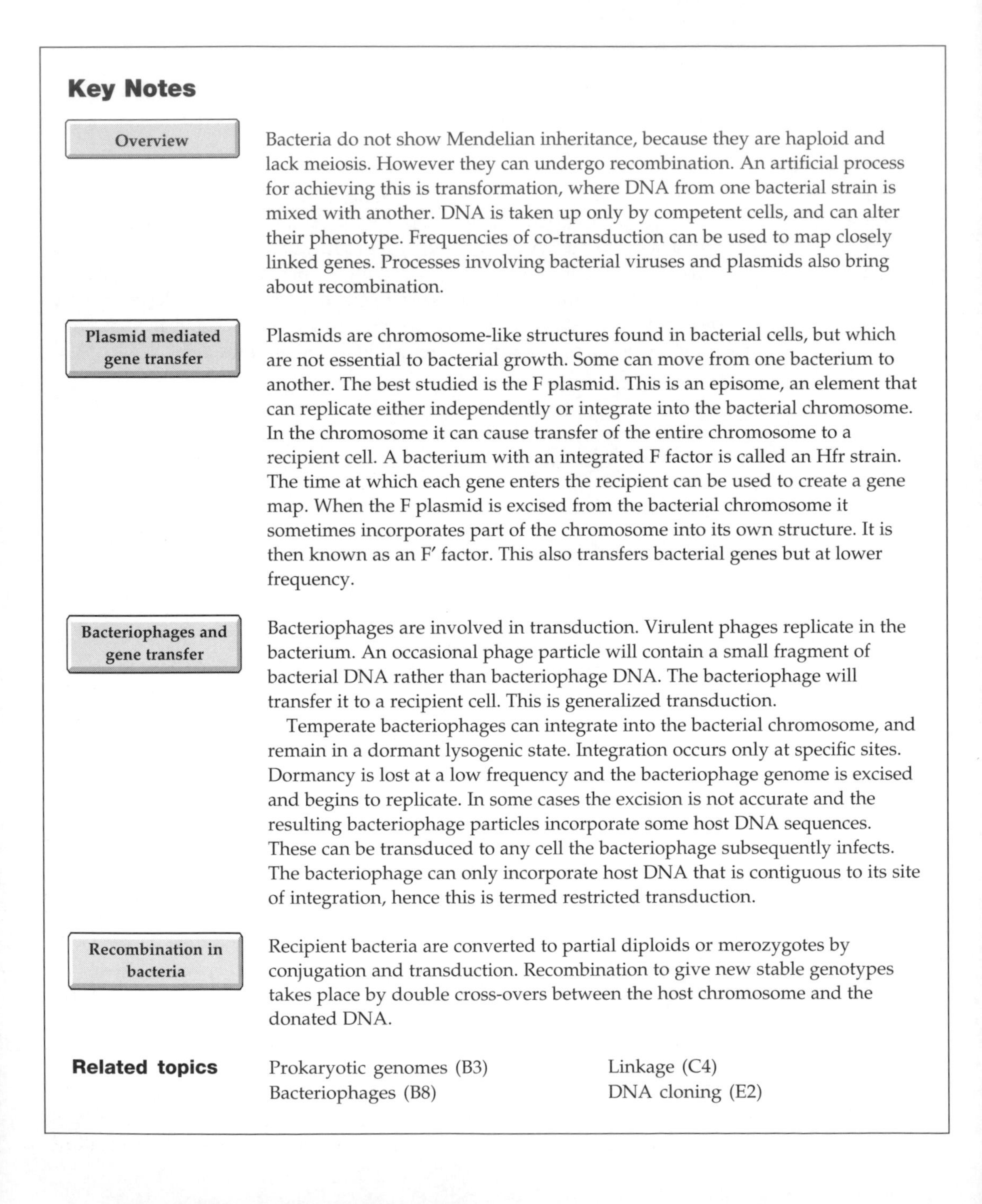

Key Notes

Overview

Bacteria do not show Mendelian inheritance, because they are haploid and lack meiosis. However they can undergo recombination. An artificial process for achieving this is transformation, where DNA from one bacterial strain is mixed with another. DNA is taken up only by competent cells, and can alter their phenotype. Frequencies of co-transduction can be used to map closely linked genes. Processes involving bacterial viruses and plasmids also bring about recombination.

Plasmid mediated gene transfer

Plasmids are chromosome-like structures found in bacterial cells, but which are not essential to bacterial growth. Some can move from one bacterium to another. The best studied is the F plasmid. This is an episome, an element that can replicate either independently or integrate into the bacterial chromosome. In the chromosome it can cause transfer of the entire chromosome to a recipient cell. A bacterium with an integrated F factor is called an Hfr strain. The time at which each gene enters the recipient can be used to create a gene map. When the F plasmid is excised from the bacterial chromosome it sometimes incorporates part of the chromosome into its own structure. It is then known as an F′ factor. This also transfers bacterial genes but at lower frequency.

Bacteriophages and gene transfer

Bacteriophages are involved in transduction. Virulent phages replicate in the bacterium. An occasional phage particle will contain a small fragment of bacterial DNA rather than bacteriophage DNA. The bacteriophage will transfer it to a recipient cell. This is generalized transduction.

Temperate bacteriophages can integrate into the bacterial chromosome, and remain in a dormant lysogenic state. Integration occurs only at specific sites. Dormancy is lost at a low frequency and the bacteriophage genome is excised and begins to replicate. In some cases the excision is not accurate and the resulting bacteriophage particles incorporate some host DNA sequences. These can be transduced to any cell the bacteriophage subsequently infects. The bacteriophage can only incorporate host DNA that is contiguous to its site of integration, hence this is termed restricted transduction.

Recombination in bacteria

Recipient bacteria are converted to partial diploids or merozygotes by conjugation and transduction. Recombination to give new stable genotypes takes place by double cross-overs between the host chromosome and the donated DNA.

Related topics

Prokaryotic genomes (B3)
Bacteriophages (B8)
Linkage (C4)
DNA cloning (E2)

Overview

Bacteria are prokaryotes and do not show the conventional Mendelian genetics of eukaryotes. This is primarily because they are haploid and lack any cell division process that is equivalent to meiosis. However, this does not mean that they do not show genetic recombination or transfer of genes between individuals; they have a number of mechanisms that allow gene transfer, some of these are prokaryote-specific but some have an equivalent in eukaryote cells.

It is possible to transfer DNA extracted from one bacterial culture to recipient bacteria *in vitro*. This is a process known as **transformation**, or transfection. Recipient bacteria are treated so as to increase the proportion of cells that are **competent** to take up exogenous DNA. These are then mixed with purified DNA and those that take up the DNA are selected as clones on agar plates. This can be useful in constructing strains of bacteria with specific genetic markers or in moving genes between species. It can also be used to identify and map linked genes. This is achieved by comparing the frequencies of transformation for individual genes with the frequency at which they show co-transformation.

If gene *A* transforms recipient bacteria with a frequency of 10^{-4} and gene *B* with a frequency of 4×10^{-5}, the predicted frequency of bacteria co-transformed with both genes would be 4×10^{-9} (the product of the two individual frequencies). This assumes that each transformation event is due to uptake of separate fragments of DNA carrying either gene *A* or *B*. If however, the two genes are close together on the chromosome of the bacteria used as donor for the DNA in the transformation experiment, then a proportion of the DNA molecules will carry both genes and so the frequency of co-transformation will be increased. This process only allows identification of closely linked genes, but if one gene can be shown to be linked to two other genes in separate experiments, then all three genes must lie close together on the bacterial chromosome even if the outer two are too far apart to show co-transformation. Thus a gene map can be built up in a manner similar to the linkage maps in eukaryotes where recombination frequencies are measured.

Transformation, although it may possibly occur in the natural environment of bacteria, is basically an artificial system for the mapping and manipulation of genes in the laboratory. However natural processes in prokaryote organisms can also be utilized in mapping bacterial genes. These depend on either the action of bacterial viruses (**bacteriophages**), or chromosome-like structures known as **plasmids**.

Plasmid mediated gene transfer

Plasmids are found in many bacterial species. They are composed of double-stranded DNA, code for proteins and are analogous to the bacterial chromosome itself. However plasmids can be removed (cured) from bacteria without harm to the host cells and so it is clear that they are nonessential structures. In many cases plasmids can move from one bacterium to another, and even between bacteria of different species. The transfer process is directed by genes carried on the plasmids. The ability of plasmids to move between bacteria can be utilized to study transfer of genes from the bacterial chromosome.

The first plasmid to be shown to move between bacteria was the F plasmid. This plasmid carries genes that direct the construction of pili, structures formed between bacteria that contain the plasmid, F^+, and those which lack the plasmid, F^-, through which the F plasmid can be transferred. The plasmid is duplicated by the process of rolling circle replication (see Topic A9) as it is transferred. This means that the donor cell retains a copy of the plasmid, and that in a mixture of F^+ and F^- bacteria the F plasmid will eventually spread to all bacteria in the culture.

The F plasmid is sometimes known as the F, or fertility factor, and F^+ bacteria as males. This is unhelpful, as there is no parallel with the processes of sex determination in eukaryotes.

The F plasmid can exist in the bacterial cell in an autonomous state where it replicates independently of the bacterial chromosome. It can also integrate into the bacterial chromosome and replicate as part of that chromosome. We refer to plasmids that can exist in these two states as **episomes**. The process is reversible and occasionally the F plasmid is excised from the chromosome. Sometimes the excision process is aberrant resulting in variable lengths of the bacterial chromosome being incorporated into the plasmid, which is now termed an F′ (F prime). This is shown in *Fig. 1(a)*. F primes are capable of moving to F^- bacteria and in so doing bring their complement of host genes with them. This is known as **sexduction** (*Fig. 1b*). Gene transfer by this process occurs at low frequencies.

Much higher frequencies of transfer are obtained when the F plasmid is remains integrated within the chromosome. It still retains its ability to transfer itself to F^- bacterial cells but now carries the rest of the chromosome with it. The process is known as **conjugation** (*Fig. 2a*), and operates as follows: the bacterial chromosome is broken within the F factor and transferred into recipient F^- bacterium; the portion of the F plasmid which is transferred first is called

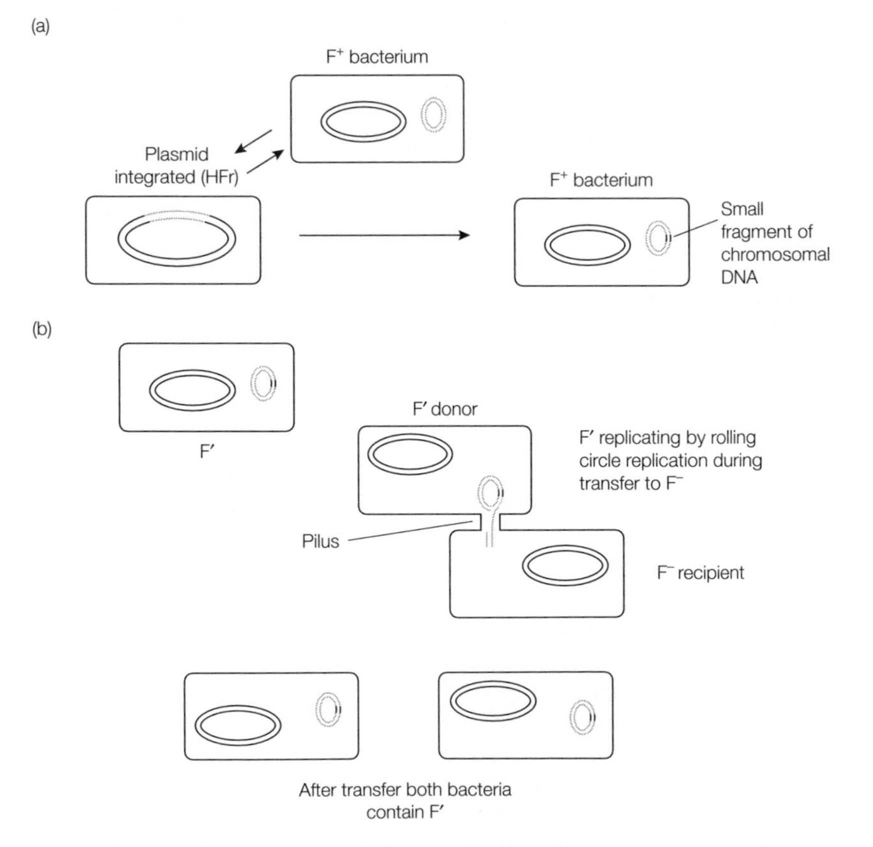

Fig. 1. (a) Relationship between F plasmid and bacterial chromosome. Plasmid DNA, shaded; chromosomal DNA, solid. (b) Transfer of F^1 plasmid from F' bacterium (donor) to F^- bacterium (recipient). This is known as sexduction.

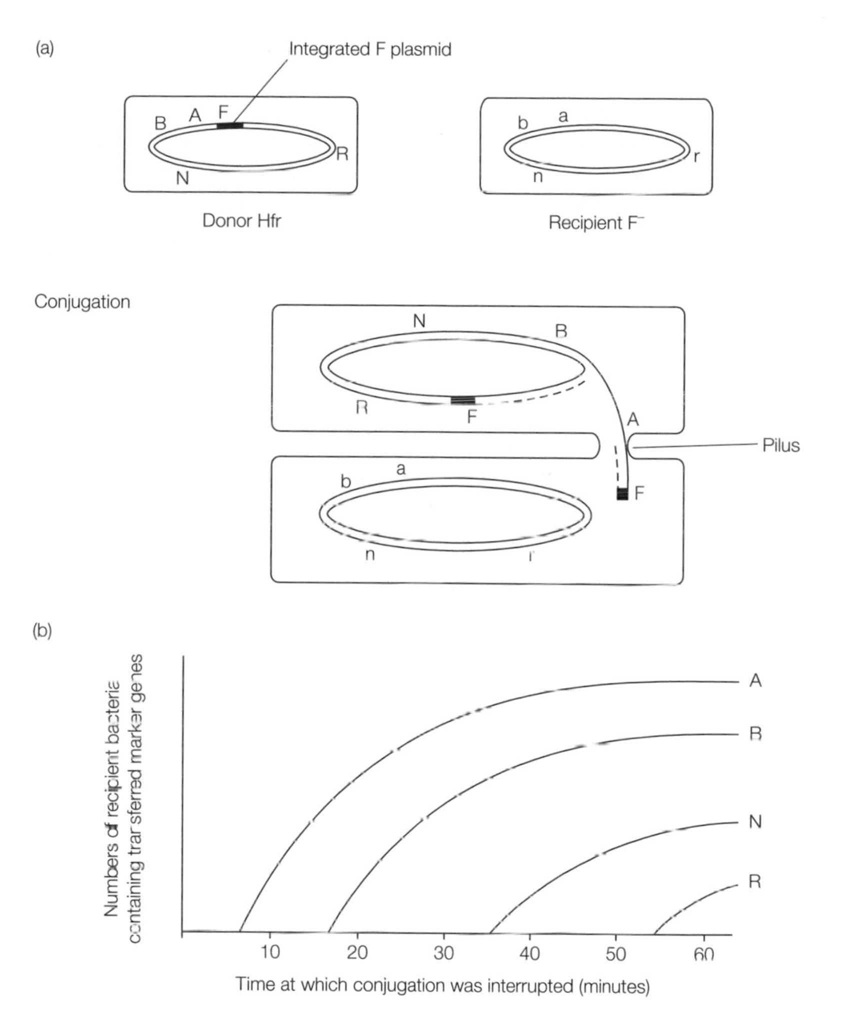

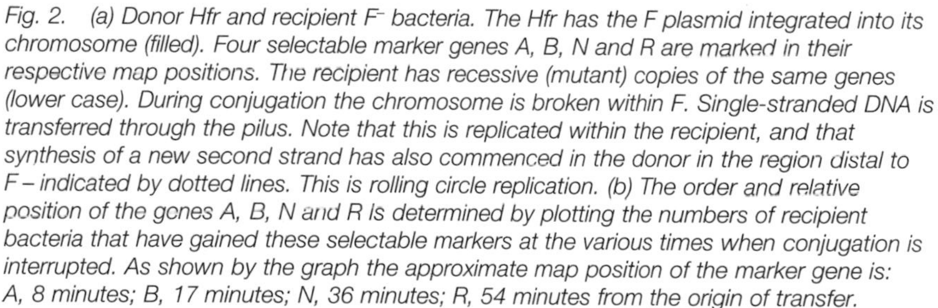

Fig. 2. (a) Donor Hfr and recipient F⁻ bacteria. The Hfr has the F plasmid integrated into its chromosome (filled). Four selectable marker genes A, B, N and R are marked in their respective map positions. The recipient has recessive (mutant) copies of the same genes (lower case). During conjugation the chromosome is broken within F. Single-stranded DNA is transferred through the pilus. Note that this is replicated within the recipient, and that synthesis of a new second strand has also commenced in the donor in the region distal to F – indicated by dotted lines. This is rolling circle replication. (b) The order and relative position of the genes A, B, N and R is determined by plotting the numbers of recipient bacteria that have gained these selectable markers at the various times when conjugation is interrupted. As shown by the graph the approximate map position of the marker gene is: A, 8 minutes; B, 17 minutes; N, 36 minutes; R, 54 minutes from the origin of transfer.

the origin of transfer; the remaining part of the F plasmid is transferred last. To move the complete chromosome into the recipient bacterium takes nearly 2 hours and this is rarely achieved because the pilus (conjugation tube) through which the chromosome moves is fragile and easily broken. However, this process results in high frequency transfer of bacterial genes and bacteria with integrated F plasmids are referred to as **HFr** (high frequency recombination) strains.

Because transfer is initiated from a fixed point and because the chromosome is transferred as a linear structure at a more or less constant rate the order and the time at which specific genes enter the F^- recipient can be utilized to construct a map of the bacterial chromosome. After mixing HFr and F^- bacteria and allowing conjugation to be initiated the process can be stopped at different times by violently agitating the culture, causing the fragile pili to rupture, and preventing further transfer of the chromosome. This is referred to as **interrupted mating**. The time at which each gene is transferred can be determined if bacteria are removed from the culture and plated out on agar containing suitable selective media at a series of time intervals during the experiment. Thus a map of bacterial genes can be constructed in map units of minutes giving the distance of each gene from the origin of transfer. An example of this is process is shown in *Fig. 2 (b)*.

It is not technically feasible to map the full length of the chromosome in this way using a single HFr strain, but different HFr strains exist, each of which has the F plasmid integrated at a unique site. These provide overlapping maps allowing the whole chromosome to be mapped and showing that the genetic map of *E. coli* is circular. This is divided into 100 units each of 1 minute.

The F plasmid is not the only plasmid capable of moving between bacteria. One other class of plasmid which has serious implications in medicine is the R plasmids. These can carry multiple antibiotic resistance genes between bacteria.

Bacteriophages and gene transfer

Under certain conditions bacteriophages (usually abbreviated to 'phages') can facilitate the transfer of genes between bacteria. The process is known as **transduction**. It depends on errors in phage replication, and is particularly useful in fine structure mapping of bacterial genes.

Phages can be divided into two classes, **virulent** or **temperate** depending on how they behave after infection of a bacterium. When a virulent phage infects a bacterium it takes over the synthetic machinery of the host and uses it as a factory for the production of new phage particles. The bacterium subsequently lyses liberating a large number of phages. Temperate phages have a choice between the **lytic** life cycle described above and an alternative **lysogenic** pathway. The latter involves integration into the host's chromosome where the phage, now termed a prophage, is dormant and replicates along with the rest of the host chromosome. An example of a temperate phage is λ. This can integrate into the bacterial chromosome only at a specific site, between the bacterial genes *gal* and *bio*. The bacterium is unaffected but is immune to further infection with the same strain of phage. Occasionally with low frequency the prophage loses its dormancy and converts to the lytic cycle, replicating phages and lysing the bacterium. The lytic phase can be induced in lysogenic bacteria by treatment with mutagens such as ultraviolet light. The mode of transfer of bacterial genes depends on the phage life cycle.

During the lytic cycle of some virulent phages errors may occur by which small fragments of the bacterial host's chromosome are randomly incorporated into new phage particles in place of phage DNA. These structures are called **transducing particles**. *Fig 3(a)* outlines the interactions between phage and bacteria and the production of transducing particles. Transducing particles may only be present in a phage stock at frequencies as low as 10^{-6}. They retain the ability to infect bacteria due to the presence of the phage coat proteins, but because they carry only bacterial DNA they simply act as vehicles for transferring DNA between bacteria. Appropriate selective media allow only bacteria that have incorporated genes in this way to grow into clones. Hence such rare

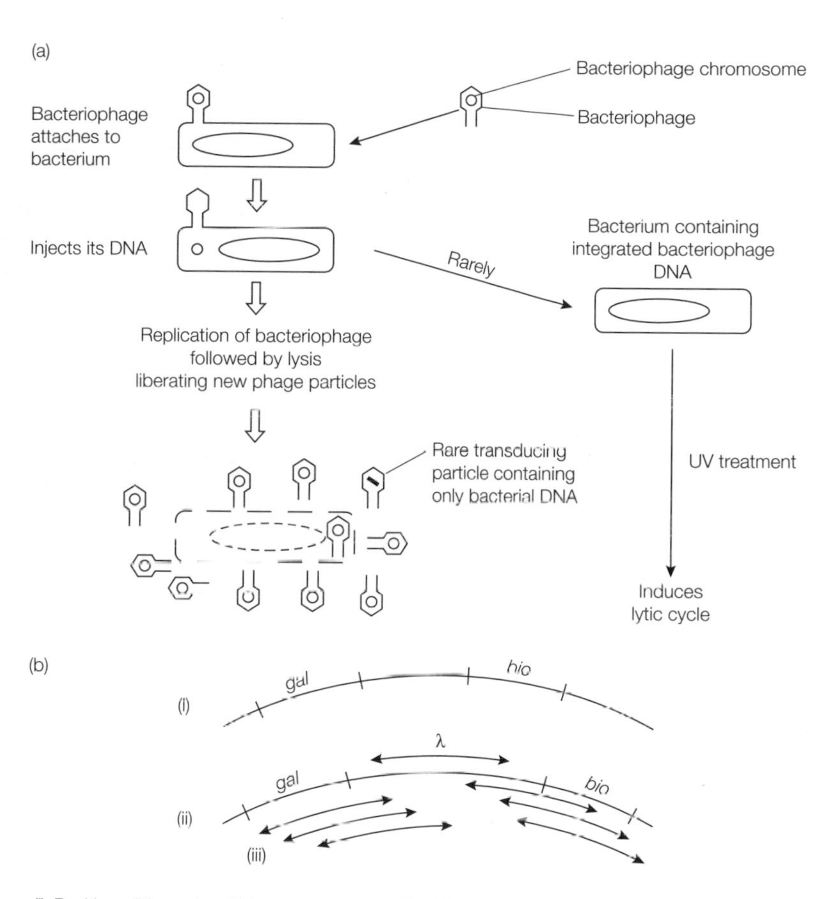

Fig. 3. Lytic infection cycle yielding rare transducing particles which allow generalized transduction. (b) Production of specialized transduction of the gal-bio *region of the* E. coli *genome.*

events can be detected and scored. Due to its fixed size the phage can only carry a small amount of DNA and only genes very close together can be **co-transduced** in this way. Frequency of co-transduction can be used for fine-scale mapping of bacterial genes. Since any fragment of the bacterial chromosome may be transduced in this manner the process is known as **generalized transduction**. Examples of phages that can be used for generalized transduction include P22 in *Salmonella* and P1 in *E. coli*.

When a lysogenic prophage loses its dormant state and begins to replicate in the host bacterium it is excised from the host DNA. This is normally an exact process in which the phage is precisely removed from the chromosome. Rarely the excision process is inaccurate, a small amount of bacterial DNA is removed, and a reciprocal fragment of the phage is left in the host chromosome. The phages resulting from such an event lack some phage genes and are defective in that they cannot carry out their full life cycle, but like the transducing particles of the virulent phages they can transfer genes between bacteria. Because temperate phage can only integrate into bacteria at one or a few sites they can obviously only transfer genes from those regions, the process is known as **specialized transduction**. Phage λ, for example, can be used for fine structure mapping of the *gal–bio* region. Specialized transduction is outlined in *Fig. 3(b)*.

Recombination in bacteria

All of the mechanisms described above relate to the moving of fragments of the bacterial chromosome between bacteria. This only allows the possibility of recombination. The recipient bacteria become partially diploid, and are termed **merozygotes**. Recombination, the production of new stable genotypes, depends on crossing-over between the host chromosome and the transferred DNA fragment. One important point must be made about this process: whereas in eukaryote recombination single cross-overs are useful in mapping genes, in merozygotes there must always be an even number of cross-overs. This is due to the bacterial chromosome being circular. A single cross-over would result in the creation of a linear chromosome, which would render the bacterium nonviable. This is illustrated in *Fig. 4*.

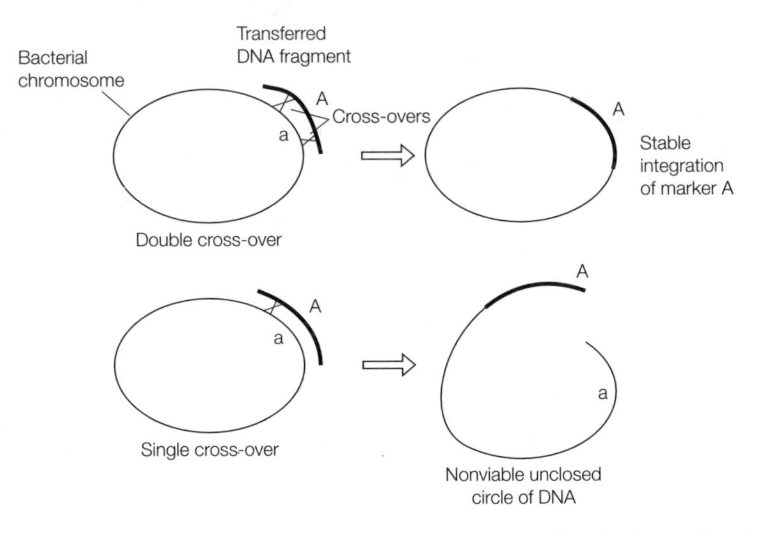

Fig. 4. The need for an even number of cross-overs to affect the integration of a marker from a merozygote.

C6 GENES IN EUKARYOTIC ORGANELLES

Key Notes

Mitochondrial and chloroplast genomes

The mitochondria and chloroplasts of eukaryotes contain their own DNA genomes. These vary considerably in size but are usually circular. They probably represent primitive prokaryote organisms that were incorporated into early eukaryotes and have co-evolved in a symbiotic relationship. The organelles have their own ribosomes and synthesize some of their own proteins, but others are encoded by nuclear genes.

Maternal inheritance

Pollen and sperm rarely contribute cytoplasmic organelles to the zygote. Hence genes carried by mitochondria or chloroplasts can only be inherited on the maternal side. In plants the most common examples of maternal inheritance involve chloroplasts that fail to synthesize chlorophyll and hence form white regions in leaves. In yeast, mitochondrial mutants can inhibit aerobic respiration. Some inherited human diseases are caused by mitochondrial mutants, and damage to mitochondrial genomes is associated with age.

Maternal inheritance in evolution

In animals mitochondrial DNA has a higher mutation rate than nuclear DNA. This makes it useful for relatively short-term evolutionary studies of divergence within populations. This has been used to show that the current human population is probably descended from a single female. Despite the high mutation rate almost all mitochondria in one individual are identical, homoplasmy. This is thought to be because only a small number of mitochondria are transferred from one generation to the next.

Maternal effects

In some cases the phenotype of an individual is affected by the genotype of the mother. This can mask the true genotype of the individual. This is different from maternal inheritance. A good example is shell coiling in water snails.

Related topics Basic Mendelian genetics (C1) More Mendelian genetics (C2)

Mitochondrial and chloroplast genomes

Eukaryote cells contain more than one genome. Most genetic studies concentrate on the nuclear genome, but the mitochondria and chloroplasts carried in the cytoplasm also contain their own DNA. It is widely accepted that these organelles represent the ancestors of prokaryote organisms that were 'captured' by primitive eukaryotes and have since lived in a symbiotic relationship. The organelle genomes differ greatly from the nuclear genome in that they resemble their prokaryote ancestors. The mitochondrial and chloroplast genomes of several species have been characterized and completely sequenced. In most plant species chloroplast genomes are circular, and range from 50 to 200 kb in size. Many copies of the genome can be found in any one chloroplast. Mitochondrial genomes are also circular but vary greatly in size between different species. Human mitochondria contain 16.6 kb of DNA, whereas the yeast mitochondrial genome is approximately five times larger. Plants have mitochondrial genomes of about 1 Mb.

Both mitochondria and chloroplasts carry their own ribosomal RNA (see Topic A6) and transfer RNA genes (see Topic A7), and their ribosomes are similar in size to prokaryotic ribosomes. Some of the proteins found in these organelles are synthesized on their own ribosomes from mRNA transcribed from the organelle genome, but most proteins are coded by nuclear genes and are imported from the cytoplasm. This dependence on host genes gives an indication of the large amount of co-evolution that has taken place between the organelle and the eukaryote cell.

Maternal inheritance

Although there are some exceptions, most plant pollen does not transfer chloroplasts or mitochondria to the zygote. In animals, sperm do not contribute mitochondria. For this reason any phenotypes coded for by organelle genomes show a pattern of inheritance in which a phenotype can only be inherited from the mother, this is known as **maternal inheritance**. There are numerous examples of this.

Variegated plants, showing white patches or stripes on the leaves, are often due to the presence of a proportion of chloroplasts that have a mutation which results in loss of the ability to form chlorophyll. At each cell division the chloroplasts segregate randomly between the daughter cells, and occasionally a daughter cell will be formed that contains only mutant chloroplasts. All of its descendants will lack chlorophyll, producing a white stripe in grasses or a white patch in broad-leaved plants. The frequency of such an event will depend on the numbers of chloroplasts per cell and the ratio of mutant to normal chloroplasts. In certain species such as the Four O'Clock plant, *Mirabilia jalapa*, cuttings can be taken from green stems, white stems, or variegated stems. These cuttings will grow on to form flowers which produce seed. Plants derived from the green and white cuttings breed true, but the seeds produced on variegated cuttings can give green, white or variegated plants. This is because the ova that develop into seeds after fertilization can carry all normal, all mutant or a mixture of chloroplasts (see *Fig. 1*).

A commercially important example of cytoplasmic inheritance in plants involves male-sterility in maize. Here mutant genes derived from the maternal cytoplasm prevent pollen production. This can be utilized in the production of F1 hybrid seed (see Topic C7). In mixed stands of male-sterile plants of one variety and normal plants of another variety, all seed produced on male-sterile plants must have been fertilized by pollen from the other variety, and hence must be hybrid. There is an interaction between the cytoplasmic gene for male-sterility and nuclear genes. Such interactions are commonly found in examples of cytoplasmic inheritance.

Cytoplasmic inheritance is also observed in fungi. Mutations in the mitochondrial DNA of two species have been associated with reduced growth rates. In the Baker's yeast *S. cerevisiae* these are referred to as 'petite' mutants. Two forms of mitochondrial petite mutants are known. Neutral petites are recessive and when normal yeast are crossed with petite yeast the resulting ascospores have normal growth rates. This is because each cell contains many mitochon-

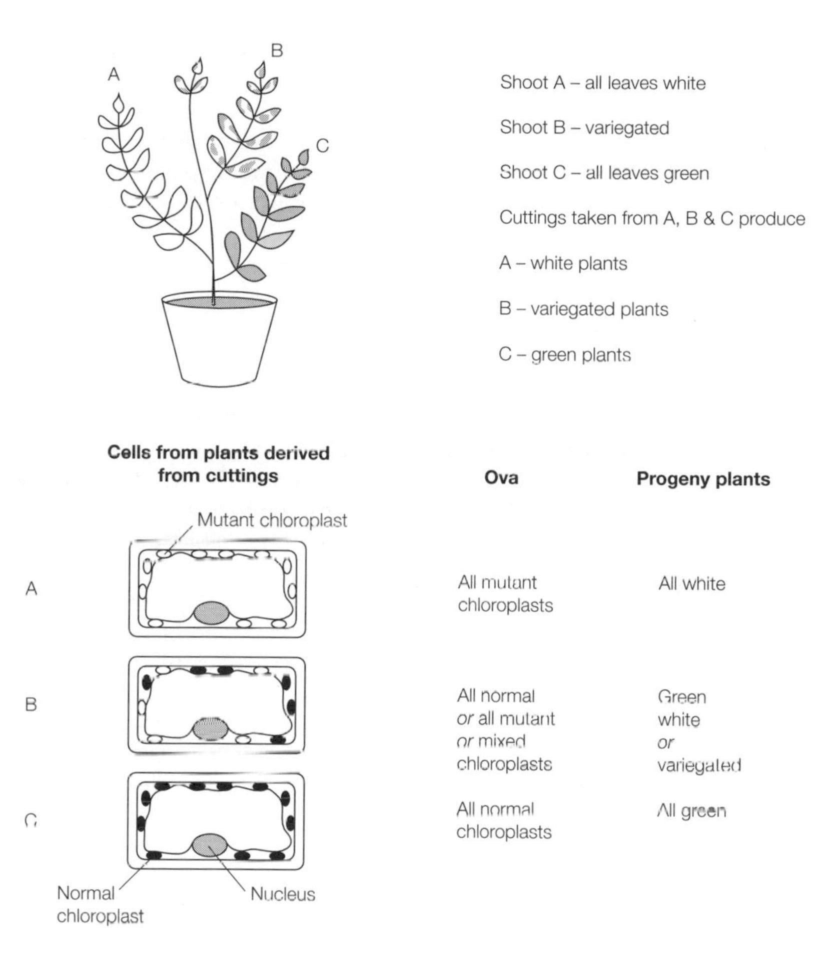

Fig. 1. The inheritance of chloroplast phenotypes in variegated plants.

dria, and although the hybrid contains a mixture of mutant and normal mitochondria, it is highly improbable that at any cell division the distribution of mitochondria between daughter cells would result in a cell containing only mutant mitochondria. Other petite mutants, suppressor petites, are not recessive and when crossed with normal yeast produce normal and petite progeny in nonMendelian ratios (see Topics B1 and B2). Petite mutants are incapable of supporting aerobic respiration. The precise defect differs between mutants, but frequently this involves major deletions of mitochondrial DNA. Suppressor petites appear to be able to induce changes in the DNA of normal mitochondria in the same yeast cell.

Mutations in mitochondrial DNA can lead to antibiotic resistance in animal cells grown in culture, and certain inherited human disorders show a pattern of maternal inheritance and can be traced to mutations in mitochondrial DNA. An example of one such disease is Leber's hereditary optic atrophy. Here, as in other examples, nuclear genes interact with the expression of defective mitochondrial genotypes; there is growing evidence that an accumulation of damage to mitochondrial DNA may play a part in the aging process.

Maternal inheritance in evolution

Mitochondrial DNA sequences have proved extremely useful in evolutionary studies, particularly in humans and other animals. There are several reasons for this. The rate of mutation in animal mitochondrial DNA is high compared with that in nuclear DNA, but despite this the mitochondria of any one individual usually show no variation (homoplasmy). This observed lack of variation within an individual is difficult to reconcile with high mutation rates, but may be connected with selection of small numbers of mitochondria to form the mitochondrial pool of oocytes. The maternal mode of inheritance would then ensure homoplasmy within the somatic tissue of any offspring. It is worth noting that plant mitochondria behave very differently and evolve slowly. The rapid mutation and lack of recombination allow changes in mitochondrial DNA to be used to follow divergence in populations. By examining the diversity of mitochondrial DNA sequences in the current human population it has been estimated that the entire human race has descended from a single woman who lived approximately 200 000 years ago. This estimate is, however, controversial and depends on a number of assumptions relating to constancy of mutation rate and sizes of populations (see Topic D9).

Maternal effects

Some confusion can occur between the maternal inheritance of genes encoded in organelle genomes and situations where the phenotype of progeny is affected by its mother, to the extent where its own genotype is masked. This can occur to a greater or lesser degree in many organisms because of maternal contribution of protein or RNA to the embryo. An extreme example of this is displayed by the pattern of shell coiling in the water snail *Limnae*. In this species a single gene with two alleles determines whether the snail shell follows a right-handed (dextral) or left-handed (sinistral) coil. There are two alleles, *D* for dextral, and *d* for sinistral coiling, dextral being dominant over sinistral. The water snail is a hermaphrodite and can either self-fertilize or mate with other snails. When a *D/D* dextral snail is mated with a *d/d* sinistral snail the progeny all have the same shell coiling pattern as the mother. If these F1 individuals are allowed to self-fertilize, the progeny are all dextral. A further round of self-fertilization produces an F3 generation with a ratio of 3 dextral : 1 sinistral. Note that the expected Mendelian ratios appear one generation later than would be expected in the case of conventional Mendelian genetics. Although this has some similarities with cytoplasmic maternal inheritance it is clear that the phenotype does not consistently follow that of the female parent.

Why is the true ratio delayed for a generation? The development of the snail is predetermined by the mother. Its influence directs the manner in which the shell coils. Hence the phenotype of each snail reflects the genotype of its mother (or single parent in a self-fertilization) rather than its own genotype. This is the best known example of maternal effect, but other examples are found in insects where the phenotype of the young of the next generation resembles the mother, but this is lost with age. In the case of the snail the coiling of the shell is probably not reversible.

C7 QUANTITATIVE INHERITANCE

Key Notes

Quantitative traits

Most classic genetic traits are discontinuous, different genotypes produce quite different phenotypes with no overlap (e.g. brown or blue eyes). Quantitative traits such as height vary on a continuous distribution. All values of height within a range are possible. They are controlled by the cumulative effects of many genetic loci, and also by the effects of differences in the environments experienced by individuals. Quantitative traits are also called multifactorial, polygenic or multilocus.

There is also a newly recognized category which varies discontinuously, but is controlled quantitatively (continuously). A combination of genetic and environmental factors cause some individuals to cross a threshold from one phenotypic state to another. Examples are diabetes and cancer where genotype affects risk, but a combination of genotype and environment push some individuals over the threshold from health to disease.

Measuring continuous variation

Continuously variable characters are described by their mean and variance, usually assuming a normal distribution (x_i = individual height measurements of n individuals).

$$\text{mean} = \bar{x} = \frac{\sum(x_i)}{n}$$

$$\text{variance } s^2 = V = \frac{\sum(x_i - \bar{x})^2}{(n-1)}$$

Components of variation

The total phenotypic variance is equal to the sum of the variance due to all causes, namely the environment (E), gene–environment interactions (GE), additive genetic effect (GA), dominant genetic effects (GD), and epistatic interactions (GI) between different genes (loci) respectively as in the formula:

$$V_{\text{total}} = V_{\text{ph}} = V_{\text{E}} + V_{\text{GE}} + V_{\text{GA}} + V_{\text{GD}} + V_{\text{GI}}$$

This is often simplified by putting all the genetic components together:

$$V_{\text{ph}} = V_{\text{E}} + V_{\text{G}}$$

Broad sense heritability

Broad sense heritability (H_{B}^2 also written H_{B}, H^2 or h_{B}^2) is the proportion of total phenotypic variance in a population sample that is due to genetic effects.

$$H_{\text{B}}^2 = \frac{V_{\text{G}}}{V_{\text{ph}}} = \frac{V_{\text{G}}}{(V_{\text{E}} + V_{\text{G}})}$$

Narrow sense heritability

Narrow sense heritability (H_{N}^2, also h^2 or H_{N}) is the component of variation caused by alleles or genes with additive effects. It is passed from parent to offspring, each parent contributing 0.5 of the offspring's character:

$$H_{\text{N}}^2 = \frac{V_{\text{A}}}{V_{\text{ph}}}$$

Measuring heritability

(i) Controlled breeding. Environmental effects can be measured by comparing inbred (homozygous) lines in crop plants or monozygotic twins in humans. The extra variance between unrelated individuals in the same environment must be the broad sense heritability. Controlled crosses can be used to produce genetically uniform F1 and variable F2 and backcross generations. By making the correct comparisons, the environmental and additive, dominant and epistatic genetic contributions to the range of phenotypes can be calculated. The F2 generation contains all possible genotypes, and can be used to calculate broad sense heritability.

$$H_B^2 = \frac{(V_{ph \cdot F2} - V_E)}{V_{ph \cdot F2}}$$

Similar calculations can be performed in humans by suitable comparisons between people with different degrees of relatedness, and adoptive relatives.
(ii) Realized heritability. Realized heritability is the response of a population to selection, and can be calculated after a selective breeding program. It is the ratio of change in character to selection differential:

$$H = \frac{(M_O - M_T)}{(M_P - M_T)} = \frac{(34 - 30)}{(40 - 30)} = \frac{4}{10} = 0.4$$

where M_T is original mean, M_O is the mean in the next (offspring) generation after selection, and M_P is the mean of the individuals selected to be parents.
(iii) Offspring – parent regression. A graph of offspring values (y) against the average (mid-point value) of their two parents (x) for a particular trait (characteristic) will have a slope (b) equal to the narrow sense heritability of that trait. Put simply, the correlation between a characteristic in parents and offspring reveals how much the offspring inherit that characteristic from their parents. In humans they also share their family environment, which confuses the measurement. The formula for a straight line graph is:

$$y = a + bx \text{ where } a \text{ is a constant and } b = \text{slope}$$

so

$$H_N^2 = b$$

Limitations of heritability

Heritability changes as conditions change because it is the ratio of genetic to total variance. High environmental variance causes lower heritability than low environmental variance, although genetic effects stay constant. Heritability only means something in the population and conditions where it is measured because numerical values cannot be compared meaningfully between different populations or conditions.

Quantitative trait loci

Quantitative trait loci are locatable genetic markers (typically variable number tandem repeats) which are closely linked (physically nearby in the DNA) to the genes affecting interesting character traits. The correlation between the markers and the trait is used to find the genetic locations of the genes controlling the trait.

Human studies

Monozygotic twins are genetically identical, so all their differences should be due to their different environments. Dizygotic twins are only 50% genetically identical, and can be used as a comparison. Twins adopted separately or together at birth indicate the environmental difference between families.
All the components of genetic and environmental variation in a trait can be

calculated by examining the correlations between monozygotic (identical) and dizygotic twins and sibs to each other and to natural and adoptive parents and sibs. Concordant events affect both twins of a pair (discordant events affect only one twin) and can be used to examine heritability of qualitative traits. For any particular trait (e.g. schizophrenia), the extent that concordance between monozygotic twins (80%) exceeds concordance between dizygotic twins (13%) is a measure of the effect of the 50% genetic difference between the dizygotic twins.

Human quantitative traits

Many interesting human traits are quantitative, the most intensely studied being intelligence and personality. Genetic loci that contribute to variation in these traits have been identified. The advantage of knowing the comparative effects of the environment and genetics (nature and nurture) is that resources can be directed where they are most beneficial, however there is a risk of the information being distorted and employed politically for racial or class-biased motives.

Related topics

Basic Mendelian genetics (C1)
More Mendelian genetics (C2)
Genetic diversity (D4)

Quantitative traits

Classic Mendelian traits that are controlled by a single gene (locus) with two alleles have two or three distinctly different phenotypes (see Topic C1). If one functional allele is sufficient for full activity the dominant phenotype is produced in homozygotes and heterozygotes, and the recessive phenotype is seen in homozygotes where the function is missing (*Fig. 1a*). If one functional allele is not sufficient, the heterozygote will be intermediate in character, but may still be distinct from the homozygotes; for example when red and white homozygous plants are crossed the heterozygous offspring are pink (*Fig. 1b*). In both cases, each allele makes a large contribution to the phenotype. The environment causes small differences between individuals with the same genotype, but there are clear differences between the phenotypes (e.g. tall vs. short), and the variation in this character is said to be **discontinuous**. Most genes that

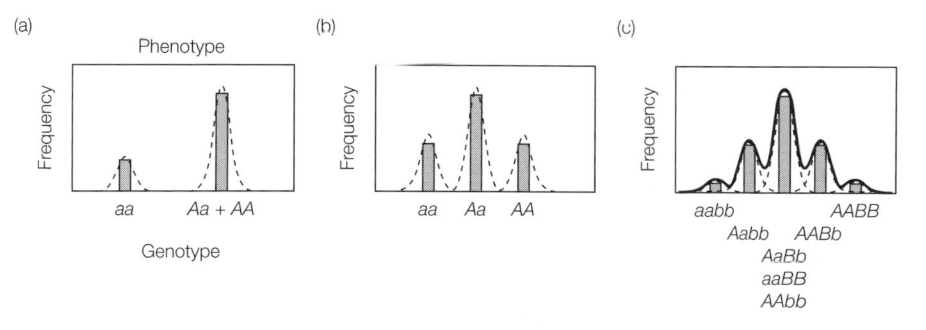

Fig. 1. (a) A single gene cross with a dominant allele gives a 3:1 ratio with discrete phenotypes (bars) broadened by environmental effects. (b) If the alleles have additive effects the heterozygotes are intermediate and the phenotypes less distinct. (c) Two genes with additive alleles give five phenotypic classes in a 1:4:6:4:1 ratio. The blurring effect of environmentally induced variation blends these into a continuous variation.

have been studied individually fall into this category. They have a large and obvious effect on the phenotype otherwise we would not notice when they were mutated.

Most genes do not have such obvious effects, instead they contribute a little to a characteristic. For example, the height of humans is not usually controlled by a single gene (although there is a dominant dwarfing allele). Most people are between 125 cm and 200 cm tall, but can be anywhere within that range. When two genes control a characteristic equally, each with two alleles and no dominance, there are five classes in an F2 ratio of 1 : 4 : 6 : 4 : 1 (*Fig. 1c*). With some superimposed environmental variations, these classes overlap. Such variation is said to be **continuous** because any intermediate value is now possible. With more genes and more environmental variation to spread the peaks, the classes for each genotype disappear into a single broad peak. These characteristics are **quantitative** because each individual must be measured, they cannot be classified qualitatively as tall or short. The genetic control of human height is clear. Children tend to be about as tall as their parents, and some races are tall (e.g. the Masai in Africa), some short (e.g. African pigmies). However, starved children do not grow as tall as their well-fed brothers and sisters.

The **environment** also plays a part; chronic malnutrition (protein or vitamin deficiency) or starvation in childhood limits growth to a height below the genetic capacity. Other continuous traits in humans include skin color, tendency to heart disease or diabetes, and intelligence and personality. These are said to be **multifactorial** or **polygenic** because many genes (loci, positions on the DNA) contribute to their control (hence the term **multilocus control**). Continuously variable traits include yield and quality of all agricultural products (e.g. milk, meat, fruit and grain). It is important to know how genes and the environment contribute to the phenotype in order to maximize production by selective breeding to improve the genome, and by modifying farming practice to improve the environment.

The biochemical basis of this is easiest to see in the control of color in wheat. Some strains have white kernels, others red. Each functional allele is transcribed into mRNA which is translated into protein that contributes to the production of red pigment. The more functional alleles there are coding mRNA, the more enzyme and pigment are produced, so such alleles and genetic loci are said to be **additive** in their effect. Kernel color in wheat is controlled by three genes (loci) each with two alleles, and crossing extreme parents (red *AABBCC* × white

	ABC	ABc	AbC	aBC	Abc	aBc	abC	abc
ABC	6	5	5	5	4	4	4	3
ABc	5	4	4	4	3	3	3	2
AbC	5	4	4	4	3	3	3	2
aBC	5	4	4	4	3	3	3	2
Abc	4	3	3	3	2	2	2	1
aBc	4	3	3	3	2	2	2	1
abC	4	3	3	3	2	2	2	1
abc	3	2	2	2	1	1	1	0

Fig. 2. A Punnett square showing the seven color phenotypes produced by three gene loci with additive effects.

aabbcc) will produce seven F2 classes with 0, 1, 2, 3, 4, 5 or 6 dominant alleles respectively in the ratio of 1 : 6 : 15 : 20 : 15 : 6 : 1. Only 1/64 will be the deepest red *AABBCC* class, and 1/64 white *aabbcc* like the parental generation. The Punnett square for this is shown in *Fig. 2*. Environmental effects such as temperature (which changes enzyme controlled reaction rates) sunlight and nutrition (which change the levels of enzymes and substrate for reactions) cause individual grains of wheat to vary within each genotype, blurring the difference between adjacent genotypic classes.

There is another category of trait where the phenotype varies discontinuously (qualitatively) but the control is quantitative under the influence of many genes as well as environmental effects. Examples are diabetes and cancer, where there are known alleles which contribute to risk, and a combination of genotype, environment, and chance events push some people over the threshold from one phenotype to another – from normal health to abnormal disease.

Measuring continuous variation

The distribution of a continuously variable character such as human height is described by its **mean** and **variance**. The mean (or average) is the sum (Σ) of the heights (x_i) of all the people measured divided by the number (n) of these people. **Mean** is usually written as $\bar{x}$ pronounced exbar:

$$\text{mean} = \bar{x} = \frac{\Sigma(x_i)}{n}$$

The spread of the distribution is called the **variance** (written s^2 or V). This is the sum of the squares of the difference (deviation) between each individual measurement (x_i) and the mean $\bar{x}$, all divided by $(n - 1)$.

$$s^2 = V = \frac{\sum(x_i - \bar{x})^2}{(n - 1)}$$

Variance is a squared value so it is always positive, but its units are also squared (e.g. cm^2, kg^2). The square root of variance, s, called the **standard deviation**, is often used instead, to restore the original units. Biological systems usually

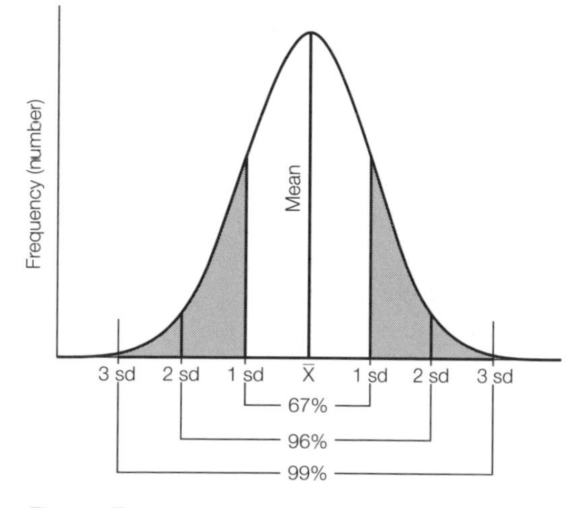

Fig. 3. The normal distribution, defined by its mean (the peak) and the breadth, or spread, measured in standard deviations (sd).

vary in a way that fits approximately to a normal distribution, an idealized mathematical distribution that gives the classic 'bell shaped curve' (*Fig. 3*). This fit is the basis for most parametrical (number-based) statistics, but should not and must not be taken for granted. If the distribution is normal, then the expected range of results can be calculated. A range from the mean of $\pm s$ includes 68.3% of all individuals, and the mean $\pm$ 1.96s includes 95% of individual measurements.

Components of variation

The total amount of variance in a phenotypic character V_{ph} is the sum of the variance caused by all the separate factors. The variance due to the **environment** V_E is the nongenetic component, and is caused by individual organisms not experiencing exactly the same conditions. Another component is the **genotype environment interaction** V_{GE}. For example, some people get fatter than others when well-fed, or starve faster when food is scarce. This is difficult to measure, but may be important (e.g. some wheat varieties increase their yield when given more nitrogen fertilizer, others do not respond to extra nitrogen, but still do well in poor soil).

There are three purely genetic components. The **additive genetic component** V_A is due to additive alleles (those with no dominance) so the *Aa* phenotype is half way between the *AA* and *aa* phenotypes. The **dominant genetic component** V_D is due to dominant alleles. Heterozygotes express these, but are equally likely to pass the recessive allele to offspring. Expression depends upon the allele received from the other parent, so this variation does not correlate well between parents and offspring. The third genetic component is **interactions between genes** V_{GI} at different loci (epistasis, see Topic C2). For example *AA* may be expressed differently according to whether alleles *BB*, *Bb* or *bb* are present at another locus. These genes are likely to be inherited independently, so combinations in the parents are broken up at gametogenesis. New combinations are made in the offspring, so interactions are not passed intact from parents to offspring.

These components can be determined accurately in controlled breeding experiments between inbred lines. The total variation is therefore:

$$V_{total} = V_{ph} = V_E + V_{GE} + V_{GA} + V_{GD} + V_{GI}$$

This is usually simplified by putting all the genetic components together and calling them genetic variation V_G or broad sense heritability:

$$V_{ph} = V_E + V_G$$

Broad sense heritability

Broad sense heritability (H^2_B also written H_B H^2 or h^2_B) is the contribution that genetic differences make to the variation in a particular character in a particular population in a particular environment. (The square sign indicates variance, but some authors prefer the *B* subscript for 'broad'). Broad sense heritability is the ratio of the genetic contribution to variation to the total variation, therefore:

$$H^2_B = \frac{V_G}{V_{ph}} = \frac{V_G}{(V_E + V_G)}$$

Narrow sense heritability

Narrow sense heritability (H^2_N also h^2 or H_N) is a useful measure for plant and animal breeders. It is the additive component of genetic variance, V_{GA} or simple

V_A, and is passed on reliably to offspring, allowing the response to a program of artificial selection to be predicted. In the example of wheat color (above) the number of alleles for red predicts the final color:

$$H_N^2 = \frac{V_A}{V_{ph}}$$

The situation is quite different when dominance and genetic interactions affect the phenotype because they cannot be passed on intact.

Measuring heritability

(i) Controlled breeding

Inbred strains become homozygous (see Topic C11) and so should have no heritable genetic variation. The phenotypic variation V_{ph} then equals V_E. If two such strains (parental one, P1, and parental two, P2) are crossed, the F1 will again be genetically uniform and so $V_{ph} = V_E$. We now have three estimates of V_E:

$$V_E = \frac{(V_{EP1} + V_{EP2} + V_{EF1})}{3}$$

The F2 will show the full range of genotypes and will demonstrate all the genetic components of variation (V_G). This can be calculated by subtracting the value of V_E obtained above from $V_{ph.F2}$:

$$V_G = V_{ph\cdot F2} - V_E$$

$$H_B^2 = \frac{(V_{ph\cdot F2} - V_E)}{V_{ph\cdot F2}}$$

If all the genetic effects are additive, the mean value of the F1 and F2 should be midway between the means of the parent strains. Any deviation from this pattern indicates that dominance and/or epistatic interactions are occurring. The genetic variance of the F2, V_G, can be partitioned into V_{GA}, V_{GD} and V_{GI} although these components may not all be present in a particular case. The F1 can be crossed to the parental strains producing two backcross strains (B1 and B2). The contribution of additive and dominant effects to the variance in each (F2, B1 and B2) can be calculated and values for V_A and V_D produced from the three simultaneous equations. More complicated procedures are required to study humans.

(ii) Realized heritability

It is also possible to calculate realized heritability retrospectively from the results of selection. Suppose that a population of tomatoes has a mean fruit size M_T of 30 g, and plants with large fruits are selected, mean size $M_P = 40$ g, as parents to breed from. The selection differential $(M_T - M_P)$ is 10 g. If the next generation (offspring) mean, $M_O = 34$ g, then:

$$\text{heritability} = \frac{\text{gain}}{\text{selection differential}}$$

$$H = \frac{(M_O - M_T)}{M_P - M_T} = \frac{(34 - 30)}{(40 - 30)} = \frac{4}{10} = 0.4$$

(iii) Offspring–parent regression

Heritability can be calculated by comparisons of relatives (in humans this is the only method) and it is possible to estimate narrow sense heritability from such comparisons by calculating a parent–offspring regression coefficient. If all variation is additive, offspring will have a phenotype midway between their parent's phenotypes. Thus if a graph is plotted of offspring phenotype (y) against the midpoint between the parents (x) it will give a perfect straight line with a gradient (slope) of 1 if all the variance is due to additive genetic effects. A graph of one parent against offspring or brothers/sisters against each other will have a regression of 0.5 if heritability is one, because they are related by 0.5. Environmental effects will cause scatter in the points and reduce the regression. When all the variation is environmental there is no correspondence between parents and children ($H^2_N = 0$). Points are scattered randomly across the graph and so the slope is zero. The formula for a straight line (linear) graph is $y = a + bx$ where b = slope, and so for regression of parental-midpoint against offspring:

$$H^2_N = b$$

The null hypothesis is that there is no correspondence so the slope $b = 0$. The value of b is calculated by regression analysis, and its statistical significance is determined by a Student's t-test. If $b > 1$ then there must be hybrid vigor (also called heterosis, overdominance) such as might occur in a breeding program.

A simple alternative is to use a calculator to determine a correlation coefficient r between mid-parent and offspring which will indicate H^2_N if offspring and midparent values follow the same distribution (have the same mean and variance). The principal difference between the tests is that a correlation coefficient does not assume any cause and effect relationship, whereas regression analysis assumes that the parent's genotype influences the offspring.

Limitations of heritability

Heritability is relative to the total variation. If a crop (for example, tomatoes) is raised under very variable conditions the environment will generate a high variability, and heritability will be low. If the same tomatoes are raised in a carefully controlled environment, there will be less phenotypic variation, and the heritability will be higher, although the genetic control is just the same. **Heritability is only numerically meaningful in the conditions where it is measured**. If heritability is high, breeding to improve strains will improve yield; if heritability is low, it may be more productive to change or control the environment. Within genetically uniform strains and self-pollinating species such as wheat, characters are not heritable at all because although they are still genetically controlled there is no genetic variation.

Quantitative trait loci

The many genes responsible for polygenic inheritance of particular characteristics are scattered around the genome. Their positions are known as quantitative trait loci (QTL). It is useful to know where they are for both medical and agricultural reasons. In the case of disease susceptibility, it is useful to identify the individual genes so that their normal function can be identified, and attempts made to design corrective medical treatments. In the case of animal and plant breeding it would be useful to identify young individuals with favorable alleles without waiting for their expression at maturity. Those with an unfavorable genotype could be removed earlier from selective breeding programs, while potentially high quality types could be cloned immediately.

The procedure in breeding situations is to take inbred lines that differ in the trait of interest, and also vary for markers [typically variable number tandem repeats (VNTRs) see Topics B4, F4, F5 and F6] at numerous probe sites. They are crossed and both the F2 progeny and later generations are examined for the desired trait and for the variations at the probe sites. If the presence of the trait correlates with inheritance of a particular marker allele, it is likely that one or more genes affecting the trait is located on the DNA close to that marker.

The same procedure can be followed in human families, particularly for disease susceptibility loci, but is more complicated and difficult because the family sizes are smaller (see Topics E1 and E2). The usual case is that one or two genes cause most of the variation, and there are increasingly more genes with smaller effects. Genes that contribute 5% or less to the variation in a trait are very difficult to find. The most important single genetic factor in insulin dependent diabetes is the allele of the promoter of the gene coding for insulin itself, followed by several other loci of decreasing importance.

Human studies

Monozygotic twins are genetically identical clones, so the only variation between them should be due to environmental effects. (This is not quite true for females who may differ in the extent to which they inactivated their two X chromosomes.) Comparison of the variation between monozygotic twins reared together gives the 'within twins' environmental effect, while comparison of twins separated and adopted at birth gives the 'between family' environmental effect on variance. **Dizygotic twins** are simply sibs (sisters and brothers) conceived and born together. The extra variation between normal sibs compared to the variation between dizygotic twins is the variation due to being born separately. Twins adopted together or separately at birth can be correlated to each other and to both adoptive and natural parents and adoptive and natural sibs to study the environmental and genetic contribution to their phenotype. Natural children of the adoptive parents serve as controls. The correlation of adopted children to the average phenotype of their natural parents is a measure of narrow sense heritability.

Given various means of estimating the environmental components of phenotypic variation, the remaining variation must be caused by genetic differences, that is the broad sense heritability. It is important to remember that families also share environments, making the separation of genetic and environmental effects difficult. Children adopted at birth are used as controls for this because they share genes but not environment. The environmental effects begin with the position of implantation in the womb and continue during intrauterine development. Twins and sibs must share the same mother, so they also share antenatal environmental effects as well as genes. Having obtained estimates of broad and narrow sense heritability the difference ($H^2_B - H^2_N$) may be attributed to genetic dominance effects and genetic interactions.

It is apparent that estimates of heritability for human characters are less accurate than those produced in plant breeding experiments because the samples available are smaller and they may not be fully representative of the population. They may also be compiled by combining data obtained in different groups, which is not strictly allowable.

Concordant traits occur in both twins of a pair, and can be used to estimate heritability for qualitative traits. The difference in frequency of concordance between monozygotic and dizygotic twins is taken as a measure or indication of the genetic component of a trait. A high concordance in monozygotic compared with dizygotic twins is found for manic depression (80% : 20%), schizophrenia (80% : 13%) and blood pressure (63% : 36%), but only 95% : 87% for measles. This suggests high genetic variability for schizophrenia and low genetic variability in susceptibility to catching measles.

Human quantitative traits

Most human traits are quantitative. A simple example is skin color. The data suggest that there are at least four important genes, perhaps six, with 8–12 additive alleles and 9–13 phenotypic classes. More interesting traits are intelligence and personality. It is useful to know relatively how much the environment (nurture–family and social upbringing) and genetic inheritance (nature) contribute to variation in intelligence. If heritability is high (variation is mainly genetic) then increasing expenditure on education will not make everyone a genius. The more and less able groups will require different types of education, and the weaker might benefit more from help than the geniuses who may need help less. If heritability is low (environmental effects high) it would be more productive to spend money to bring the environment of the disadvantaged up to the standards of the advantaged. Some recent studies put heritability of IQ above 0.6, often around 0.8. What does this mean? Remember that heritability is the **ratio** of genetic variation to total variation. If IQ tests are carried out on a group of children from, say, a middle class background, where all their families have a high regard for education, and they all have similar opportunities, then the environmental factors may be low, so heritability will be high. If the whole population of a city is included, from slums to rich neighborhoods, there will be much more environmental variation, so heritability may be lower while in fact the genes involved are the same in both tests! One research group found that a particular allele of the gene for insulin-like growth factor 2 receptor is more frequent (about 30%) in 'gifted' children than in normal children (16%). Notice that nearly everyone with this allele is normal.

Personality and behavior certainly have genetic components, because they depend to a large extent upon the activity of neurotransmitters and their receptors. This is why psychoactive drugs work. There are over 100 neurotransmitters identified so far, and most have several receptors, so large genetic variation is to be expected. A serotonin receptor gene has two alleles of its promoter, a high activity and low activity version. People with the high activity allele are slightly more anxious on average than people with the low activity allele. Some comparisons of twins and siblings suggest a broad sense heritability of 30–45% (and narrow sense heritability of 10–25%) for scores in psychometric tests of 'agreeableness', 'conscientiousness', 'extrovertism', 'neuroticism' and 'openness'. The decisions taken as a result of genetic data are political. If people are genetically disadvantaged do you help them, prevent them from having children, or kill them? The misuse of human genetics for racial or socio-political causes has led to protests against the publication of controversial data. The scientific response is usually that knowledge is useful if it is used correctly, and cannot be worse than the ignorance which led to prejudice, witch-hunts and genocide in the past.

C8 SEX DETERMINATION

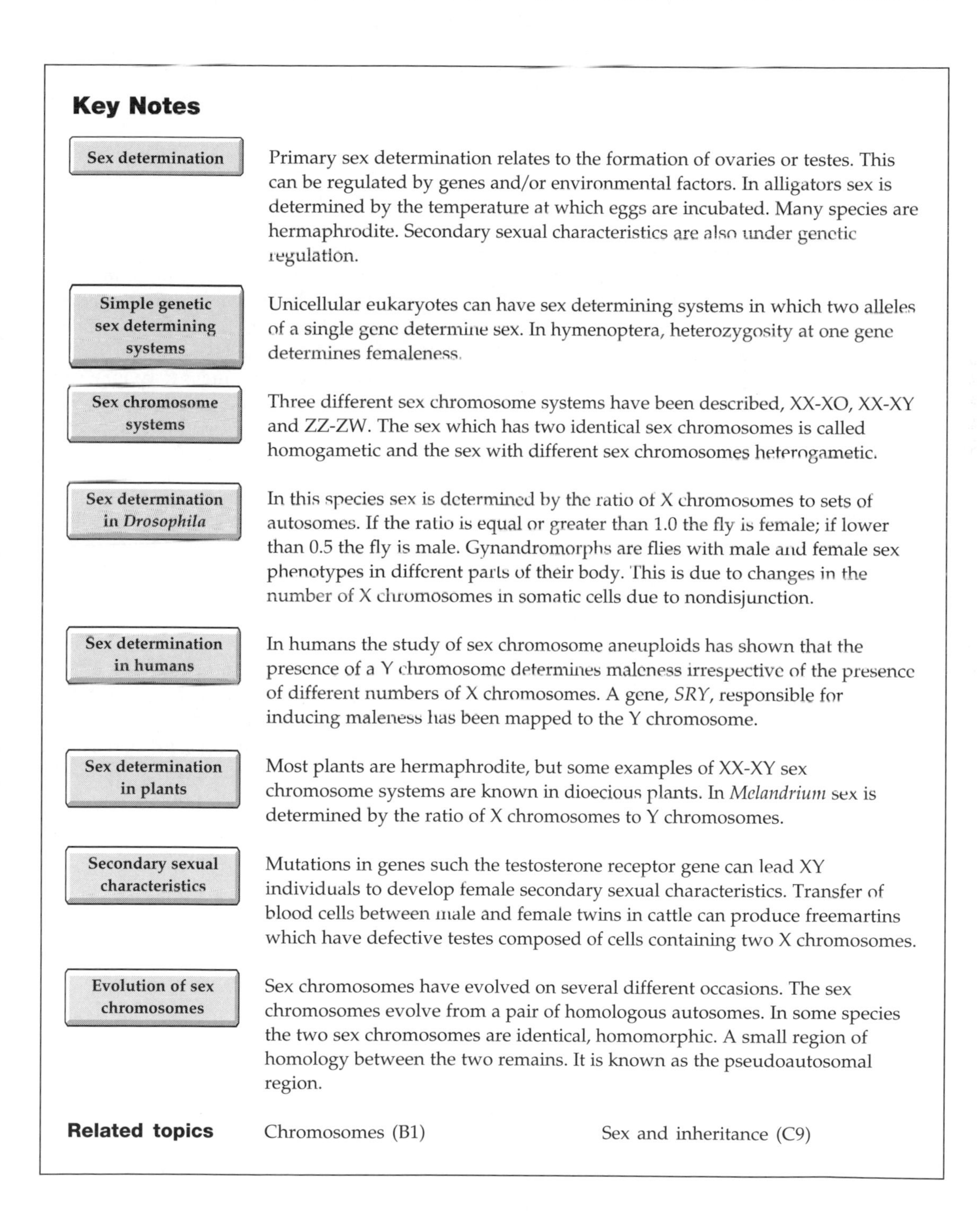

Key Notes

Sex determination

Primary sex determination relates to the formation of ovaries or testes. This can be regulated by genes and/or environmental factors. In alligators sex is determined by the temperature at which eggs are incubated. Many species are hermaphrodite. Secondary sexual characteristics are also under genetic regulation.

Simple genetic sex determining systems

Unicellular eukaryotes can have sex determining systems in which two alleles of a single gene determine sex. In hymenoptera, heterozygosity at one gene determines femaleness.

Sex chromosome systems

Three different sex chromosome systems have been described, XX-XO, XX-XY and ZZ-ZW. The sex which has two identical sex chromosomes is called homogametic and the sex with different sex chromosomes heterogametic.

Sex determination in *Drosophila*

In this species sex is determined by the ratio of X chromosomes to sets of autosomes. If the ratio is equal or greater than 1.0 the fly is female; if lower than 0.5 the fly is male. Gynandromorphs are flies with male and female sex phenotypes in different parts of their body. This is due to changes in the number of X chromosomes in somatic cells due to nondisjunction.

Sex determination in humans

In humans the study of sex chromosome aneuploids has shown that the presence of a Y chromosome determines maleness irrespective of the presence of different numbers of X chromosomes. A gene, *SRY*, responsible for inducing maleness has been mapped to the Y chromosome.

Sex determination in plants

Most plants are hermaphrodite, but some examples of XX-XY sex chromosome systems are known in dioecious plants. In *Melandrium* sex is determined by the ratio of X chromosomes to Y chromosomes.

Secondary sexual characteristics

Mutations in genes such the testosterone receptor gene can lead XY individuals to develop female secondary sexual characteristics. Transfer of blood cells between male and female twins in cattle can produce freemartins which have defective testes composed of cells containing two X chromosomes.

Evolution of sex chromosomes

Sex chromosomes have evolved on several different occasions. The sex chromosomes evolve from a pair of homologous autosomes. In some species the two sex chromosomes are identical, homomorphic. A small region of homology between the two remains. It is known as the pseudoautosomal region.

Related topics

Chromosomes (B1)

Sex and inheritance (C9)

Sex determination

The sex of an individual can be determined at several levels. This topic is principally concerned with primary sex determination, which relates to whether an individual develops testes or ovaries. Secondary sexual characteristics – forms of development associated with one sex or the other – such as feather-coloring, presence of horns or manes, are also under genetic control.

Unicellular organisms can have simple systems for sex determination, but multi-cellular species differ greatly in the strategies they employ to generate the male and female gametes that are necessary for sexual reproduction. In many instances a single individual may have male and female reproductive organs. Such individuals are known as **hermaphrodites**. This is common in many invertebrates and in plant species. Hermaphrodites can be of both sexes simultaneously or may change from one sex to the other. The importance of genes in determining sex in such species is clearly less than in most higher organisms, however there is evidence that genes can regulate the timing and extent of different sexual phases in some hermaphrodite invertebrate species.

Hermaphroditism, in animals, is not entirely limited to invertebrates. Several fish species such as bass undergo sex-reversal, often under environmental or hormonal influence. Domestic fowl can also occasionally undergo spontaneous sex-reversal. The most clear-cut example of environmentally determined sex is found in alligators where the temperature at which eggs are incubated determines the sex of the individual.

Simple genetic sex determining systems

Probably the simplest sex determining mechanism is found in yeast. Two alleles of a single gene determine mating type. In *S. cerevisiae* a gene on chromosome 3 known as *MAT* has two alleles *a* and α. For most of its life cycle *S. cerevisiae* is haploid (has only one set of chromosomes), and the yeast cell will carry either the *a* or α allele. This determines mating type. Only yeast cells of opposite mating types can fuse to form diploids which undergo meiosis and release new haploid spores. Thus the *MAT* gene can be regarded as an early type of sex-determining system. A similar single gene system is responsible for sex determination in unicellular algae species such as *Chlamydomonas.*

The hymenoptera (ants, bees and wasps) have an unusual method of sex determination. Male bees (drones) develop from eggs that were not fertilized and are haploid. Female bees (workers and queens) develop from fertilized eggs and are diploid. It was thought that the difference in ploidy was the sex determining factor. However, the mechanism depends on a gene with multiple alleles. If this gene is heterozygous the fly will be female. Haploid drones cannot be heterozygous at any gene since they have only one copy of each chromosome. Intensive inbreeding results in high levels of homozygosity and in highly inbred bee stocks diploid males have been detected.

Sex chromosome systems

In many species, sex determining genes are associated with specific chromosomes known as sex chromosomes. Several different sex chromosome systems are known:

- **XX-XO system**. This is found in many insect species. Females contain a pair of chromosomes known as X chromosomes. Males have only one X chromosome. This is the case in grasshoppers, and the bug *Protenor* and is sometimes known as the **Protenor system**.

- **XX-XY**. This is found in mammals and also in certain insects including *Drosophila* (the fruit fly). Here females have two copies of the X chromosome and males have an X and a Y chromosome.
- **ZZ-ZW.** This is essentially the reverse of the XX-XY system, where the female is ZW and the male ZZ. It is found in birds, lepidoptera (butterflies) and snakes.

The terms **homogametic** and **heterogametic** are used to describe these systems. Homogametic means that with respect to sex chromosomes gametes are all identical. For instance, in the XX-XY system females are the homogametic sex as all gametes will carry one X chromosome. In the ZZ-ZW system the female is the heterogametic sex as two classes of gametes containing either Z or W as the sole sex chromosome are found in equal numbers.

Although two species may share the same sex chromosome system, this does not mean the genes which determine sex operate in similar ways. To illustrate this point three different examples of the XX-XY sex chromosome system are compared below.

Sex determination in *Drosophila*

Sex is determined in *Drosophila* by the ratio of X chromosomes to sets of autosomes (sets of autosomes simply refers to the ploidy of the fly). When the ratio is 1.0 or greater flies are female. When it is 0.5 or less flies are male. Intermediate values give rise to intersex flies. Some typical examples are given in *Table 1*.

Extreme ratios such as 0.33 and 1.5 give rise to flies that are called metamales or metafemales. Although clearly of their respective sex these flies are poorly developed and have a shortened life-span.

The fact that sex determination is a result of a balance of X chromosomes and autosomes suggests that genes that cause female development are clustered on the X chromosome and genes for maleness on the autosomes. One important point to note concerns the Y chromosome. The data above indicate that it has no role in sex determination in *Drosophila*. This is correct, but although flies that lack a Y chromosome may be male, they are infertile because a gene on the Y chromosome is essential for the development of functional sperm.

A similar genetic balance mechanism regulates sex determination in other species such as the nematode *Caenorhabditis elegans* (round worm). However this is slightly more complex as male and hermaphrodite individuals exist in this species.

One significant feature of sex determination in *Drosophila* is the presence of abnormal flies known as **gynandromorphs**. These are the result of non-disjunction (see Topic B1) in the somatic cells of the flies. If this results in a change in the number of X chromosomes in a cell the X : autosome ratio will

Table 1. Ratio of X chromosomes to sets of autosomes, and sex determination in Drosophila

Number of X chromosomes (X)	Number of sets of autosomes (A)	X : A ratio	Sex
3	2	1.5	Female
3	3	1.0	Female
2	2	1.0	Female
2	3	0.67	Intersex
1	2	0.5	Male
1	3	0.33	Male

be changed and may affect the sex of the cell. This can occur because in flies sex is determined autonomously in every cell. As the cell continues to divide, its descendants will form a patch of cells (clone) which, depending on their position in the organism, may differentiate to form structures of the opposite sex to that of the rest of the fly. In the most extreme case, loss of an X chromosome in the first division after fertilization can result in a fly which develops bilaterally into two halves, one male and the other female. This type of event is not found in mammals where the production of secondary sexual characters is determined hormonally.

Sex determination in humans

Sex determination in humans is typical of the process in other mammalian species, and although the sex chromosomes are XX and XY, the genetic basis of sex determination differs markedly from that described for *Drosophila*. Our understanding of this subject comes from the study of sex chromosome aneuploids (see Topic B1). Aneuploidy of sex chromosomes arises more frequently than for autosomes because very few genes are present on the Y chromosome, and due to the phenomenon of X chromosome inactivation (see Topic B1), only one X chromosome is expressed in any cell. Hence alterations to the numbers of sex chromosomes have less effect on viability than do changes to the autosomes. The sex of several sex chromosome aneuploids is given in *Table 2*. Not all of these sex chromosome configurations result in fertile individuals and individuals with the more extreme deviations from normal suffer severe mental retardation.

From these examples it is clear that at the chromosomal level the presence of a Y chromosome is the factor which determines maleness in humans. During early embryonic development the presence of a Y chromosome causes the undifferentiated gonad to grow more rapidly and subsequently to develop into testes. In birds, where the ZZ-ZW sex determining system is essentially the reverse of the XX-XY, the presence of a W chromosome induces the development of ovaries from undifferentiated gonads. A specific gene, *SRY*, mapping to the Y chromosome in both humans and mice has been isolated that is responsible for the switch from female to male development in embryos (see Topic C9).

Table 2. Relationship between sex chromosome numbers and sex determination in humans

Sex chromosomes	Chromosome number	Sex
X	45	Female
XXX	47	Female
XXXX	48	Female
XXXXX	49	Female
XYY	47	Male
XXY	47	Male
XXXY	48	Male

Sex determination in plants

Most angiosperms are hermaphrodite, flowers contain both male and female organs. However, in some species male and female flowers are borne on separate plants (**dioecious**). One plant genus in which the chromosomal basis of sex determination has been worked out is *Melandrium* (Campion). Here the ratio of X : Y chromosomes is the important factor in determining whether plants produce male or female flowers. This implies that genes on both of the sex chromosomes interact to produce the sex phenotype. Studies of plants that had radiation-induced deletions in X or Y chromosomes showed that the Y chromosome contains regions that repress female development and induce male development, whereas the X chromosome has regions that stimulate development of female flowers.

Secondary sexual characteristics

As noted at the start of this topic the production of secondary sexual characteristics is also under genetic control. This is well illustrated by the gene *Tfm* in mammals. This codes for a protein that acts as the receptor for the male-specific steroid hormone, testosterone, and is expressed in both males and females. Testosterone is produced only in the testes and is responsible for secondary sexual characteristics in males. Mutant alleles of this gene are responsible for a syndrome known as testicular feminization (androgen insensitivity). Here, the presence of a Y chromosome causes testes to form and these produce testosterone. However, the hormone has no effect on target cells because they lack functional testosterone receptors. Individuals with this syndrome develop as infertile females. Their gonads are testes but these remain internal, and their sex chromosomes are XY.

A different misalignment of primary and secondary sexual differentiation occurs in **freemartins**. These are sheep, goats or cattle that develop as infertile females, but have defective internal testes. Cells of the testes and other organs have two X chromosomes, but the blood contains some cells that have X and Y sex chromosomes. This abnormal development is only found in females that have been a member of a pair of mixed sex twins. The embryonic blood supplies of twins in these species are fused *in utero* and hence XY blood cells and hormones can enter the circulation of the female twin. This is sufficient to force the gonads to develop into testes, and to block partially normal female development even though the animal is genetically XX. The male twin is not affected by the presence of XX cells in its blood system.

These two examples should indicate the complex interactions that can occur between the genetic and hormonal determinants of developmental processes.

Evolution of sex chromosomes

It is generally considered that sex chromosomes evolved from a pair of homologous autosomes. This process must have taken place a number of different times during evolution. One of the best examples is found in snakes. Sex chromosomes have been studied in primitive and highly evolved snake species. In primitive species the two sex chromosomes appear identical (**homomorphic**). The two chromosomes are only differentiated by the fact that the W chromosome replicates late in S phase (see Topic B2). In more advanced species the W chromosome becomes reduced in size, heterochromatic and clearly different from the Z. The pair of sex chromosomes retain only a small region of homology. This is known as the **pseudoautosomal region** and is required to allow the two chromosomes to pair and segregate accurately at meiosis (see Topics C3 and C9). In vertebrates the sex chromosome which is limited to the heterogametic sex, (Y or W) is generally found to carry few genes and to accumulate large amounts of satellite DNA sequences and constitutive heterochromatin (see Topics B1 and B4).

C9 Sex and inheritance

Key Notes

Sex-linked inheritance

Recessive alleles of genes mapping to the X chromosome are not expressed in heterozygous female mammals but will be expressed in males because males have only one X chromosome. Males transmit the recessive allele to their daughters, where it is not expressed. They are referred to as carrier females. The daughters, in turn, transmit the allele to half of their sons, where it is re-expressed. Some genes are on the Y chromosome and are passed directly from father to son. This is known as holandric inheritance. A small region of homology exists between the X and Y chromosomes. Genes in this region, the pseudoautosomal region, do not show sex-linked inheritance.

Sex-limited and sex-influenced traits

Sex-limited traits are inherited traits caused by a single gene which are expressed only in one sex. Sex-influenced traits are those which are observed more frequently in one sex than in the other. This can be caused by dominance relationships being different in the two sexes.

Related topic

Sex determination (C8)

Sex-linked inheritance

Due to the fact that, in mammals, the X chromosome is present in only one copy in males and in two copies in females, genes which map to this chromosome show a particular pattern of inheritance that differs from the normal expectations for Mendelian inheritance (see Topics B1 and B2). This is known as **sex linkage**.

Recessive alleles are not expressed in heterozygous females. The male has only one copy of the X chromosome (**hemizygous**), and hence recessive alleles present are expressed. A useful example is colorblindness in humans. This is due to a recessive allele of a gene which maps to the X chromosome. The normal allele is denoted *CB* and the mutant allele responsible for colorblindness *cb*. The possible genotypes are shown in *Table 1.*

The *cb* allele is relatively rare in human populations, with approximately one male in 40 being affected. Colorblind females must have two *cb* alleles and thus occur at a much lower frequency, approximately 1 / 1600 (the square of the frequency in males).

A colorblind man must have inherited the *cb* allele with his X chromosome from his mother. If she had normal vision then she must have been a heterozygote *CB / cb*. A colorblind man cannot transmit his *cb* allele to his son as, by definition, his son must inherit a Y chromosome from his father. By the same rule he must pass *cb* to all of his daughters. Any daughter that inherits the *cb* allele from her father and a normal allele from her mother will be a carrier and transmit the syndrome to, on average, half of her sons. A daughter of a carrier and a colorblind father will have a 50% chance of inheriting the syndrome (*Fig. 1*).

Hence a phenotype present in a male disappears in the next generation, and then reappears in his grandsons. All affected males, except for new mutants,

Table 1. Possible genotypes at the colorblind locus

Normal male	*CB*/Y
Colorblind male	*cb*/Y
Colorblind female	*cb/cb*
Normal female	*CB/CB* or *CB/cb*

Y denotes the Y chromosome. The *CB/cb* female is a carrier for the syndrome.

must have inherited the allele in question from their mother. These are the hallmarks of sex-linked inheritance.

Duchenne muscular dystrophy and hemophilia are other examples of sex-linked conditions in humans. In the case of hemophilia, family records show that Queen Victoria had a mutant allele for hemophilia and transmitted the disease to some of her sons and many of the European royal houses through marriage of her daughters. As there is no evidence of hemophilia in her ancestors, Queen Victoria must have inherited a mutation that arose in the germ cells of one of her parents.

Sex linkage is complicated to some extent by the process of X-inactivation (see Topic B1) in female mammals. This means that approximately half of the cells of a carrier female will express the allele responsible for the syndrome. This can easily be shown, in examples such as glucose-6-phosphate deficiency or Lesch–Nyhan syndrome, by the analysis of single cells. However this is insufficient to affect the overall phenotype.

Sex linkage is not displayed by genes which map to a small segment of the X chromosome, the **pseudoautosomal region**, the part of the X chromosome which pairs with the Y chromosome at meiosis. In humans this is found at the tip of the short arm of the X chromosome. Because there is homology between the X and Y chromosomes in this region, crossing-over can occur and alleles can switch between X and Y. Genes mapping to this region show the same inheritance pattern as genes on autosomes.

Any genes resident on the Y chromosome are obviously passed directly from father to son. There have been some controversial examples of this including traits

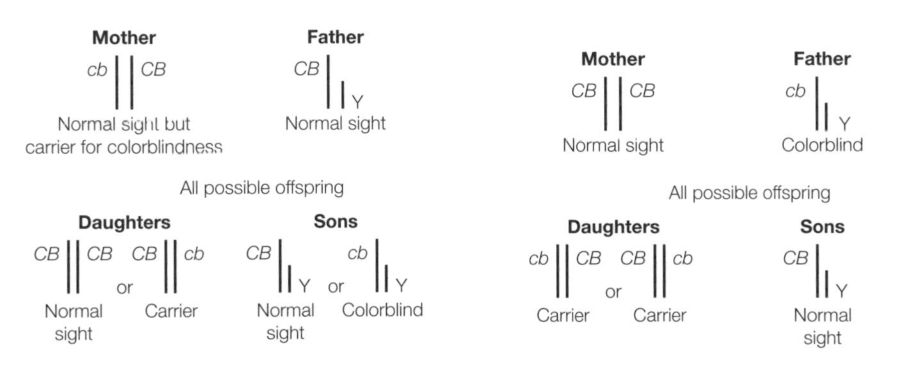

Fig. 1. The inheritance of colorblindness. X chromosomes are depicted as the long vertical lines, Y chromosomes as the short vertical lines.

such as 'porcupine man' and webbing of toes. A gene for hairs on the outer rim of the ear appears to show this pattern of **holandric inheritance**. Recently it has been shown that the sex-determining gene *SRY* and a gene for a minor histocompatibility antigen, *H-Y*, map to the Y chromosome, and hence are passed directly from father to son.

Sex-limited and sex-influenced traits

These terms relate to situations where the phenotype produced by a specific genotype is altered because of the sex of an individual. The two terms are easily confused, and care must be taken to differentiate between them.

Alleles of sex-limited genes will be expressed only in one sex. One example of this concerns mutant alleles of the breast cancer susceptibility gene *BRCA1*, which are dominant and cause breast cancer in females but not in males. In contrast to this, a second breast cancer susceptibility gene, *BRCA2*, causes breast cancer in both males and females, and is thus clearly not sex limited.

The difference between sex-limited and sex-influenced genes is subtle but important. In the former a phenotype is restricted to one sex, but in the latter the same phenotype will occur in both sexes but is more common in one. A good example of this is inherited pattern baldness. This is the form of baldness where hair loss spreads out from the crown of the head, and is controlled by a single gene with two alleles *B* and *b*. Homozygous *BB* individuals show premature pattern baldness, and *bb* homozygotes do not. The phenotype of the heterozygote, *Bb*, differs between males and females. In males the *B* allele is dominant and heterozygotes are bald, but in females it is recessive and no hair loss occurs. This is set out in *Table 2*. The dominance of such an allele is clearly influenced by the hormone balance of the individual. In this context it is interesting to note that this gene is also associated with polycystic ovarian disease in females.

Sex-limited and sex-influenced genes are autosomal and the genotypes follow normal Mendelian patterns of inheritance, but the phenotypes are altered by the hormonal environment. In contrast, the pattern of inheritance of sex-linked genes is due to the inheritance of genotypes caused by the genes being located on sex chromosomes.

Table 2. The expression of pattern baldness genotypes in male and female humans

Genotype	Female phenotype	Male phenotype
Bb	Pattern baldness	Pattern baldness
Bb	Normal hair	Pattern baldness
bb	Normal hair	Normal hair

C10 SOMATIC CELL FUSION

Key Notes

Linkage groups and chromosomes

Linkage studies allow genes to be placed in linkage groups. This gives no indication on which chromosome, or chromosome region these groups of genes are located. Originally methods to perform this depended on showing linkage between a gene and a chromosomal aberration. An example of this was the assigning of Duffy blood group to human chromosome 1. These methods have been superseded by procedures based on segregation of genes and chromosomes from somatic cell hybrids.

Tissue culture and cell fusion

Cells from most species of eukaryotes can be grown in culture in a manner resembling that used for microorganisms. Cultured cells can be made to fuse together by treatment with certain viruses or chemicals. The fused cells can go on to produce clones of somatic cell hybrids. In interspecific somatic cell hybrids one set of chromosomes is usually preferentially lost. Human chromosomes are lost from hybrids with either mouse or Chinese hamster cells.

Assigning genes to chromosomes

Clones are isolated from human × rodent somatic cell hybrids. Each clone contains a unique set of human chromosomes. A panel of clones is used in which each human chromosome can be individually identified. Each clone is screened for the presence or absence of human genes. Genes that show a concordant pattern of segregation with only a single human chromosome must be located on that chromosome. Genes that are assigned to the same chromosome are said to be syntenic. These do not necessarily show genetic linkage.

Constructing a fine scale map

To map genes to subregions of chromosomes human cells carrying translocations are used as parents for interspecific somatic cell hybrids. Depending on the segregation pattern of a human gene it can be assigned to one side or other of the translocation breakpoint. Other methods are available to produce hybrids containing only small parts of human chromosomes.

Related topics Linkage (C4) The human genome project (F5)

Linkage groups and chromosomes

All of the methods of gene mapping described (see Topic C4) have allowed genes to be ordered into linear linkage groups, but do not permit either individual genes or linkage groups to be assigned to specific chromosomes or chromosomal regions. To achieve this important step a combination between gene mapping and cytogenetic methodologies is required. In some cases this can be achieved by making use of cytogenetic polymorphisms that occur in natural populations. Translocations which cause the linkage relationships of groups of genes to be altered can be used for this purpose. More important are alterations in the morphology of a chromosome that allow its presence or

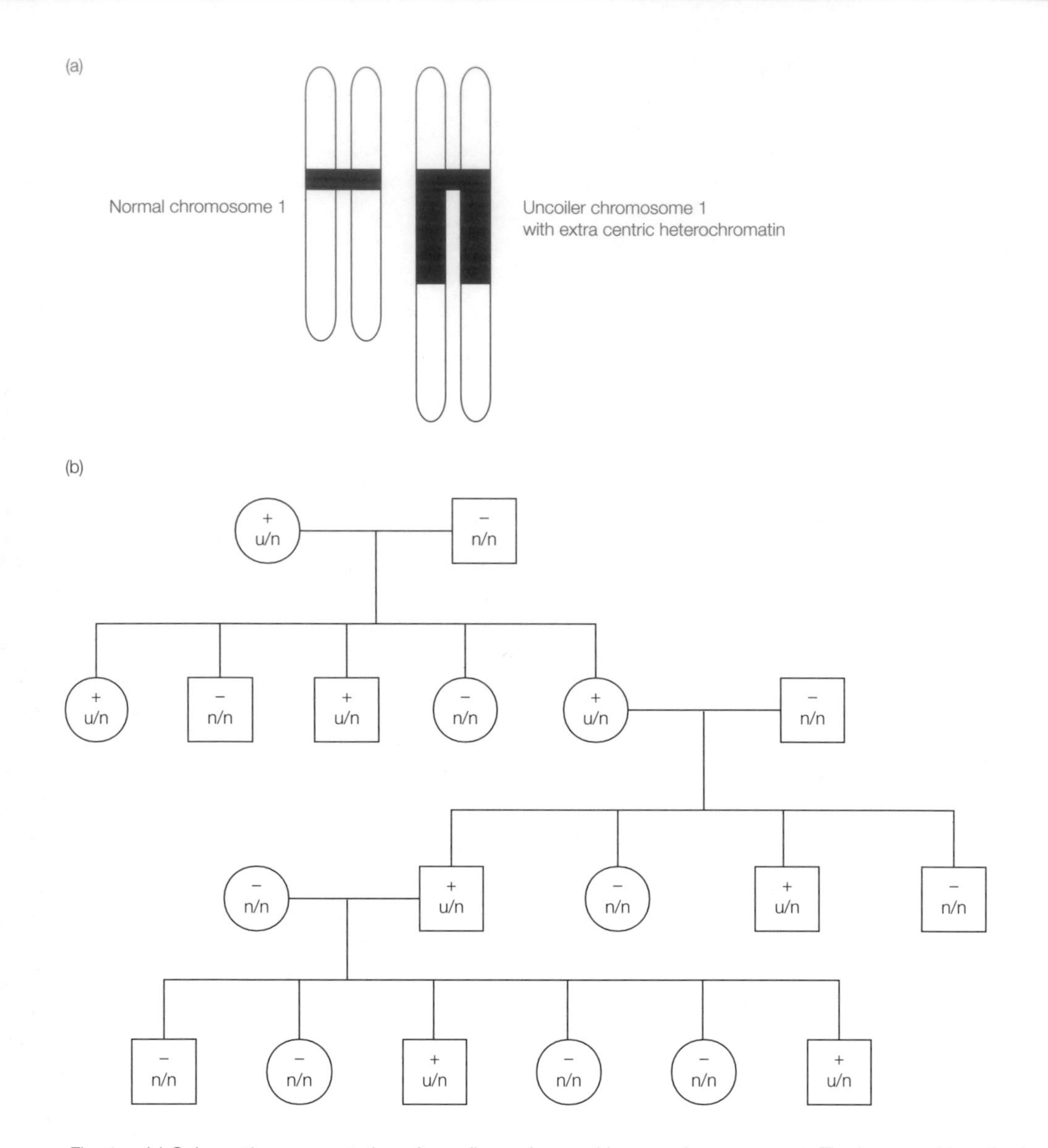

Fig. 1. (a) Schematic representation of uncoiler and normal human chromosome 1. The increased length of the uncoiler chromosome is due entirely to extra centric heterochromatin. (b) Members of a hypothetical family segregating Duffy blood group and the uncoiler version of chromosome 1. The great-grandmother is heterozygous for both markers. The Duffy blood group consistently co-segregates with the uncoiler version of the chromosome and shows no recombination. Circles, females; squares, males. +, individual is positive for Duffy blood group; –, individual is negative for Duffy blood group; u, uncoiler version of chromosome 1; n, normal chromosome 1.

absence to be correlated with the presence of absence of a phenotype governed by different alleles of a single gene. An example of this facilitated the first gene to be mapped to a human autosome in 1969 (*Fig. 1*). The morphology of chromosome 1 is known to vary within the human population due to changes in centric heterochromatin (see Topic B1). The chromosome carrying the longer centromere is known as 'uncoiler', and it was noted that in an extended

American family this version of chromosome 1 was only found in family members who were positive for the Duffy blood group. This gene has two alleles, one positive and the other negative for this blood group. Investigation of other unrelated families, showing segregation for this gene and the 'uncoiler' chromosome confirmed the association between the Duffy blood group and chromosome 1, and the gene was subsequently mapped close to the centromere.

Clearly this approach depends to a large extent on luck and is not suitable for systematic mapping of genes to chromosomes, particularly in species such as humans where the chromosome number is high and the frequency of chromosomal rearrangements in fertile individuals is low.

The technique which has been employed most successfully to position genes on chromosomes owes its origins to a remarkable observation made by Weiss and Ephrussi in 1967 when they were carrying out experiments in which animal cells grown in tissue culture were fused to each other.

Tissue culture and cell fusion

Cells of most animal and plant species can be adapted to grow in culture in a manner analogous to the conventional cultivation of microorganisms. For animal cells this involves aseptically disaggregating tissue and placing cells in a medium containing amino acids, vitamins, an energy source (usually glucose) and a supplement of animal serum. Cultures of normal human cells have a limited life-span with an upper limit of approximately 50 cell divisions. Importantly these cells retain their normal diploid chromosome complement. Cultures of rodent cells often alter their properties *in vitro* and go on to grow indefinitely in culture.

Cell fusion is a process which occurs naturally in situations such as sperm–egg fusion, or in the fusion of myoblasts to form myotubes. It is also a result of infection by viruses such as measles or mumps. The initial experiments which fused cells grown in culture used inactivated viruses, but these have since been replaced by chemical agents such as polyethylene glycol. In culture, fusion can be carried out between cells of the same species or of different species; there have even been examples of plant and animal cells being fused together. The initial product of the fusion is a heterokaryon, a cell containing more than one genetically different nucleus. Most of these fail to grow but a proportion go on to produce viable cultures of hybrid cells. A number of selection systems have been developed to isolate hybrid cells from unfused cells in the culture but these need not be described here.

Weiss and Ephrussi found that although hybrids made between cells of the same species showed stability of chromosome numbers, hybrids made between mouse and human cells were chromosomally unstable and chromosome numbers decreased as the hybrid cell proliferated. This drop in chromosome number was due to preferential loss of human chromosomes from the cells at cell division. Early after fusion human chromosomes were lost rapidly but as the number of human chromosomes present in each cell approached zero the rate of loss dropped and the remaining human chromosomes were retained with relative stability. A similar pattern is found in human × Chinese hamster hybrids, but here human chromosomes are lost much more rapidly. Although the processes that control this phenomenon are still unknown, loss of one set of chromosomes from interspecific somatic cell hybrids is commonly observed. *Table 1* shows which sets of chromosomes are lost in somatic cell hybrids made between different species.

Table 1. Pattern of loss of parental chromosomes in somatic cell hybrids

Interspecific hybrid	Chromosomes lost
Human × mouse	Human
Human × hamster	Human
Rat × mouse	Stable
Hamster × mouse	Mouse
Monkey × hamster	Monkey
Pig × hamster	Pig
Sheep × hamster	Sheep

Assigning genes to chromosomes

As human × rodent somatic cell hybrids lose (segregate) human chromosomes, and because each human chromosome appears to be lost independently, these hybrids can be used to genetically fractionate the human karyotype. After the period when human chromosomes are lost rapidly, each cell in a hybrid cell line will contain a small number of human chromosomes. If single cells of the hybrid cell line are isolated and allowed to grow on (cloned) at this stage most resulting clones will contain discrete groups of human chromosomes. The chromosomes present can be identified by cytogenetic analysis and a panel of clones chosen in which each human chromosome can be uniquely identified.

Table 2. Clones 1–11 represent clones of mouse × human somatic cell hybrids, each with a unique set of human chromosomes. The presence of any specific human chromosome in a hybrid clone is indicated by 'p' and its absence by 'a'. The presence of the human insulin gene as determined by Southern blotting is denoted by 'i' and its absence by 'n'

Chromosome	Clone 1	Clone 2	Clone 3	Clone 4	Clone 5	Clone 6	Clone 7	Clone 8	Clone 9	Clone 10	Clone 11
1	p	a	a	a	p	a	a	a	a	a	a
2	a	a	a	a	a	a	a	a	p	a	a
3	a	a	a	a	a	a	p	a	a	a	p
4	a	a	a	p	a	p	a	a	a	p	a
5	a	a	a	p	a	a	a	a	a	a	a
6	p	a	p	a	a	a	a	a	a	p	a
7	a	a	p	a	a	p	a	a	a	a	a
8	a	p	a	p	p	a	a	a	p	a	a
9	a	a	a	a	a	a	a	p	a	a	a
10	a	a	a	a	a	a	p	a	p	a	p
11	p	p	a	a	p	a	a	p	p	a	a
12	a	a	a	a	a	a	a	a	a	p	a
13	a	a	p	a	p	a	a	a	a	a	a
14	a	a	a	p	a	p	a	a	a	a	a
15	p	a	a	a	a	a	a	a	a	a	a
16	a	a	a	a	a	a	a	a	p	p	a
17	p	p	a	a	a	a	a	a	a	a	a
18	a	a	p	a	a	a	a	p	a	a	a
19	a	a	p	a	a	a	a	a	a	a	a
20	p	a	a	a	a	a	p	a	a	a	a
21	a	a	a	a	p	a	p	a	a	a	a
22	a	a	a	a	a	p	a	a	a	a	a
X	a	p	p	a	a	a	a	a	a	a	a
Insulin	i	i	n	n	i	n	n	i	i	n	n

Such a panel is shown in *Table 2*. The panel can be used to assign human genes to specific chromosomes by analyzing hybrid clones for the presence or absence of particular human genes and correlating this information with the human chromosomes present. The assignment of the human insulin gene to chromosome 11 is shown in the example in *Table 2*. The presence or absence of the human insulin gene can be determined for each of the hybrid clones by Southern blot hybridizations (see Topic E4), using a cloned human insulin gene as the probe. More recently PCR (see Topic E2) has been used to recognize the presence of human genes. As shown, the human insulin gene is present in clones 1, 2, 5, 8 and 9. It is absent from clones 3, 4, 6, 7 and 11. The only human chromosome which shows the identical pattern of presence and absence is chromosome 11. This is referred to as **concordant segregation** and is strong evidence that the insulin gene is present on chromosome 11. When a gene and a chromosome do not segregate together this is referred to a **discordant segregation**. If two genes show concordant segregation this means that they are on the same chromosome. They are said to be **syntenic**. Note that syntenic genes may be far apart on the same chromosome and need not show genetic linkage to each other (see Topic C4).

Where a gene which is assigned to a chromosome has been shown, by recombination analysis, to be part of a linkage group all other members of the linkage group are also, by definition, assigned to the same chromosome. This technology made a huge impact on human genetics in the period 1970–1990. The same procedures could also be applied to mouse genes using mouse × hamster somatic cell hybrids from which mouse chromosomes are lost.

However this is only part of the story. Although it was a major step to assign genes to specific chromosomes this did not give any information as to where on a chromosome a gene was located. Further refinements of this technology were required to improve the mapping of genes.

Constructing a fine scale map

Regionalization of genes on chromosomes is achieved using a variety of related techniques in which interspecific hybrids carrying only a fragment of a human chromosome were studied.

The simplest way to achieve this was to make hybrids using human cells that came from an individual carrying a known translocation. For example, if the human cells used in the fusion carried a balanced translocation between chromosomes 4 and X in which the short arm of the X chromosome had been translocated onto chromosome 4, analysis of hybrid clones for genes on the X chromosome would give different segregation patterns. Those on the short arm would have concordant segregation with genes on chromosome 4 whereas those on the long arm would be retained or lost in parallel with the truncated X chromosome. This allows genes to be regionalized to one arm or other of the X chromosome. By use of a large number of specific translocations between different chromosomes genes have been assigned to subchromosomal regions. Translocation breakpoints have been used to divide chromosome 21 into 19 subdivisions suitable for gene mapping.

Other approaches to producing fragments of human chromosomes in interspecific hybrids include chromosome mediated gene transfer, in which purified metaphase chromosomes are transferred to cells. The chromosomes are damaged in this process and only small fragments are incorporated into the hybrids. It is also possible to take hybrids carrying a single human chromosome and isolate clones in which fragments of this chromosome have been lost.

Genes which are also lost in this process must map to the deleted chromosomal regions.

Somatic cell fusion has proved a major tool in the mapping of genes in humans and a few other species. It must be remembered, however, that this technique is limited to genes for which a molecular probe, such as a cDNA clone, or sufficient DNA sequence for PCR, is available. Genes responsible for inherited syndromes where the basic biochemical defect has not been identified cannot be mapped in this way.

C11 Inbreeding

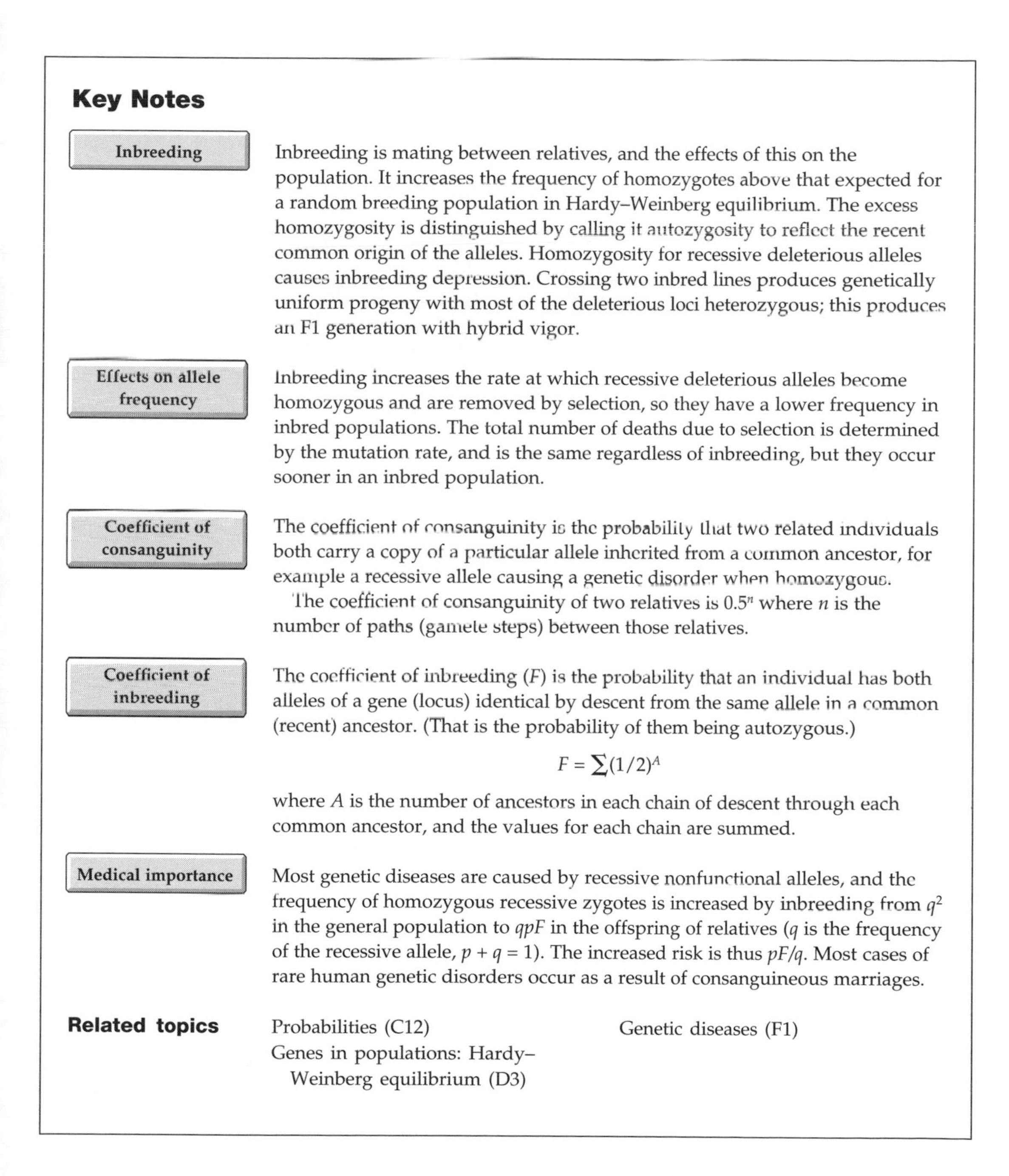

Key Notes

Inbreeding

Inbreeding is mating between relatives, and the effects of this on the population. It increases the frequency of homozygotes above that expected for a random breeding population in Hardy–Weinberg equilibrium. The excess homozygosity is distinguished by calling it autozygosity to reflect the recent common origin of the alleles. Homozygosity for recessive deleterious alleles causes inbreeding depression. Crossing two inbred lines produces genetically uniform progeny with most of the deleterious loci heterozygous; this produces an F1 generation with hybrid vigor.

Effects on allele frequency

Inbreeding increases the rate at which recessive deleterious alleles become homozygous and are removed by selection, so they have a lower frequency in inbred populations. The total number of deaths due to selection is determined by the mutation rate, and is the same regardless of inbreeding, but they occur sooner in an inbred population.

Coefficient of consanguinity

The coefficient of consanguinity is the probability that two related individuals both carry a copy of a particular allele inherited from a common ancestor, for example a recessive allele causing a genetic disorder when homozygous.

The coefficient of consanguinity of two relatives is 0.5^n where n is the number of paths (gamete steps) between those relatives.

Coefficient of inbreeding

The coefficient of inbreeding (F) is the probability that an individual has both alleles of a gene (locus) identical by descent from the same allele in a common (recent) ancestor. (That is the probability of them being autozygous.)

$$F = \sum(1/2)^A$$

where A is the number of ancestors in each chain of descent through each common ancestor, and the values for each chain are summed.

Medical importance

Most genetic diseases are caused by recessive nonfunctional alleles, and the frequency of homozygous recessive zygotes is increased by inbreeding from q^2 in the general population to qpF in the offspring of relatives (q is the frequency of the recessive allele, $p + q = 1$). The increased risk is thus pF/q. Most cases of rare human genetic disorders occur as a result of consanguineous marriages.

Related topics

Probabilities (C12)
Genes in populations: Hardy–Weinberg equilibrium (D3)
Genetic diseases (F1)

Inbreeding

Inbreeding is the effect of matings between relatives, or in extreme cases self-fertilization by hermaphrodites. This is quite common in plants. The effect of inbreeding is to increase homozygosity (*Fig. 1*). Under self-fertilization conditions the proportion of the population which is heterozygous halves at each generation, because half of the combinations of gametes from heterozygotes form homozygous zygotes. (As $Aa \times Aa$ gives a $1AA:2Aa:1aa$ ratio in Mendelian monohybrid crosses.) The danger in this is that deleterious recessive alleles which are present in a low frequency in the whole population will become homozygous in inbred offspring. This causes a decrease in vigor known as **inbreeding depression**. This is not a problem for plants which normally inbreed (e.g. wheat and peas) because selection has nearly eliminated deleterious alleles. In fact, inbreeding helps farmers because when a cultivar is homozygous offspring are all genetically identical to their parent. The crop will be predictable and consistent from year to year, and all the plants will have the same characteristics (e.g. size, ripening time, quality). In organisms that normally outbreed, inbreeding depression is a serious problem. It is necessary to cross two inbred lines to make genetically uniform but fully heterozygous F1 plants. The restored quality is known as **hybrid vigor** but it is not passed on to the F2 generation, which shows the full range of phenotypes including the phenotypes of both parents. Inbred maize is very weak, and four inbred lines are used to make two F1 strains which are then crossed to produce sufficient double hybrid seed for commercial production.

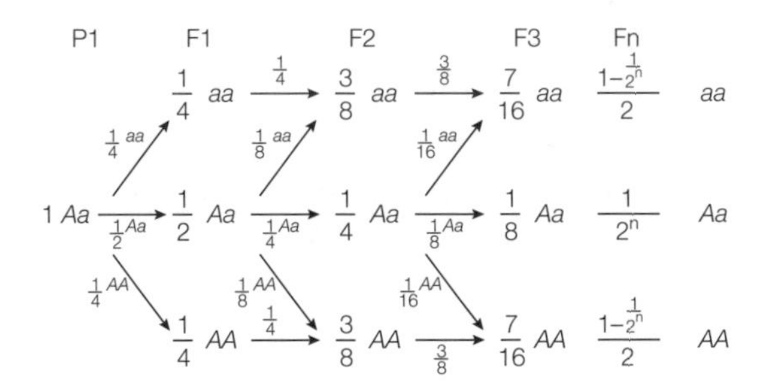

Fig. 1. Increase in homozygosity with successive rounds of self-fertilization (extreme inbreeding).

Inbreeding produces **excess homozygosity**. Many loci are homozygous in most organisms, simply because there is a limited amount of genetic variation. Alleles with a frequency of q will be homozygous with a frequency of q^2 in a randomly breeding population (see Topic D3). Most specific mutations only arose once, and to some extent everyone is descended from a common ancestor if their family trees are traced back far enough. Inbreeding is specifically about mating between comparatively close relatives. This puts the frequency of homozygotes in the population above the level expected from random mating at the Hardy–Weinberg equilibrium. We distinguish this by saying that the inbreeding coefficient is the probability of **autozygosity,** a special case of homozygosity, where both alleles of a locus have a recent common ancestor.

Effects on allele frequency

In an inbred population, deleterious recessive alleles become homozygous and are subject to selection more quickly than they would be in a large outbreeding population. They are removed more quickly after they are formed by mutation, and do not survive so long in the population. For this reason deleterious recessive alleles are present at a lower frequency in inbred populations than in outbred populations. The total number of deaths is not changed by inbreeding because deaths are due to mutations, not inbreeding. Every new deleterious mutation must be removed by a death. Mutations just become homozygous and are removed sooner in an inbred population.

Coefficient of consanguinity

Consanguineous marriages are marriages between relatives. The coefficient of consanguinity is the probability that two relatives will both have alleles which are identical by descent from a specific allele in a common ancestor. The use of this is that their consanguinity equals the inbreeding coefficient of their offspring, but it only works for a specific allele. This is medically useful if an identified common ancestor (say a grandparent) is known to have carried a particular recessive deleterious allele. It then predicts the chance of descendants (e.g. grandchildren, who are cousins) producing children who will be affected by the genetic disorder because they are homozygous for that allele.

Coefficient of consanguinity is calculated as follows. Each gamete carries half the alleles from the parent that produced it, so children have half their alleles in common with their father and half in common with their mother. What is the probability that a son and daughter (brother and sister) each carry a copy of the same allele? We can calculate this using *Fig. 2*. The probability of son C getting allele *z* from his father A is 1/2 (0.5) and daughter D has the same probability of getting *z*, so the probability of them both getting allele *z* is $(1/2)^2 = 1/4$. This is their coefficient of consanguinity. (Note that they also have a probability (1/4) of both inheriting the other allele *Z* so the probabilities of them both having copies of the same unspecified allele are actually 1/2.) If we continue down the family tree, there is a 1/2 probability of C carrying *z*, and if he does, a probability of 1/2 of passing it (*z*) to A's grandson E, so E has $1/2 \times 1/2 = 1/4$ chance of getting *z*. The consanguinity of nephew E and aunt D is $1/4 \times 1/2 = 1/8$ (= 0.125). (It is the same for uncle C and niece F.) Similarly there is a $1/2 \times 1/2 - 1/4$ probability of D having and passing on a copy of *z* to granddaughter F. The consanguinity of first cousins E and F is there-

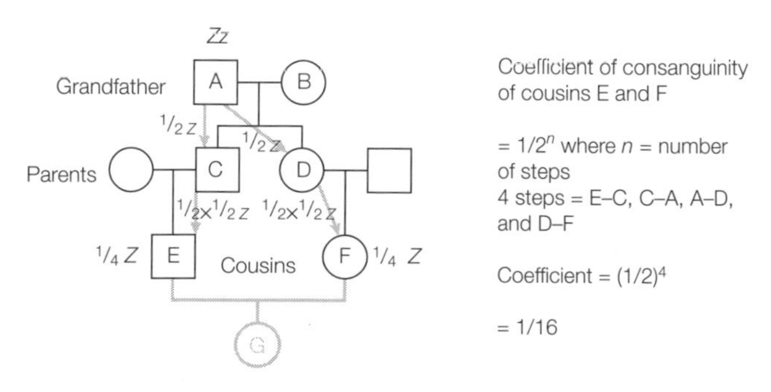

Fig. 2. The coefficient of consanguinity of cousins E and F is the probability of them both inheriting a copy of the same allele (say z) from one particular common parent (in this example their grandfather, A) and is 1/16, the same as the inbreeding coefficient of their child G. (Circles = females, squares = males.)

fore = $1/4 \times 1/4 = (1/2)^4 = 1/16$ or 0.0625. This calculation can be summarized as **path analysis**. The formula is:

$$\text{coefficient of consanguinity} = 0.5^n$$

where n is the number of paths (gamete steps) between them. For E and F (*Fig. 2*) these paths are E–C, C–A, A–D, and D–F, which equals four paths or steps. (They could be written E–C–A–D–F.)

Coefficient of inbreeding

The coefficient of inbreeding (F) is the probability that an individual has both alleles of a gene identical by descent from the same allele in a common (recent) ancestor. The inbreeding coefficient can be calculated by following the paths of descent and calculating the probability of the same allele being inherited from both parents. *Fig. 3a* shows the paths for the product of a brother–sister mating. Consider a particular allele z_1 in the grandfather A. There is a probability of 1/2 that it will be in the gamete contributed to the father C, and a probability of 1/2 that it will be passed on again to the child I (sib), so the child has a probability of $(1/2)^2$ of getting z_1 from his father. On the other side of the pedigree he also has the same probability of getting z_1 from his mother D. If he gets z_1 from both parents he will be homozygous. The probability of getting z_1 from both parents and so being homozygous is:

$$(1/2)^2_{\text{(father)}} \times (1/2)^2_{\text{(mother)}} = (1/2)^4 = 1/16$$

However there are four alleles in the grandparents (z_1, z_2, z_3, z_4) and I (sib) is equally likely to be homozygous for any of them, so the probability of him being homozygous for any allele at this locus (his coefficient of inbreeding) is $F = 4 \times (1/2)^4 = 4 \times 1/16 = 1/4$. (This is numerically the same as the coefficient of consanguinity of the parents C and D.) Following this to the offspring of a cousin marriage (I $_{\text{(cous)}}$ *Fig. 3b*) we can see that here there is another step in each side of her ancestry, so the chances of her getting any particular allele are reduced by $(1/2)^2$ and the inbreeding coefficient is $4 \times (1/2)^6 = (1/2)^4 = 1/16$.

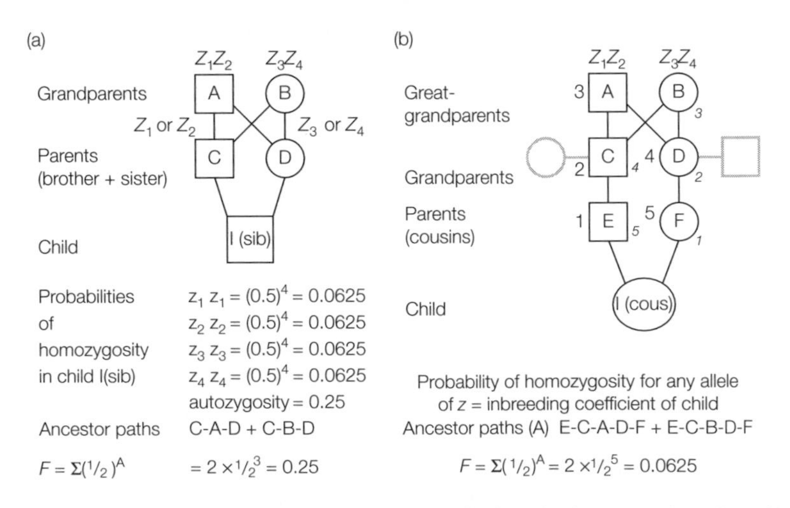

Fig. 3. (a) Inbreeding coefficient F *of progeny (I) of mating between sister D and brother C (sibs). (b) Inbreeding coefficient* F *of a mating between cousins. Normal numbers to left show chain of ancestors via A, small italics to right show ancestors via B. (Circles = females, squares = males.)*

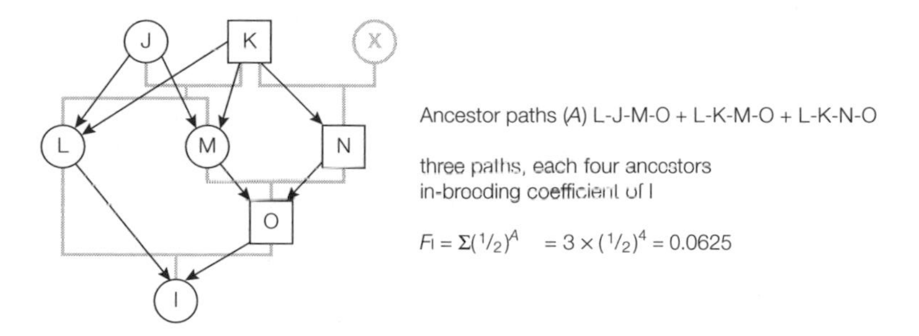

Fig. 4. A more complicated pattern of inheritance from a breeding program. J and K produce L and M; K also sires N. Half sibs (N × M) produce O, which is mated to L, producing I, with an inbreeding coefficient of 3/16. Female X does not contribute to inbreeding. (Gray lines are family pedigree, black arrows are paths of alleles. Circles = females, squares = males.)

A formula for calculating F is:

$$\text{inbreeding coefficient} = F = \Sigma(1/2)^A$$

where A is the number of ancestors in each chain of descent through each common ancestor, and the values are summed for each chain. For the example of a brother–sister marriage ($I_{(sib)}$ *Fig. 3a*) the paths are C–A–D and C–B–D, so $A = 3$ for two paths and $F_{(Isib)} = (1/2)^3 + (1/2)^3 = 1/4$. In the case of a cousin marriage ($I_{(cous)}$ *Fig. 3b*) there are five ancestors via A: E–C–A–D–F, and there are another five via B: E–C–B–D–F. Thus $A = 5$ for each chain and

$$F_{(cous)} = (1/2)^5 + (1/2)^5 = 1/32 + 1/32 = 1/16$$

A more complicated example that might arise in a breeding program is shown in *Fig. 4*. Each ancestor in each chain must be used once only in each chain. There are three chains, each with four ancestors: L–J–M–O and L–K–M–O, and L–K–N–O:

$$F_{(I)} = 3(1/2)^4 = 3/16$$

Medical importance

The frequency of homozygosity for rare recessive alleles in a random breeding population is given by the Hardy–Weinberg equilibrium (see Topic D3) and is q^2 where q is the frequency of the recessive allele in the population. Most genetic diseases are caused by homozygosity for rare recessive alleles. If we consider an allele with frequency $q = 0.0001$ (one in 10 000) then $q^2 = 1 \times 10^{-8}$ (one in 100 000 000), however for the child of a cousin marriage it is $qpF = 0.0001 \times 0.9999 \times 1/16 \times =$ (approximately) 6.25×10^{-6} (one in 160 000). The frequency of homozygotes is increased 625 times compared with the general population. The increase equals pF/q. The majority of cases of the rarest human genetic diseases occur as a result of consanguineous marriages. Studies suggest that marriages between cousins have at least two times, and up to 10 times the rates of miscarriage and neonatal mortality as marriages between unrelated parents.

C12 PROBABILITIES

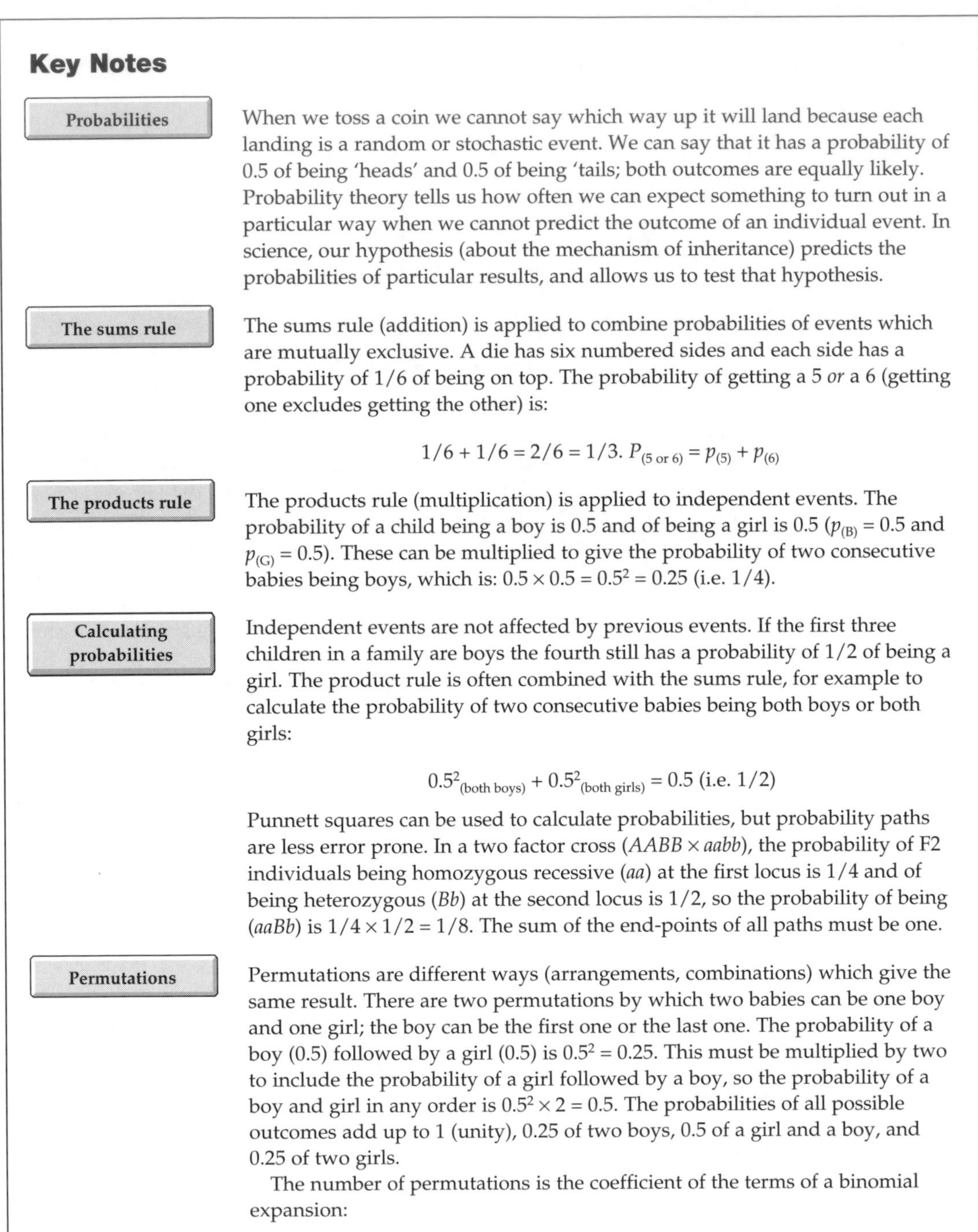

Key Notes

Probabilities

When we toss a coin we cannot say which way up it will land because each landing is a random or stochastic event. We can say that it has a probability of 0.5 of being 'heads' and 0.5 of being 'tails; both outcomes are equally likely. Probability theory tells us how often we can expect something to turn out in a particular way when we cannot predict the outcome of an individual event. In science, our hypothesis (about the mechanism of inheritance) predicts the probabilities of particular results, and allows us to test that hypothesis.

The sums rule

The sums rule (addition) is applied to combine probabilities of events which are mutually exclusive. A die has six numbered sides and each side has a probability of 1/6 of being on top. The probability of getting a 5 *or* a 6 (getting one excludes getting the other) is:

$$1/6 + 1/6 = 2/6 = 1/3.\ P_{(5\ \text{or}\ 6)} = p_{(5)} + p_{(6)}$$

The products rule

The products rule (multiplication) is applied to independent events. The probability of a child being a boy is 0.5 and of being a girl is 0.5 ($p_{(B)} = 0.5$ and $p_{(G)} = 0.5$). These can be multiplied to give the probability of two consecutive babies being boys, which is: $0.5 \times 0.5 = 0.5^2 = 0.25$ (i.e. 1/4).

Calculating probabilities

Independent events are not affected by previous events. If the first three children in a family are boys the fourth still has a probability of 1/2 of being a girl. The product rule is often combined with the sums rule, for example to calculate the probability of two consecutive babies being both boys or both girls:

$$0.5^2_{\text{(both boys)}} + 0.5^2_{\text{(both girls)}} = 0.5 \text{ (i.e. } 1/2)$$

Punnett squares can be used to calculate probabilities, but probability paths are less error prone. In a two factor cross ($AABB \times aabb$), the probability of F2 individuals being homozygous recessive (*aa*) at the first locus is 1/4 and of being heterozygous (*Bb*) at the second locus is 1/2, so the probability of being (*aaBb*) is $1/4 \times 1/2 = 1/8$. The sum of the end-points of all paths must be one.

Permutations

Permutations are different ways (arrangements, combinations) which give the same result. There are two permutations by which two babies can be one boy and one girl; the boy can be the first one or the last one. The probability of a boy (0.5) followed by a girl (0.5) is $0.5^2 = 0.25$. This must be multiplied by two to include the probability of a girl followed by a boy, so the probability of a boy and girl in any order is $0.5^2 \times 2 = 0.5$. The probabilities of all possible outcomes add up to 1 (unity), 0.25 of two boys, 0.5 of a girl and a boy, and 0.25 of two girls.

The number of permutations is the coefficient of the terms of a binomial expansion:

$$P = (n!/s!t!)$$

where n = total number of trials (e.g. babies), s = number of one outcome (e.g. a boy) and t = number of other outcome (e.g. a girl) and $s + t = n$. The '!' means factorial, the product of all integers down to 1, so $3! = 3 \times 2 \times 1 = 6$. If the probability of the first outcome is p and of the second is q (where $p + q = 1$) then the probability of s first events and t second events in n trials (where $n = s + t$) is:

$$P_{(\text{s and t})} = (n!/s!t!)p^s q^t$$

The coefficient $(n!/s!t!)$ is the number or permutation, and $p^s q^t$ is the probability of a specific sequence of events. With n events, each with two possibilities, there are $(n + 1)$ numerical categories: if there are three children, zero, one, two or three could be boys. The coefficients (number of ways of getting) for each category can also be read from the $(n + 1)$ row of Pascal's triangle. When there are more than two alternative outcomes for each event (e.g. six numbers on a die) a multinomial expansion must be used.

Related topics

Basic Mendelian genetics (C1)
More Mendelian genetics (C2)
Tests for goodness of fit: chi-square and exact tests (C13)
Genes in populations: Hardy–Weinberg equilibrium (D3)

Probabilities

In genetics it is often necessary to calculate the expected frequency and numbers of particular genotypes. For example in predicting the outcome of Mendelian crosses (see Topics C1 and C2) or predicting the frequency of particular genotypes in a population for comparison with Hardy–Weinberg equilibrium (see Topic D3). The scientific method requires that predictions are made so that they can be tested (see Topic C13).

We know that the probability of a tossed coin falling heads up is 0.5, but we cannot predict what the result of tossing a coin once will be, because each toss is unique and is effectively a **random** or **stochastic** event. Similarly we may know the proportion of flowers that will be pink from a particular cross, but we cannot predict which color each individual plant will produce, we must examine it to see. **Probability theory** allows us to calculate what we expect to get.

The probability (P) that something will happen is the number of times it *does* happen (a) divided by the total number of times it could *possibly* have happened (n).

$$P = a/n$$

P can be measured directly or deduced from the nature of the event. We can assume that the probability of a coin falling heads up is 0.5, or of getting a six when we throw a die is 1/6 (because a die is equally likely to fall with 1, 2, 3, 4, 5 or 6 uppermost). **The sum of all the probabilities must add up to 1** because **something** must happen; one of the six sides *must* be uppermost. If something is certain to happen it has a probability of one (often called unity), if it can never happen it has a probability of 0 (nil or zero).

The sums rule

The sums rule is **applied to mutually exclusive events**. The die can give *either* a 5 *or* a 6, but not both. What is the probability of getting a 5 or a 6?

Each has a probability of 1/6. These are added to give the probability of getting both:

$$P_{(5 \text{ or } 6)} = 1/6 + 1/6 = 1/3$$

A genetic example would be a cross where we expected to get 1/4 red : 2/4 pink : 1/4 white flowers. The probability of any particular flower being red or pink is 1/4 + 2/4 = 3/4

The products rule

The products rule is **applied to independent events and events happening in a specific order**. What is the probability of getting two sixes if we throw two dice? The fall of one die will not affect the fall of the other. The probability of the first being 6 is 1/6, and that will not affect how the second one falls, it also has a probability of 1/6, so in 1/6 of our throws the first die gives a 6, and in 1/6 of these the second die also gives a 6, so:

$$P_{(6 \text{ and } 6)} = 1/6 \times 1/6 = 1/36$$

A genetic example is a marriage between two people heterozygous for a recessive genetic disease, say phenylketonuria (PKU). If they have three children, what is the probability (chance) of all three being affected? The probability of a child being homozygous for the recessive allele is 1/4, therefore:

$$P_{(\text{all 3 PKU})} = 1/4 \times 1/4 \times 1/4 = (1/4)^3 = 1/64$$

Calculating probabilities

Independent events are not affected by the outcome of previous events. Consider the PKU family above. $P_{(\text{all 3 PKU})} = 1/64$ is only true while they have no children. If they already have two children, and both children have PKU, the probability of the third child having PKU is not changed, it is still 1/4 because the third child is an independent event. Similarly if you throw 10 heads consecutively with a coin, the next toss is still equally likely to be heads or tails.

The sums rule and products rule are often combined. The probability of two children being a boy and a girl is the probability of the first being a boy multiplied by the probability of the second being a girl (1/2 × 1/2 = 1/4) summed with the probability of a girl first followed by a boy (also 1/4), so the boy–girl combination in any order is 1/4 + 1/4 = 1/2. The segregation of each different genetic locus in a cross is an independent event (Mendel's second law). The easiest way to calculate the frequency of combinations of independent events (genoype or phenotype) is to make a probability path. They are simpler than Punnett squares, and less likely to lead to mistakes. For example, suppose we consider a three-factor cross with Mendel's peas. If we cross a strain with the dominant characters round yellow gray-coated seeds (genotype *RRYYGG*) with a strain carrying the recessive characters wrinkled green white-coat (genotype *rryygg*), the F1 (genotype *RrYyGg*) will produce eight different gamete genotypes, and a Punnett's square will have 64 cells in it. It is easier to work out the F2 probabilities directly. What is the probability (i.e. what is the proportion) of an F2 seed being round, green and gray-coated? Of the total, 3/4 will be round (round is dominant), 1/4 will be green (green is recessive) and 3/4 will be gray-coated (dominant) so we can construct a probability path (*Fig. 1*). The answer is 3/4 round × 1/4 green × 3/4 gray = 9/64 round green gray-coated. The probabilities of all the possible phenotypes add up to 64/64 = 1, as they must if our math is correct. The same principle can be applied to genotypes.

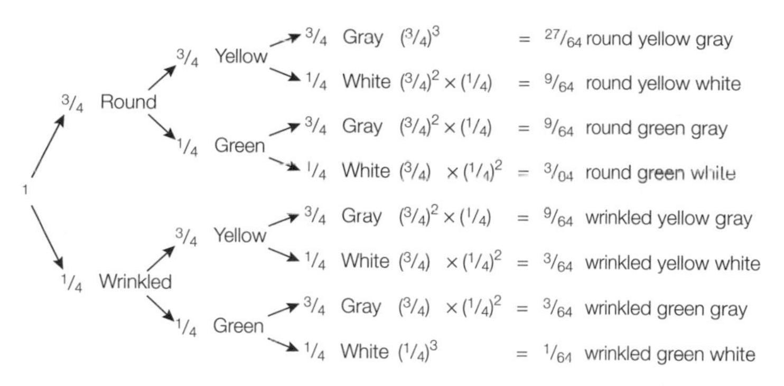

Fig. 1. Probability paths for phenotype frequencies in the F2 of a Mendelian trihybrid cross.

What will be the frequency of *RryyGg* genotypes in the F2? The frequency of *Rr* will be 1/2, of *yy* will be 1/4 and of *Gg* will be 1/2, so the frequency of *RryyGg* will be $1/2 \times 1/4 \times 1/2 = 1/16$ (or 4/64 in a Punnett square).

Permutations

Permutations are different ways of doing the same thing. There is only one way that tossing two dice can give two sixes, both dice must fall 6-up. The same is true if we specify the order of events (e.g. getting a 3 then a 6 from two throws: $P_{(3\text{ then }6)} = 1/6 \times 1/6 = 1/36$). However, if we wanted a 3 and a 6, this could happen in two ways, the first die 3 and the second die 6, or the first die 6 and the second die 3. These are two permutations or arrangements:

$$P_{(3\text{ and }6)} = 1/6 \times 1/6 \times 2 = 1/18$$

Boys and girls are equally frequent ($P = 1/2$), so the probability of two children both being boys is $(1/2)^2 = 1/4$, the probability of both being girls is similarly 1/4, but the probability of one being a boy and the other a girl is $1/2 \times 1/2 \times 2 = 1/2$ because they could occur in the order boy–girl or girl–boy. These are two permutations (also called **combinations** or **arrangements**). In the case of two parents heterozygous for PKU who have three children, what is the probability that they will have one affected child and two normal children? The affected child could be the first, second or third born, so there are three permutations.

Probability of one PKU affected child (A) and two normal (N)

$$= P(\text{A then N then N}) + P(\text{N then A then N}) + P(\text{N then N then A})$$

$$= (1/4 \times 3/4 \times 3/4) + (3/4 \times 1/4 \times 3/4) + (3/4 \times 3/4 \times 1/4)$$

$$= 3[(3/4)^2 \times 1/4] = 27/64$$

The number of permutations (combinations) is given by the formula for the terms of the **binomial expansion**: $(p + q)^n$ where p is the probability of one outcome (say a normal child, 3/4) and q is the probability of the other outcome (a child being affected by PKU, 1/4) and n is the number of trials (total number of children). Note that $(p + q) = 1$. If there are n children, the probability of s children being normal and t children being affected is:

$$P = (n!/s!t!)p^s q^t$$

The number of permutations is given by the part $n!/s!t!$ of the equation and is called the **coefficient**, whereas $p^s q^t$ is the probability for each particular ordered occurrence. Note that $(p + q) = 1$ and $(s + t) = n$. The symbol ! as in $n!$ is called **factorial**, (pronounced n-factorial) and means the product of all the integers from n down to 1 (e.g. $5! = 5 \times 4 \times 3 \times 2 \times 1 = 120$). $0! = 1$, the same way that $n^0 = 1$. If we recalculate the probability of one child in three having PKU ($t = 1$) and two being normal ($s = 2$) we get:

$$P = (n!/s!t!)p^s q^t$$

$$P = (3!/2!1!) \times (3/4)^2 \times (1/4)^1$$

$$= [(3 \times 2 \times 1)/ ((2 \times 1) \times 1)] \times 9/16 \times 1/4$$

$$= 3 \times 9/64 = 27/64$$

This gives the same result as the direct calculation above, but has the advantage that it is much easier for complicated combinations, such as the chance of having three affected children in a family of seven ($n = 7$, $s = 4$, $t = 3$).

Pascal's triangle has all the coefficients (number of permutations) of the terms of the binomial expansion (*Fig.* 2) and can be used by those who find the formula difficult. Each row starts and ends with 1, and each coefficient is the sum of the two adjacent terms in the row above. In the three-children example above ($n = 3$) we use the fourth ($n + 1$) row of Pascal's triangle. We could have 0, 1, 2 or 3 children affected. The fourth row of Pascal's triangle shows that there is one way of getting zero, three ways of getting one affected / two normal, three ways of getting two affected / one normal, and one way of getting three affected (1 : 3 : 3 : 1). It is easy to generalize upwards to three or more alternatives using the general **multinomial expansion**. If the probabilities of flowers being red, pink or white are p, q and r respectively, the probability of getting s red, t pink and u white flowers is:

$$P = (n!/s!t!u!)p^s q^t r^u$$

where the total number of flowers $= n = s + t + u$. In fact this can be extended to any number of classes.

Pascal's triangle

Row		Total
1	1	1
2	1 1	2
3	1 2 1	4
4	1 3 3 1	8
5	1 4 6 4 1	16
6	1 5 10 10 5 1	32
7	1 6 15 20 15 6 1	64
8	1 7 21 35 35 21 7 1	128
9	1 8 28 56 70 56 28 8 1	256

Fig. 2. Pascal's triangle shows the number of permutations (coefficients) of binomial expansions.

C13 TESTS FOR GOODNESS OF FIT: CHI-SQUARE AND EXACT TESTS

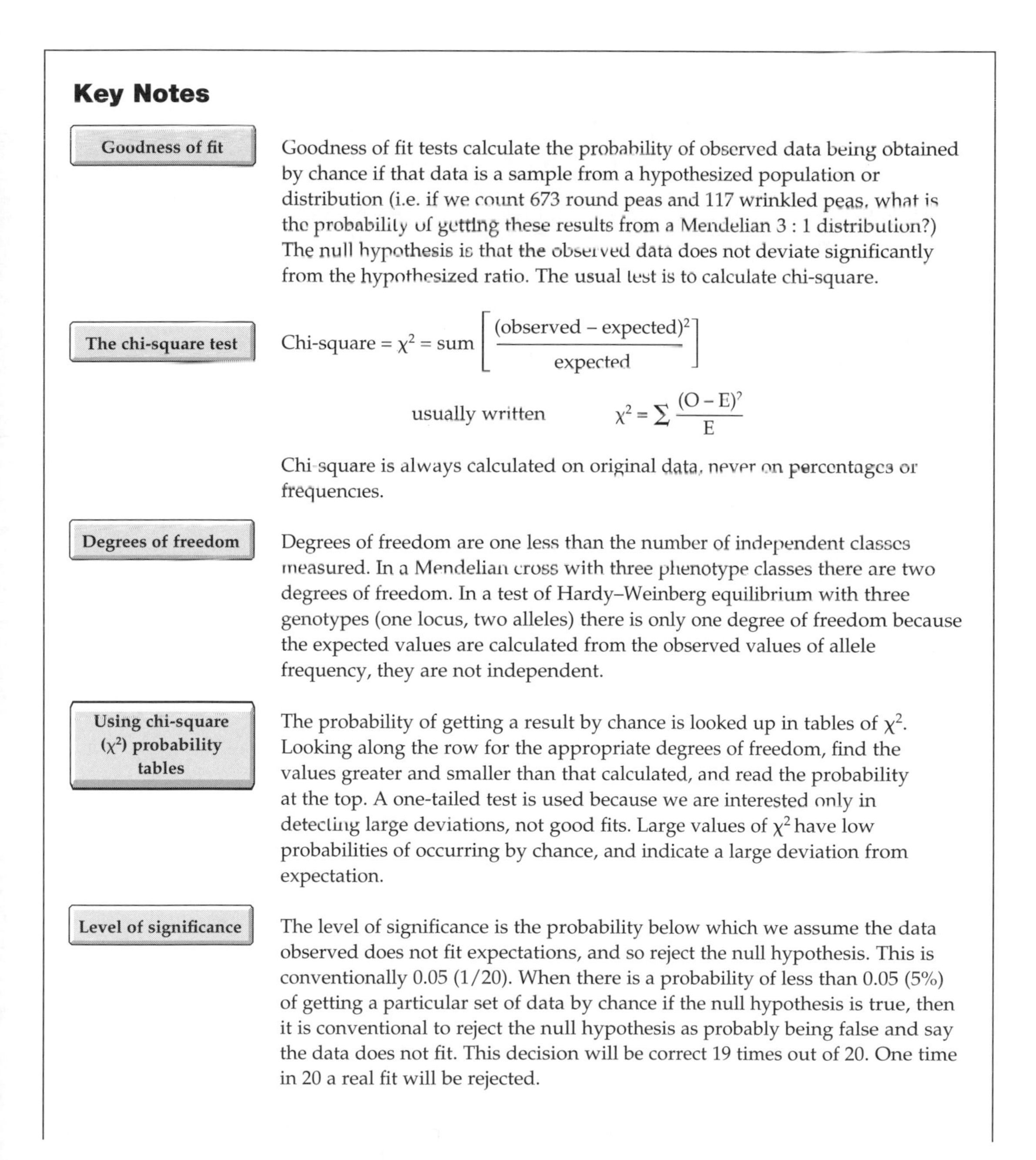

Key Notes

Goodness of fit

Goodness of fit tests calculate the probability of observed data being obtained by chance if that data is a sample from a hypothesized population or distribution (i.e. if we count 673 round peas and 117 wrinkled peas, what is the probability of getting these results from a Mendelian 3 : 1 distribution?) The null hypothesis is that the observed data does not deviate significantly from the hypothesized ratio. The usual test is to calculate chi-square.

The chi-square test

$$\text{Chi-square} = \chi^2 = \text{sum}\left[\frac{(\text{observed} - \text{expected})^2}{\text{expected}}\right]$$

usually written $\chi^2 = \sum \frac{(O - E)^2}{E}$

Chi-square is always calculated on original data, never on percentages or frequencies.

Degrees of freedom

Degrees of freedom are one less than the number of independent classes measured. In a Mendelian cross with three phenotype classes there are two degrees of freedom. In a test of Hardy–Weinberg equilibrium with three genotypes (one locus, two alleles) there is only one degree of freedom because the expected values are calculated from the observed values of allele frequency, they are not independent.

Using chi-square (χ^2) probability tables

The probability of getting a result by chance is looked up in tables of χ^2. Looking along the row for the appropriate degrees of freedom, find the values greater and smaller than that calculated, and read the probability at the top. A one-tailed test is used because we are interested only in detecting large deviations, not good fits. Large values of χ^2 have low probabilities of occurring by chance, and indicate a large deviation from expectation.

Level of significance

The level of significance is the probability below which we assume the data observed does not fit expectations, and so reject the null hypothesis. This is conventionally 0.05 (1/20). When there is a probability of less than 0.05 (5%) of getting a particular set of data by chance if the null hypothesis is true, then it is conventional to reject the null hypothesis as probably being false and say the data does not fit. This decision will be correct 19 times out of 20. One time in 20 a real fit will be rejected.

Yates' correction

If the expected number in a class is five or less, the numerical value of the difference between observed and expected is reduced (moved closer to zero) by 0.5 before it is squared. This is to allow for the fact that only integers (whole numbers) can be counted, whereas we may expect fractions.

Fisher's exact test

Fisher's exact test is simply to work out the exact probability of obtaining the observed result, plus the probability of obtaining an even more extreme result. A man is known to be heterozygous for phenylketonuria but his wife's genotype is unknow. The null hypothesis is that she is hetereozygous, in which case each child will have a 3/4 chance of being normal. If they have 11 children all normal, with a probability of $(3/4)^{11} = 0.04223$ then we can reject the nul hypothesis with a significance level of 4.22%. That is the exact chance of us being wrong, and her having so many normal children if she is heterozygous. (If they have one affected child, she is also heterozygous.)

Related topics

Basic Mendelian genetics (C1)
More Mendelian genetics (C2)
Probabilities (C12)
Genes in populations: Hardy–Weinberg equilibrium (D3)
Genetics in forensic science (F5)

Goodness of fit

It is often necessary to discover whether observed numbers fit an expected hypothesis. For example Mendelian genetics suggest a 3 : 1 ratio in the F2 generation of a monohybrid cross (see Topic C1). Suppose we cross a red flower with a white flower, interbreed the F1 (which are all red) and observe a ratio of 63 red : 37 white in the F2. Is 63 : 37 a good fit for a 3 : 1 ratio allowing for chance error in sampling? Chi-square (χ^2) is a statistical test which tells us the probability of obtaining the observed result by chance if the null hypothesis is true. It does not tell us whether the null hypothesis is actually true. The null hypothesis is that the data fits the ratio we suggest, and this is what is tested. We must have a null hypothesis before we can do any meaningful test. Our null hypothesis is that the results (63 : 37) do fit a 3 : 1 ratio.

The chi-square test

Chi-square (χ^2) = sum[(observed – expected)^2/expected]

usually written $\chi^2 = \sum \frac{(O - E)^2}{E}$

χ^2 is always calculated from original data, never from percentages, frequencies or proportions.

The χ^2 for comparing the 63 : 37 ratio above is calculated in *Table 1*. The expected numbers are calculated by multiplying the total by the predicted frequency (e.g. $100 \times 1/4$ white). The deviation, or difference, between the observed and expected number is squared to remove negative numbers, and divided by the expected value to undo the squaring and standardize the numbers. This is repeated for each class (the red class and the white class in this example) and the values are added (summed) to give the overall value of χ^2. Notice that the value of χ^2 increases when the deviations from expected are large, so large values of χ^2 lead us to reject the null hypothesis. The data does not fit if χ^2 is large. A perfect fit gives a χ^2 of zero. We are not interested in identifying unusually close fits, so we use a **one tailed test** which tells us the

Table 1. Chi-square calculations

Null hypothesis (predicted ratio)	Observed			
3 : 1	Observed	(100 total)	63 Red	37 White
	Expected		100 × 3/4	100 × 1/4
			75	25
	Observed – expected		–12	12
	(Observed – expected)^2		144	144
	(Observed – expected)^2/expected		144/75 = 1.92	144/25 = 5.76
$\chi^2 = 1.92 + 5.76 = 7.68$ Degrees of freedom = 2 – 1 = 1 d.f. From *Table 2* probability (p)(1 d.f.) $0.01 > p > 0.005$				
9 : 7	Observed	(100 total)	63 Red	37 White
	Expected		100 × 9/16	100 × 7/16
			56.25	43.75
	Observed – expected		6.75	–6.75
	(Observed – expected)^2		45.5625	45.5625
	(Observed – expected)^2/expected		45.5625/56.25 = 0.81	45.5625/43.75 = 1.04
$\chi^2 = 0.81 + 1.04 = 1.85$ Degrees of freedom = 2 – 1 = 1 d.f. Probability (p)(1 d.f.) $0.20 > p > 0.05$				

probability of getting a large deviation from the expected. The example (*Table 1*) gives a χ^2 value of 7.68. What does this mean? We need another piece of information before we can look this up in statistical tables. We need to know the degrees of freedom.

Degrees of freedom

Degrees of freedom are one less than the number of classes. They tell us something about the number of independent numbers we have, and this relates to the usefulness of our data. In this case we have two classes, red and white. All the plants that are not red must be white. When we have counted the red ones, the number of white ones is fixed so we have only one degree of freedom. If we had red, pink and white flowers – three classes – there would be two degrees of freedom; when two classes had been counted, the third would be fixed. When testing Hardy–Weinberg equilibrium with one locus and two alleles (three genotypes) there is only one degree of freedom. This is because the expected numbers are calculated from the observed allele frequencies, they are not independent (whereas a 3 : 1 ratio prediction is independent of the observed results). The only independent variable is the frequency of one allele (e.g. p), because this fixes q ($p + q = 1$) and $2pq$.

The number of classes has a practical effect on the value of χ^2 because an extra value is added into the calculation for each extra class. Therefore χ^2 gets bigger as the degrees of freedom get bigger, and we must look at a different line in the table of probabilities.

Using χ^2 probability tables

Now we can look up the level of significance of our result. Go to *Table 2*, and follow the line for 1 degree of freedom (top line) to find the nearest values of χ^2 above and below our value. We can see that the value of 6.6349 is exceeded with a probability of 0.01 (that is 1%, or 1/100) and 7.87944 is exceeded with a

Table 2. Values of the χ^2 *distribution exceeded with probability* P

P d.f.	0.995	0.975	0.050	0.025	0.010	0.005	0.001
1	$392704 \cdot 10^{-10}$	$982069 \cdot 10^{-9}$	3.84146	5.02389	6.63490	7.87944	10.828
2	0.0100251	0.0506356	5.99146	7.37776	9.21034	10.5966	13.816
3	0.0717218	0.215795	7.81473	9.34840	11.3449	12.8382	16.266
4	0.206989	0.484419	9.48773	11.1433	13.2767	14.8603	18.467
5	0.411742	0.831212	11.0705	12.8325	15.0863	16.7496	20.515
6	0.675727	1.23734	12.5916	14.4494	16.8119	18.5476	22.458
7	0.989256	1.68987	14.0671	16.0128	18.4753	20.2777	24.322
8	1.34441	2.17973	15.5073	17.5345	20.0902	21.9550	26.125
9	1.73493	2.70039	16.9190	19.0228	21.6660	23.5894	27.877

probability of 0.005 (i.e. 0.5%, = 1/200). Our value, 7.68, is between these. This means that if we repeated the experiment 100 times we would not expect to get this big a deviation by chance once. This suggests that our data does not fit the expectation of a 3 : 1 ratio.

Level of significance

The conventionally accepted significance level is 5% (0.05); if the deviation is so large that the probability of it occurring by chance is less than 5%, we reject the null hypothesis because the data differs **significantly** from the predictions of the null hypothesis. So if χ^2 exceeds the value which has a 5% (i.e. 1/20) probability of occurring by chance, we will be correct to reject the null hypothesis 19 times out of 20. Note that one time in 20 we will reject the null hypothesis when it really is correct. Statistics do not 'prove' anything, they tell us the probability of certain things happening if our particular assumptions are correct. In this case we reject the null hypothesis that the results fit a 3 : 1 ratio because a 63 : 37 ratio will only occur by chance on less than one occasion in 100.

The highest applicable value of χ^2 is normally quoted to emphasize the quality of a conclusion. For example, if the probability of obtaining a particular value of χ^2 by chance is less than 0.01 it is normal to quote this value (1%) rather than the conventional 5%. The 1% significance means we will only be wrong in rejecting the null hypothesis one in 100 experiments.

There is another possible explanation for the 63 : 37 ratio which is that there are two genes that differ between the original red- and white-flowered strains. If these code for enzymes that work in the same biochemical pathway they will both be needed to produce red pigment. The new null hypothesis is that the ratio will be 9 : 7 (see Topic C2). The χ^2 based on this null hypothesis is also calculated in *Table 1*, and is 1.85. *Table 2* tells us that the value of χ^2 will be less than 1.64 in 0.20 (one in five) of our experiments, and greater than 3.84 in 0.05 (5% or one in 20) of our experiments. Our result, 1.85, has a probability of occurring by chance between these values. Since a deviation this great or greater will occur by chance in more than one experiment in 20, we will accept the null hypothesis that the data fit a 9 : 7 ratio and assume that two genes are involved in producing the red color. There is a small chance that we are wrong, but we have no better hypothesis to test at present.

Yates' correction

Yates' correction should be applied in χ^2 calculations in any class where the value is less than five. χ^2 is a continuous distribution (it expects fractional values) but we count in integers (whole numbers) – you cannot grow half a pea. So if we expect 3.5 white flowers, the nearest we can get is three or four, a minimum error of 0.5. Yates' correction is applied by reducing the numerical value of (observed – expected) by 0.5 if the value expected is five or less. If observed minus expected = –1.6 then it is adjusted to –1.1; if it is 2.7 it is adjusted to 2.2, before it is squared.

Fisher's exact test

Fisher's exact test can be applied when there are only a few events, or when some or all of the classes are rare, and only small numbers are expected, because it gives an exact probability for any observed result. The probability of the particular outcome, and all more deviant outcomes, can be calculated from the probabilities of each individual event. These can be summed to give the probability of getting a deviation as great as, or greater than, the observed deviation by chance (see Topic C12). If this probability is less than 0.05 we reject the null hypothesis. For example if the null hypothesis is that babies are equally likely to be boys or girls (each has a probability of 0.5) the probability of two consecutive births both producing girls is exactly $0.5^2 = 0.25$. Thus if two consecutive children are both girls, this is not significantly different from the expected 1 : 1 ratio. The probability of getting six girls consecutively is $0.5^6 = 0.015\,625$ so (if we had no other information) a family with six children, all girls, would lead us to reject the null hypothesis that boys and girls are equally probable with a significance level of 1.5625%. Consider a more complex example: a pair of short-haired dogs are mated and produce four puppies, three long-haired and one short haired. The parents must be heterozygous, and short hair must be dominant, so our null hypothesis is that the data does not deviate from the expected ratio of three short-haired to one long-haired. What is the probability of getting this particular reversed result by chance? What is the probability of getting a deviation this great or greater by chance? The numbers are much too small to use χ^2. The greatest deviation would be four long-haired puppies, so we start there. *Table 3* shows the individual and cumulative probabilities of all the possible combinations of long-haired and short-haired puppies in a litter of four. The probability of getting our family of three long-haired and one short-haired or the more extreme case of four long-haired puppies is 0.050 781 25, slightly greater than one in 20, so we have no reason to reject the null hypothesis. Even if all the puppies were long-haired, (probability 0.0039, one in 256) we really should repeat the experiment because a sample of four is not big enough to justify profound conclusions. Fisher's exact test is also used to check fit of data to the Hardy–Weinberg equilibrium when some alleles and genotypes are rare. An important application is to estimate the degree of inbreeding and linkage (association of alleles at different loci) in the human population to validate assumptions about genotype frequencies for forensic work (see

Table 3. Exact test for hair length in puppies

Hair length		Individual probability x permutations	Probabilities of specific family	Cumulative probabilities
Long (n)	Short (n)			
4	0	$1/4 \times 1/4 \times 1/4 \times 1/4$	0.00390625	0.00390625
3	1	$1/4 \times 1/4 \times 1/4 \times 3/4 \times 4$	0.046875	0.05078125
2	2	$1/4 \times 1/4 \times 3/4 \times 3/4 \times 6$	0.2109375	0.26171875
1	3	$1/4 \times 3/4 \times 3/4 \times 3/4 \times 4$	0.421875	0.68359375
0	4	$3/4 \times 3/4 \times 3/4 \times 3/4$	0.31640625	1

Topic F5). Some alleles are so rare that particular combinations will not be found when a few thousand people are genotyped. There are too many combinations for the exact probabilities of them all to be calculated. The exact test is simulated by using a computer to put together many thousands of random combinations of the alleles at their observed frequency. The genotype frequencies from the simulations are counted to give an approximation of the exact probability of obtaining very rare genotypes assuming Hardy–Weinberg equilibrium.

D1 INTRODUCTION

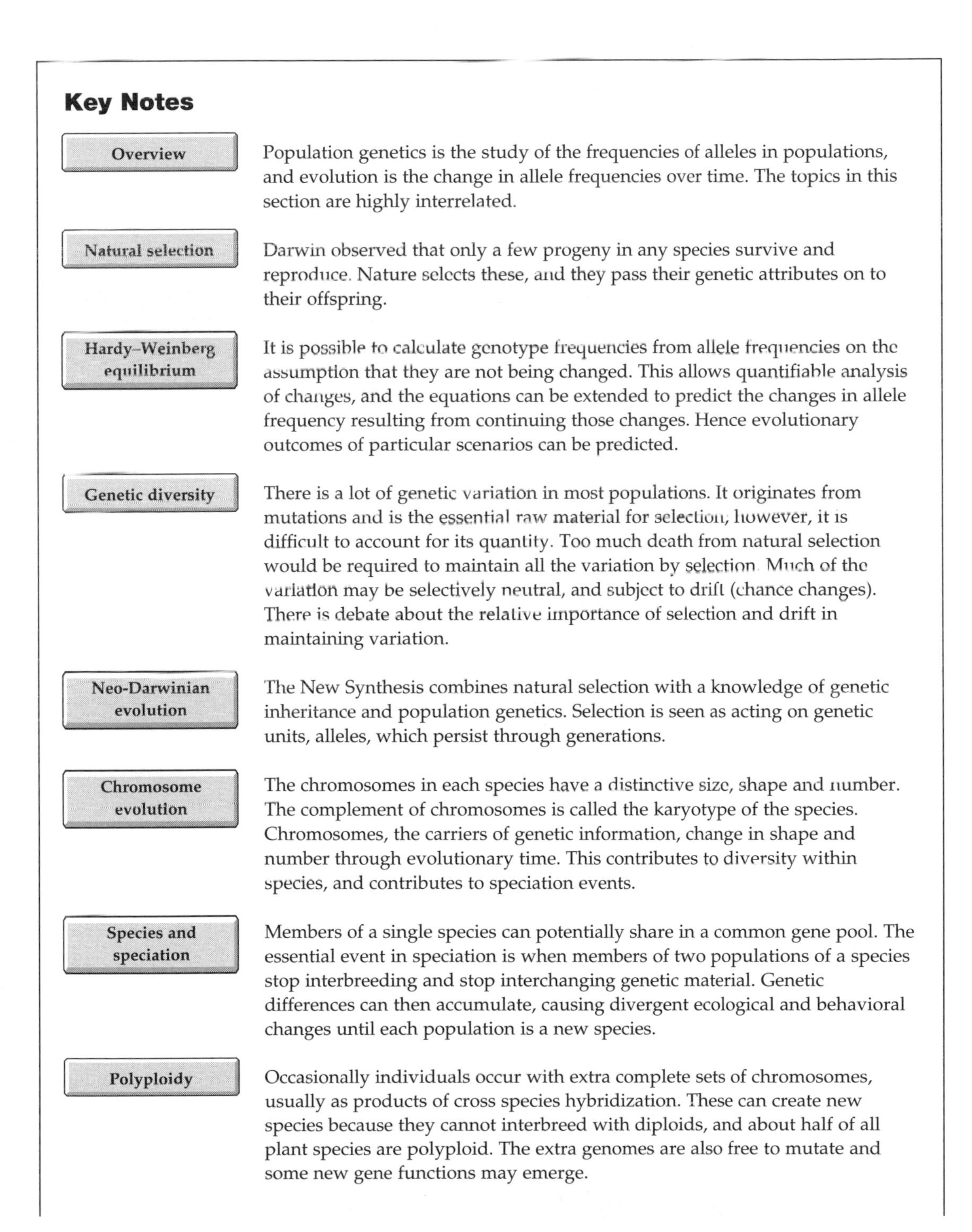

Key Notes

Overview

Population genetics is the study of the frequencies of alleles in populations, and evolution is the change in allele frequencies over time. The topics in this section are highly interrelated.

Natural selection

Darwin observed that only a few progeny in any species survive and reproduce. Nature selects these, and they pass their genetic attributes on to their offspring.

Hardy–Weinberg equilibrium

It is possible to calculate genotype frequencies from allele frequencies on the assumption that they are not being changed. This allows quantifiable analysis of changes, and the equations can be extended to predict the changes in allele frequency resulting from continuing those changes. Hence evolutionary outcomes of particular scenarios can be predicted.

Genetic diversity

There is a lot of genetic variation in most populations. It originates from mutations and is the essential raw material for selection; however, it is difficult to account for its quantity. Too much death from natural selection would be required to maintain all the variation by selection. Much of the variation may be selectively neutral, and subject to drift (chance changes). There is debate about the relative importance of selection and drift in maintaining variation.

Neo-Darwinian evolution

The New Synthesis combines natural selection with a knowledge of genetic inheritance and population genetics. Selection is seen as acting on genetic units, alleles, which persist through generations.

Chromosome evolution

The chromosomes in each species have a distinctive size, shape and number. The complement of chromosomes is called the karyotype of the species. Chromosomes, the carriers of genetic information, change in shape and number through evolutionary time. This contributes to diversity within species, and contributes to speciation events.

Species and speciation

Members of a single species can potentially share in a common gene pool. The essential event in speciation is when members of two populations of a species stop interbreeding and stop interchanging genetic material. Genetic differences can then accumulate, causing divergent ecological and behavioral changes until each population is a new species.

Polyploidy

Occasionally individuals occur with extra complete sets of chromosomes, usually as products of cross species hybridization. These can create new species because they cannot interbreed with diploids, and about half of all plant species are polyploid. The extra genomes are also free to mutate and some new gene functions may emerge.

Evolution and populations

Comparison of DNA sequence and allele frequency allow retrospective study of the separation and divergence of populations and species.

Related topics

Evolution by natural selection (D2)
Genes in populations. Hardy–Weinberg equilibrium (D3)
Genetic diversity (D4)
Neo-Darwinian evolution: selection acting on alleles (D5)
Chromosome changes in evolution (D6)
Species and speciation (D7)
Polyploidy (D8)
Evolution (D9)

Overview

Population genetics and evolution are inextricably linked. Population genetics is the quantitative study of the frequencies of alleles and genotypes in populations, whereas evolution is the change in those frequencies over time. Factors which change allele frequencies are the factors which cause evolution. The topics in this section all interrelate, and many texts have one or two rambling chapters on the subject. The topics here each take a key approach or concept as a focus for learning, but the interactions are clearly indicated. The order of presentation is also problematic, because it is often necessary to understand two things together, but something must come first. It is important to remember that many topics in this section are the same subject examined from different perspectives. The topics, and their place in this section, are outlined below

Natural selection

Natural selection (see Topic D2) examines Darwin's original observations and deductions. These are the foundation for our understanding of factors which cause adaptive genetic changes in populations in response to their environment, and these changes over time are evolution. Any habitat can only support a limited population of any species, far less than the potential for the population to reproduce. As individuals compete to reproduce, the most successful pass on to their offspring any genetic variant which made them successful. Thus evolution proceeds.

Hardy–Weinberg equilibrium

Hardy and Weinberg independently developed this calculation (see Topic D3) to show that an equilibrium in genotype frequencies will occur after one generation of random mating, and those genotype frequencies can be calculated from allele frequencies. The equilibrium frequencies will be maintained from generation to generation unless some force changes the allele frequencies. Hardy and Weinberg disproved the erroneous notion that dominant characteristics become more common. The introduction of simple algebra started the science of population genetics, and allows us to quantify the evolutionary effects of different levels of selection on homozygotes and heterozygotes, or on dominant or recessive alleles. This topic is fundamental to Neo-Darwinian evolution (see Topic D5) and also to understanding the distribution of genetic disease in human populations.

Genetic diversity

Most populations of most species show extensive genetic variation between individuals (see Topic D4). The source of this variation is mutations, but why is there so much variation persisting in the population? There are two alternative explanations: natural selection and chance. Natural selection could be continually replacing alleles, so we see a snapshot of the changes, or there could be frequency-dependent selection favoring rare alleles, keeping them in the population. Both explanations require a large number of deaths by natural selection in each generation. There is so much variation that the required level of selection would probably be intolerable. The alternative view is that most of the variation does not affect fitness, so it is selectively neutral, and the frequency of the alleles involved fluctuates by chance (drift). The real explanation is some combination of these two, but arguments continue about their relative importance.

Neo-Darwinian evolution

The combination of population genetics and natural selection (the New Synthesis) allows us to examine the effect of selection on genotypes and predict the outcome (see Topic D5); it turns the study of genetic evolution into a quantitative science. We can calculate the number of generations needed for a particular level of selection on a dominant or recessive allele to lead to replacement of one allele by another. We find that recessive alleles are less affected when rare because they are mainly hidden in heterozygotes, and selection for heterozygotes preserves both alleles, whereas selection against heterozygotes removes rare alleles, even if they are fit as homozygotes. This leads us to consider selection as acting on alleles (alleles are regions of DNA) rather than on phenotypic characters, and establishes the concept of 'selfish genes' (see Topic D2) which secure their own transmission, not the fitness of the individual.

Chromosome evolution

Chromosomes carry the genetic information. Their shape and number is a characteristic of each species, its **karyotype**, and can change over time. Gametogenesis (see Topic C3) requires the accurate division of genetic material in meiosis, and this is difficult to achieve if the chromosomes do not match. There are species where heterozygosity for chromosome shape is common, parts of chromosomes being inverted, or translocated to other chromosomes. Some special genetic effects result from this. Meiotic recombination (crossing-over, see Topic C3) in a heterozygous rearrangement often produces inviable products, so recombination is prevented in the rearranged chromosome segment, and all the alleles there can co-evolve. In other cases, chromosomal differences cause sterility in hybrids, and can act as an agent towards speciation. Comparison of chromosomes between species may indicate evolutionary change.

Species and speciation

The formation of new species is the key occurrence in evolution. Members of different species do not interbreed and cannot share the same gene pool. This block on genetic exchange is the only essential requirement for speciation, and must ultimately be genetically controlled when two species meet and have an opportunity to interbreed. If two populations are separated physically by either time or space when mating, they cannot exchange genetic material. As neutral and selected changes arise, they remain within one population. Eventually the differences become so large that members of the two populations do not interbreed successfully when they do meet, and speciation has occurred.

Polyploidy

Sometimes individuals occur with three or more sets of chromosomes. They are polyploid as opposed to the normal diploid with two sets (see Topic D8). Organisms with three sets are triploid, with four sets are tetraploid, and so on: pentaploid (five), hexaploid (six), octaploid (eight). These usually arise as hybrids between two related species. They have a set of chromosomes from each parent, but they do not match. Odd number polyploids are sterile because they cannot segregate chromosomes evenly into gametes in meiosis, odd numbers are not divisible by two. However, they may accidentally double the number of chromosomes in mitosis and each chromosome can then pair with its sister; thus fertility can be restored. This is a common event in plants where 47% of angiosperms (flowering plants) are polyploid. Because their gametes only form balanced zygotes when they fuse with similar ploidy gametes, the original polyploid individual can start a new species with a functional species isolation mechanism.

Evolution and populations

Comparisons between species show us what has happened in the past. Species which share a recent common ancestor are more alike than species which diverged longer ago. Comparison of the amino acid sequence of particular proteins shows us that they change (diverge) at different rates. Proteins which interact with many other molecules must maintain their structure to keep their function, and therefore evolve very slowly. Other proteins, which are soluble and mainly interact with water, are less constrained and change much more quickly. Examination of the DNA shows that sequences which do not code for anything evolve fastest of all. In this way we can develop molecular clocks, based on the basic intrinsic mutation rate, filtered (in coding sequences) by selection against deleterious mutations. This reinforces the view that most surviving mutations are neutral. It also allows us to predict the evolutionary distances between species by comparing their DNA sequence. A similar process can be applied to populations. Particular attention is currently being given to the spread of humans from Africa during the last few hundred thousand years. On a finer scale, the spread of tribes across Europe and Asia, and into Oceania and the Americas during and after the last ice age, can also be examined, with much more emotive prospects.

D2 EVOLUTION BY NATURAL SELECTION

Key Notes

Evolution by natural selection

Natural selection is the constraint that natural conditions put on the size of populations, forcing individuals of the same species to compete for limited resources. Those types which use the resources most successfully in order to reproduce, pass on their genetic material to their offspring, and are selected for. Inefficient or unsuccessful types fail to reproduce as successfully and are selected against. The process is analogous to selection, by farmers, of the best types of stock and plants for breeding. This causes a gradual change in the genetic makeup of the population. The phenotypic characters which are selected for will become more frequent providing that they are genetically controlled. Relative fitness is the differential average reproductive success of individuals with different genotypes (combinations of alleles). Natural selection was first proposed as the cause of evolution by Charles Darwin, but he did not know how inheritance worked.

Darwin's observations and deductions

Darwin's observations were: (i) the potential numbers of descendants from any species is infinite. Many more offspring are produced than can survive; (ii) this potential population growth is prevented by limited resources; populations are relatively constant in size; (iii) there are many differences between individuals which affect their ability to survive and reproduce.

Darwin deduced that, therefore: (i) there is a struggle for the limited resources; (ii) those individuals which survive and reproduce successfully will pass on to their offspring the characteristics which helped them to survive and reproduce; (iii) Gradually, the inherited characteristics which help in the competition to reproduce will become more frequent in the population. This gradual change is evolution.

Modes of selection

(i) Stabilizing selection removes phenotypes which deviate too far from the optimum (norm). In a stable environment, change or deviation usually produces less fit phenotypes. (ii) Directional selection selects for a new optimum, and against one end of the range of phenotypes, causing the mean for the character to change towards the new optimum. This can occur after a change in environment, including competition with another species. Character displacement occurs when two species compete. Selection is greatest where their niches overlap, so each species is selected to be different from the other (e.g. one species becomes larger, the other smaller, reducing competition). (iii) Disruptive selection selects against the mean phenotype and favors two different optimum phenotypes. It can develop as a sexual difference (e.g. female sparrow hawks are larger that males and catch larger prey). Disruptive selection can work when there are two qualitatively different types, but there are theoretical problems with continuously variable characters because of the continued production of unfit intermediates.

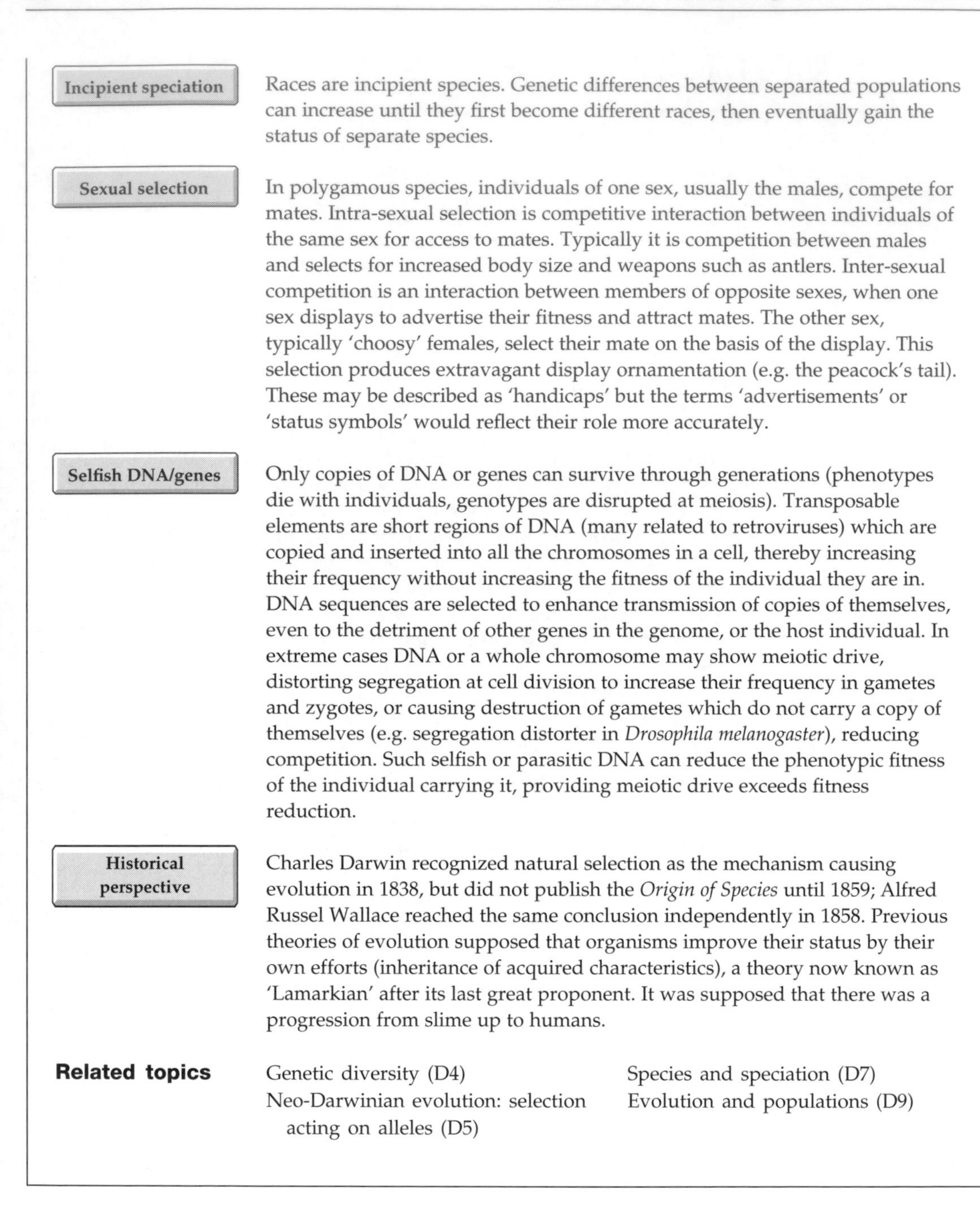

Incipient speciation

Races are incipient species. Genetic differences between separated populations can increase until they first become different races, then eventually gain the status of separate species.

Sexual selection

In polygamous species, individuals of one sex, usually the males, compete for mates. Intra-sexual selection is competitive interaction between individuals of the same sex for access to mates. Typically it is competition between males and selects for increased body size and weapons such as antlers. Inter-sexual competition is an interaction between members of opposite sexes, when one sex displays to advertise their fitness and attract mates. The other sex, typically 'choosy' females, select their mate on the basis of the display. This selection produces extravagant display ornamentation (e.g. the peacock's tail). These may be described as 'handicaps' but the terms 'advertisements' or 'status symbols' would reflect their role more accurately.

Selfish DNA/genes

Only copies of DNA or genes can survive through generations (phenotypes die with individuals, genotypes are disrupted at meiosis). Transposable elements are short regions of DNA (many related to retroviruses) which are copied and inserted into all the chromosomes in a cell, thereby increasing their frequency without increasing the fitness of the individual they are in. DNA sequences are selected to enhance transmission of copies of themselves, even to the detriment of other genes in the genome, or the host individual. In extreme cases DNA or a whole chromosome may show meiotic drive, distorting segregation at cell division to increase their frequency in gametes and zygotes, or causing destruction of gametes which do not carry a copy of themselves (e.g. segregation distorter in *Drosophila melanogaster*), reducing competition. Such selfish or parasitic DNA can reduce the phenotypic fitness of the individual carrying it, providing meiotic drive exceeds fitness reduction.

Historical perspective

Charles Darwin recognized natural selection as the mechanism causing evolution in 1838, but did not publish the *Origin of Species* until 1859; Alfred Russel Wallace reached the same conclusion independently in 1858. Previous theories of evolution supposed that organisms improve their status by their own efforts (inheritance of acquired characteristics), a theory now known as 'Lamarkian' after its last great proponent. It was supposed that there was a progression from slime up to humans.

Related topics

Genetic diversity (D4)
Neo-Darwinian evolution: selection acting on alleles (D5)
Species and speciation (D7)
Evolution and populations (D9)

Evolution by natural selection

Natural selection is the process by which the environment limits the size of populations and forces individuals to compete for resources. Some individuals inherit characteristics which help them to survive and reproduce better than individuals with other characteristics. As a result, copies of the survivors' genetic determinants (which we know as alleles of genes) for favorable characteristics become more frequent (common) in the population in future generations. Darwin made a direct analogy with selection of plants and animals by human breeders, saving individuals with favored characteristics for breeding, and destroying the rest. He used fancy pigeons as an example. Darwin did not know how inheritance worked, and like others at the time he incorrectly assumed a blending of characteristics from both parents, rather than discrete Mendelian genes. Blending should lead to a uniform type, and Darwin's main obstacle was not understanding the mechanism of inheritance, although he recognized new variant types (our mutations).

Natural selection acts on the variation already existing in populations; where there is no variation, there can be no change (see Topic C7). However, species do not become extinct because they fail to adapt. Species become extinct when their habitat is removed or changed to a state where not enough individuals with an adapted genotype remain to maintain the species. There have been many large animals which are now extinct because of predation by humans. They were well adapted and carried well adapted fleas. When the animals became extinct, so did their fleas. Both had lost their environmental niche, and had nothing else to adapt to.

Darwin's observations and deductions

Darwin made observations and deductions which may be presented simply.

Observations

Observation 1. The number of individuals in any population can increase exponentially when conditions allow them all to survive and reproduce. Any species could soon cover the entire Earth if allowed to do so.

Observation 2. This potential reproduction is always limited eventually by lack of resources. At all stages in the life cycle, different factors will limit survival or reproduction, but their relative importance may change with time, both in one life cycle, or historically over generations. Examples of limiting factors include availability of resources: food, shelter, territory, nest-sites, mates, sunlight (essential for plants), and the effects of predators, diseases and parasites.

Observation 3. Individuals are not all the same. There are differences (variations) between individuals in every species, which may, however slightly, affect their ability to compete, survive, and reproduce. Some of these variable characters are inherited.

Deductions

Deduction 1. From observations 1 and 2, there must be a 'struggle' or competition for survival and reproduction in which some individuals are less successful than others.

Deduction 2. From observation 3, some of the differences in characteristics between individuals will affect their success in the struggle for survival and reproduction. Therefore the elimination will be selective. Individuals with favorable characteristics will be more successful and produce more offspring than less successful individuals.

Deduction 3. Those characters which are genetically determined, and which assist individuals to survive and reproduce (characters which are selected for) will be more frequent in the next generation to produce a gradual change in the genetic makeup of the population over time.

Evolution only acts on the portion of variation which is controlled by genetic factors. Differences caused by environmental factors, for example differences in size due to being born where food was plentiful or scarce will affect an individual's ability to leave offspring, but these differences will not be passed through successive generations of offspring, so will not contribute to evolution.

Darwin thought that Herbert Spencer's phrase 'survival of the fittest' was not the best way to describe this process because of its misleading suggestion of physical fitness. The important aspect of fitness is the individual's relative success in leaving successful progeny, in comparison to other individuals of the same species. Individuals may adapt to changes in the environment (e.g. grow thicker hair in the cold) but this is not evolution because it does not cause a change in the genetic makeup of the population and is not passed on to the next generation. The ability to adapt is passed on, however.

Modes of selection

There are three modes of selection (*Fig. 1*).

Stabilizing selection

Stabilizing selection (*Fig. 1a*) removes individuals deviating too far from the norm, and maintains the population near the optimum. In a stable environmental niche most individuals are well adapted and new types or deviants are likely to be less fit. For example, house sparrows (*Passer domesticus*) killed in a severe snowstorm in New York were found to have wings markedly longer or shorter than the mean. Human babies with birth weights deviating furthest from 3.6 kg are more likely to die or suffer damage at or soon after birth because they are either too big to pass through the birth canal, or too small to survive. This selection is now very much reduced in countries where medical support is available.

Directional selection

Directional selection (*Fig. 1b*) favors individuals with characteristics at one end of the range found in the population. An example which changes direction with the climate relates to Darwin's finches, *Geospiza fortis*. During drought in the Galapagos islands, grasses and small plants do not grow well, but bigger deep-

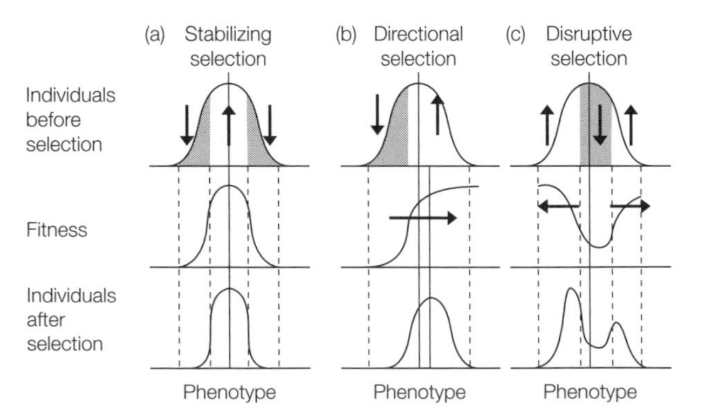

Fig. 1. Modes of selection. The shaded region is selected against. Arrows show changes in phenotype frequency expected over time. Vertical solid lines show means, dashed lines allow comparisons.

rooted plants with larger seeds still produce fruit. There is then selection which favors individual finches with larger beaks capable of eating large seeds, and selects against those with smaller beaks best suited for eating grass seeds. This is reversed in wet years when small seeds are plentiful. A common form of directional selection causes **character displacement** when two species compete. If there are two similar species of finch on one island (sympatry), one species will specialize in large seeds and evolve a large beak, the other will specialize in small seeds or insects and evolve a suitably small beak. Either species alone (allopatry) has an intermediate size beak. This effect is very common among related species.

Disruptive selection

Disruptive selection (*Fig. 1c*) selects for two distinctively different phenotypes in a population. Examples are size differences (e.g. female sparrowhawks *Accipiter nisus* are twice as large as males and catch larger prey), and behavior (e.g. some male salmon mature precociously in rivers while small without going to sea; others go to sea and get much larger, return to breed, and can aggressively displace small males and fertilize more eggs, but they have the disadvantage of longer life cycle and greater predation). Disruptive selection in a single population will work if the variation is discontinuous, with two clearly different alternative types, (e.g. sex and size in sparrowhawks). There are, however, reasons to believe that two phenotypes cannot coexist for a continuously variable trait (see Topic C7) if interbreeding between them produces hybrid offspring with unfit intermediate phenotypes. The less common of the two fit phenotypes may then be lost (see Topic D7). Disruptive selection might cause one species to evolve into two, each adapted to one of the optima, but this again requires the existence of premating isolation mechanisms to prevent continual interbreeding.

Incipient speciation

Darwin considered varieties to be incipient species. There is a continuous range between slightly different races and fully diverged species. Divergence occurs when two populations undergo different genetic change with time. This may be due to directional selection for different phenotypes in the two populations. However, even if the selection on the phenotype (e.g. bill size, optimum temperature, etc.) is in exactly the same direction in both populations, which is unlikely, a different combination of alleles may finally be selected in each population. For example, suppose two genes *A* and *B* both code for similar enzymes, and there is selection to produce less enzyme. Now *AABB* has too much enzyme, and either *A* or *B* must be lost. *AAbb* may become fixed in one population while *aaBB* becomes fixed in another. (*aabb* is lethally deficient because it has no enzyme at all.) There are also differences which arise slowly by chance (drift or mutation). When the two populations are sufficiently different in appearance, behavior, ecology or genetics, they may be considered to be separate species (see Topics D7 and D9).

Sexual selection

Selection can affects the two sexes differently, mainly because microgametes (sperm and pollen) are small and cheap to produce, so males can father many offspring from many females, while female reproduction is limited by the greater cost of producing macrogametes (eggs) and in many species, rearing young. Males have a greater potential reproduction than females, and in many species have a greater variance in success. **Intra-sexual selection** is aggressive competitive interactions between individuals of the same sex (usually between males for females) and usually selects for large size and weapons (e.g. antlers) useful for fighting. The differences are most marked in species such as elephant seals where males hold large harems. There are a few cases where females aggressively compete for males (polyandry), notably in birds nesting in exposed situations. Females of some species of lily trotter (genus *Jacana*) defend a territory containing several males, each male incubating a clutch of her eggs without her help. **Inter-sexual selection** is competition between members of one sex (usually males) to attract mates, and involves interactions between individuals of opposite sex. This often selects for ornaments for display such as peacocks' tails and bower birds' bowers, which act as status symbols advertising the surplus resources of the fittest males to the females. The females are said to be **choosy**. These costly male adornments are often called **handicaps**, but this is misleading, the human equivalent would be owning a luxury yacht, an advantage in attracting mates, and not a handicap if you can afford it. The terms **status symbol** and **advertisement** are more accurate. The males must have good health and fitness to produce the advertisement because it does cost resources. In group displays it is often not clear whether males are threatening other males or courting females, or doing both simultaneously.

Selfish DNA/genes

The genome of a sexually reproducing organism is disrupted at gametogenesis by recombination and reassortment. Individuals and their genomes are temporary; only copies of regions of the DNA pass unchanged through the generations. This leads to the concept of 'selfish DNA' and 'selfish gene' because any piece of DNA or gene which can enhance transmission of copies of itself to the next generation has a selective advantage over other DNA. The selfish DNA could be a gene for sacrificial behavior in one individual to enhance reproduction by a relative who is likely to pass on copies of the same DNA. This is called **kin selection**. It is demonstrated by worker bees, ants and wasps who help a sister to become a queen and reproduce. Males of this group (*Hymenoptera*) are haploid, so full sisters all inherit the same genes from their father and are related to each other by 0.75 of their genome, rather than the 0.5 of their genome they pass to each of their offspring. Thus genes are transmitted more successfully through sisters than through offspring. This is probably why the group has evolved so many social species. Somatic cells in all multicellular animals similarly abandon long-term reproduction to facilitate sexual reproduction of specialized germ cells (see Topic C3), which are usually genetically identical to the soma (body tissue). **Transposable elements** (see Topics B3, B4 and B9) are short DNA sequences which replicate and are inserted around the genome into all uninfected chromosomes. They are presumably 'selfish DNA'. Their transposition may be performed by proteins coded by the transposable element itself (e.g. some are similar to retroviruses), or by proteins coded by genes elsewhere in the genome. If any piece of DNA enhances its own transmission into gametes and zygotes at greater frequencies than expected by Mendelian segregation then it has a selective advantage. This is known as **meiotic drive**. In some cases this may be accompanied by a reduction in the total reproduction of the individual. For example the allele *SD* (segregation distorter) in *Drosophila* actually does this by destroying sperm not carrying copies of itself. It is then truly parasitic selfish DNA. There are also parasitic chromosomes which persist in germ line cells, but are lost from the soma where they might reduce fitness. Other parasitic chromosomes get into the egg nucleus and avoid the polar bodies.

Historical perspective

Evolution by gradual change through successive generations was already an accepted idea when **Darwin** proposed **natural selection as a mechanism for evolution** in his book *Origin of Species* in 1859. Darwin had formed the idea in September 1838, but only went public and published it in 1858 after **Alfred Russel Wallace** came to the same conclusion independently. Previous theories were based on an ancient idea of a **Great Chain of Being** with slime at the bottom and humans, angels and God at the top, all organisms striving to become more like angels. **Lamark** tried to write this formally as a scientific hypothesis in which improvements made by the efforts of parents were passed on to their offspring (**inheritance of acquired characteristics**). Evidence shows this to be incorrect; amputating the tails of mice at birth for several generations does not produce a tailless race of mice.

D3 GENES IN POPULATIONS: HARDY–WEINBERG EQUILIBRIUM

Key Notes

Introduction

Population genetics is the study of alleles of genes in populations (often called demes), and the forces which maintain or change the frequencies of particular alleles and genotypes in populations. The total genetic stock of the population is its gene pool. Individuals have a selection of alleles from that gene pool, possibly taken randomly. The Hardy–Weinberg equilibrium is a means of calculating expected genotype frequencies from allele frequencies determined in the same population (and vice versa) assuming random mating, equal reproductive success, and no effects of selection or migration affecting particular genotypes. If a population does not fit Hardy–Weinberg predictions then that is evidence of some real effect (e.g. natural selection) operating to disturb it.

Measuring allele frequency

Allele frequency is found by adding up the number of copies of each allele in a population and expressing it as a frequency. A population of N diploid individuals has $2N$ alleles. Each *Aa* heterozygote has one *A* allele, each *AA* homozygote has two *A* alleles, so:

$$\text{frequency of allele } A \text{ is } \frac{(\mathrm{n}_{Aa} + 2n_{AA})}{2N}$$

where n is the number of individuals with the respective genotype. The frequency of *A* is usually called p, the frequency of other (non*A*) alleles is denoted by q. As a check, $p + q = 1$, but this only checks your arithmetic, not whether p and q are correct.

Equilibrium genotype frequencies

Equilibrium genotype frequencies will be reached in zygotes after one round of random mating. The expected genotype frequencies are predicted by the Hardy–Weinberg equation:

$$p^2 + 2pq + p^2$$

If there are two alleles, *A* and *a*, with frequencies p and q respectively, then the expected frequency of *AA* homozygotes is p^2, of *Aa* heterozygotes is $2pq$, and of *aa* homozygotes is q^2. As a check of your arithmetic $p^2 + 2pq + q^2 = 1$. When the frequency q of an allele is low, the frequency of occurrence of homozygotes, q^2, is very low (e.g. recessive genetic diseases). Uncommon alleles (small value of q) are usually present in heterozygotes, so cannot be identified if they are recessive and are not expressed, so selection cannot act against them.

Distorting effects

A range of natural phenomena can distort genotype frequencies from those predicted by the Hardy–Weinberg equation.

(i) Selection reduces the fertility or survival of certain genotypes.
(ii) Migration replaces local individuals with immigrants from a population where the allele frequencies, and hence expected genotype frequencies, are different. If some genotypes are more likely to emigrate than others this will also change their frequency locally.
(iii) Assortative mating of either similar or dissimilar genotypes produces an excess of homozygotes or heterozygotes respectively.
(iv) Subpopulations exist. These are locally mating groups in a larger, possibly continuous population. This may be common in humans due to ethnic or class groupings, and also in organisms with low mobility, and always increases the frequency of homozygotes. In an extreme case this is inbreeding.
(v) Mutations will produce new alleles, but the rate is too low to be noticeable practically, and selection will exactly counteract this if the mutants are deleterious.
(vi) Drift (chance) will also cause small deviations from the predicted frequencies, especially in small samples, but these deviations are expected and allowed for in statistical tests.

Testing the fit

A chi-square (χ^2) test (see Topic C13) is used to compare observed genotype frequencies to expected frequencies estimated assuming Hardy–Weinberg equilibrium:

$$\chi^2 = \text{sum}\left[\frac{(\text{observed} - \text{expected})^2}{\text{expected}}\right]$$

There is one less degree of freedom than there are alleles. Once the frequency (p) of one allele (of two alleles) is known, the frequency of the other is fixed ($p + q = 1$) and so are the expected genotype frequencies. Small values of χ^2 indicate a good fit to expectation, large values of χ^2 indicate a large deviation.

Estimating allele frequency

If heterozygotes cannot be identified, the frequency of recessive alleles can still be estimated from the frequency of homozygotes by assuming Hardy–Weinberg equilibrium:

the frequency of homozygous recessives is q^2, so
the frequency of the recessive allele is q, and
the frequency of the dominant allele is $p = (1 - q)$

There is a large error associated with this method. It is difficult to measure q^2 accurately because: (i) it is often very small so a very large population must be sampled; and (ii) Hardy–Weinberg equilibrium may not apply, usually because of population structure and local inbreeding subpopulations.

Related topics

Genetic diversity (D4)
NeoDarwinian evolution (D5)
Species and speciation (D7)

Introduction

Population genetics is the study of alleles of genes in populations, and the forces which maintain or change the frequencies of particular alleles and genotypes in populations. The local interbreeding population is called a **deme** to distinguish it from a geographical population. A deme may be difficult to define, for example humans living in one area may actually marry partners chosen on criteria of religion, race, wealth or social class. Similarly, insects on trees in a forest may mate with others on the same tree, and rarely travel to the next tree to mate. The insect population may fill the forest, or each tree may have its own semi-isolated population. Natural populations are composed of many interbreeding individuals, each with a unique combination of genes and alleles, but the population shares a **gene-pool**. Evolution is the change of frequencies of alleles in the gene pool, so population genetics is of fundamental importance. Some genotypes, such as those associated with human genetic diseases, are rare, and population genetics is important for understanding why this is so, and for predicting changes in the incidence of such diseases. Most individuals have phenotypes close to the average, but some carry rare combinations of genes or alleles which cause their phenotype to deviate markedly from a typical member of the species.

There was a fear, expressed by the **eugenics movement**, that if 'defective' humans were allowed to reproduce, their defects would become more common, and the human race would degenerate. Eugenics refers to the principle of selective breeding of humans by encouraging desirable types to have more children (positive eugenics) and discouraging undesirable types from having any (negative eugenics). This has been applied in varying degrees, 'desirable' generally being interpreted to mean the ruling class or race.

Hardy and Weinberg independently published calculations showing that undesirable characteristics would not become more common, and that they were due to rare combinations of relatively common alleles. Hardy and Weinberg showed that the frequency of particular genotypes in a population of a sexually reproducing diploid species reaches equilibrium after one generation of random mating and fertilization, and will stay constant unless something changes the frequencies of alleles in the population. Hardy–Weinberg equilibrium requires random mating, random fertilization, and an absence of differential selection or migration affecting particular alleles or genotypes. (A population can be in equilibrium in the sense of being constant from generation to generation without being in Hardy–Weinberg equilibrium.)

The goodness of fit of observed genotype frequencies to frequencies expected from the Hardy–Weinberg equilibrium can be tested statistically using a chi-square (χ^2) test (see Topic C13). Deviations may reveal evidence of natural selection, assortative (like with like) or disassortative (choose unlike type) mating, migration from an unsampled population, population substructure (increasing inbreeding raising the frequency of homozygotes) or selective fertilization. (Selective fertilization can occur, for example, when pollen of the wrong compatibility type is rejected by the stigma of a female flower.) The observed frequencies will also deviate from the expected frequencies due to chance (also known as drift, sampling, stochastic, random or statistical error). This is greater in small samples and is allowed for in the statistical tests.

Measuring allele frequency

The frequency of an allele (e.g. A) in a population is the number of A alleles divided by the total number of alleles of that gene locus. In a diploid species, a population of N individuals will be produced from $2N$ gametes and contain

$2N$ alleles for each genetic locus. Suppose that there are two alleles, A and a, of a particular gene in this population. The number of A alleles is twice the number of AA homozygotes plus the number of Aa heterozygotes. (Each homozygote has two A alleles, each heterozygote has only one.) The frequency of A is the number of A alleles divided by the total, $2N$, and similarly for a. If we use n to represent number, and the suffixes A and a for alleles, and AA, Aa, and aa for genotype, then:

$$n_A = 2n_{AA} + n_{Aa}$$

If p is the frequency of allele A then:

$$p = \frac{n_A}{2N} = \frac{2n_{AA} + n_{Aa}}{2N}$$

Similarly the frequency of a is q:

$$q = \frac{n_a}{2N} = \frac{2n_{aa} + n_{Aa}}{2N}$$

Notice that all alleles must be accounted for:

$$n_a + n_A = 2N, \text{ and } p + q = 1$$

If there are more than two alleles of a gene, they can all be counted in a similar way. The frequency of the third allele is usually denoted by r.

The frequency of alleles can only be determined accurately by counting in homozygotes and heterozygotes in a sufficiently large population. It was often impossible to identify heterozygotes, but direct DNA based techniques involving polymerase chain reaction techniques (see Topic E3) are now helping to solve this problem.

Worked example

There are two alleles of the L human blood group gene on chromosome 2, L^M and L^N, (usually called M and N respectively) (*Table 1*). These alleles are co-dominant (both are expressed in heterozygotes) so they can be identified in heterozygotes by antibody tests. In a population of 1000 white Americans, 357 were MM, 485 were MN and 158 were NN. The 357 MM individuals have 714 M alleles, and the 485 MN individuals have another 485 M alleles, so the frequency of M (=p) is:

$$p = \frac{(2 \times 357) + 485}{2000} = 0.5995$$

Table 1. Measuring the frequencies of N and M in a population

	Genotype (= phenotype)			
	IMM	*MN*	*NN*	Total
Number of individuals	357	485	158	1000
Number of M allele	714	485	0	1199
Number of N alleles	0	485	316	801
Total number of alleles	715	970	316	2000

Frequency of M in population: 1199/2000 = 0.5995; frequency of N in population: 801/2000 = 0.4005

Similarly, the frequency of N =

$$q = \frac{(2 \times 158) + 485}{2000} = 0.4005$$

Equilibrium genotype frequencies

Equilibrium will be reached in a single generation. We have already assumed random mating and fertilization. The frequency with which a male gamete carrying allele A (frequency p) fuses with a female gamete carrying A (also frequency p) will be $p \times p = p^2$. Similarly the frequency with which a male gamete carrying allele a fuses with a female gamete also carrying a will be $q \times q = q^2$.

Heterozygotes (Aa) are produced by the fusion of gametes carrying different alleles. The frequency with which a male gamete carrying allele A fuses with a female carrying a will be $p \times q = pq$, and the frequency of a male gamete carrying allele a fusing with a female gamete carrying A will be $q \times p = qp = pq$, so the total frequency of Aa heterozygotes will be $2pq$.

This is all derived from the binomial expression:

$$(p + q)(p + q) = p^2 + 2pq + p^2$$

This is shown graphically for the frequency of *MM*, *MN* and *NN* bloodgroups in white Americans as calculated above (*Fig. 1a*) and for Australian aborigines (*Fig. 1b*). Notice that the allele and genotype frequencies are quite different for the two populations.

The frequency of homozygotes is the square of the appropriate allele frequency, the frequency of heterozygotes is twice the product of the two allele frequencies. This can be applied to any number of alleles. *Fig. 2* shows the frequencies of three alleles and six genotypes for the human ABO blood group gene.

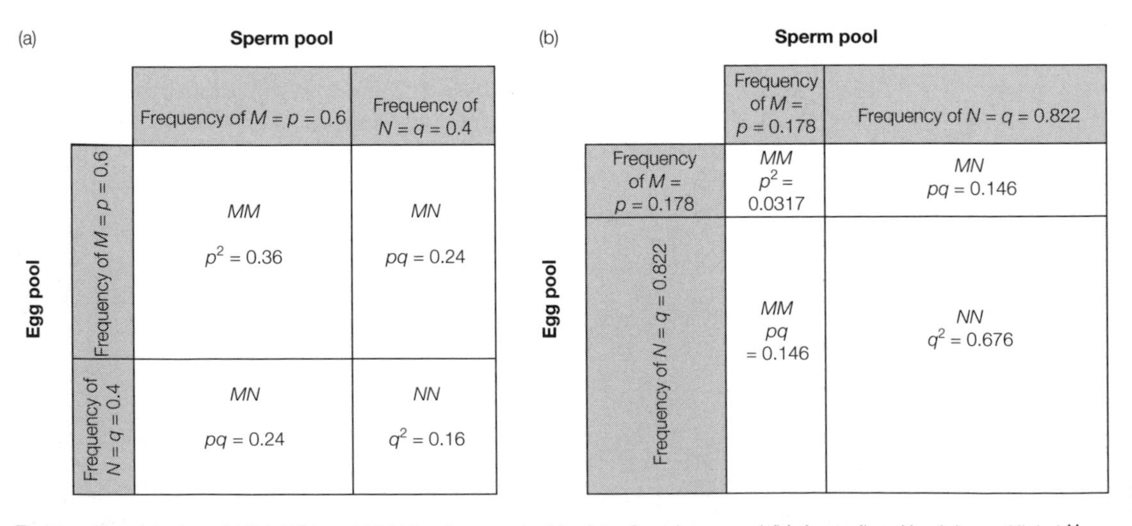

Fig. 1. Frequencies of MM, MN and NN bloodgroups in (a) white Americans and (b) Australian Aborigines. Allele L^M, frequency p, codes for production of M antigen, and L^M, frequency q, code for N antigen. These are codominant, so genotype $L^M L^M$ is phenotype M, $L^M L^N$ is phenotype MN, and $L^N L^N$ is type N.

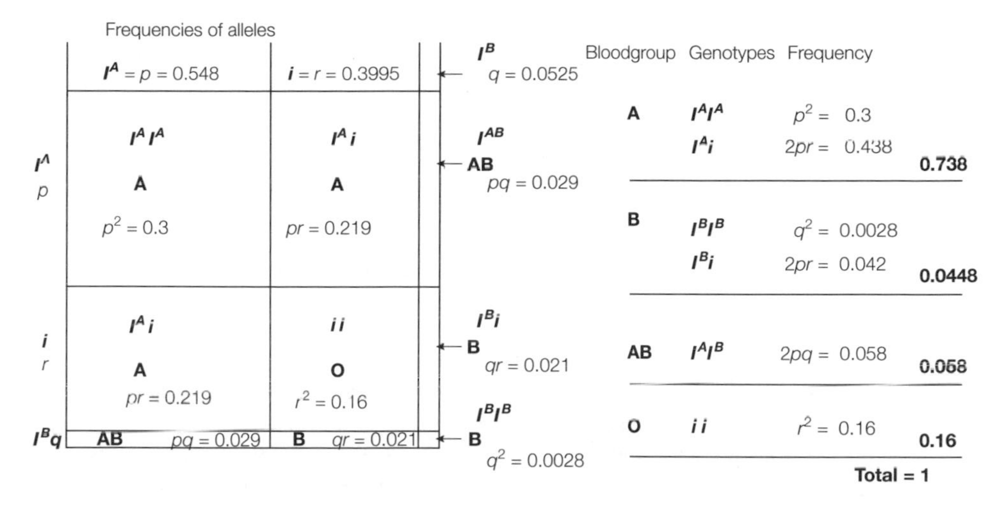

Fig. 2. Frequencies of A, B, AB and O blood groups at Hardy–Weinberg equilibrium. Frequencies of alleles are p (I^A), q (I^B) and r (i). I^A and I^B encode enzymes which produce A and B bloodgroup antigens respectively, and are codominant, whereas i is a recessive null mutation, of no effect.

Points to notice

(i) When the frequency (q) of an allele is low, the frequency of occurrence of homozygotes, q^2 is very low. Notice the rarity of *MM* homozygotes among Australian aborigines (*Fig. 1b*) and of I^BI^B genotypes (*Fig. 2*).

(ii) Uncommon alleles (small value of q) are usually present in heterozygotes, (this is the case for *M* alleles in Australian aborigines). If they are recessive they are not expressed and so cannot be identified, so neither natural selection nor eugenics programs can act against them.

Distorting effects

If a population is found not to be in Hardy–Weinberg equilibrium, there are several possible causes:

- **Selection** which affects the viability or fertility of individuals or gametes of particular genotypes in the parental generation will distort the genotype frequencies in the zygotes. Differential viability of genotypes will distort the genotype frequencies in adults, in which case the deviation from Hardy–Weinberg equilibrium may be seen to change (increase) as progressively older individuals are sampled.
- **Migration** will bring in individuals with genotypes from a gene pool different from the one sampled, and obviously introduce errors. Mixing populations always gives too many homozygotes. If different genotypes show different tendencies to migrate in or out, the deviation will be in a particular direction.
- **Assortative mating.** If individuals choose mates with similar genotypes to themselves they will produce an excess of homozygotes (*AA* × *AA* gives *AA*) while if they choose opposites there will be an excess of heterozygotes (*AA* × *aa* gives *Aa*).
- **Subpopulations** within the sample will increase the frequency of homozygotes. This can arise if the organism's range of movement and interbreeding is much smaller than the area sampled, so individuals only have a small gene pool available for mating, and they cannot mate randomly. Because of

this effect, pooling genotypes from different populations always increases the frequency of homozygotes. Social structures in human societies have the same effect. The extreme arrangement is inbreeding between members of the same family (relatives). This is demonstrated by rare human genetic disorders. For example 33% of cases of alkaptonuria and 54% of cases of microcephaly sufferers have parents who are cousins. These cases show that, in humans, the frequency of rare homozygotes is considerably increased by assortative mating with relatives. The frequency of homozygotes goes from q^2 for random mating to $pq/16$ for cousin marriages (see Topic C11).

- **Mutations** will destroy existing alleles and create new ones. This process is too slow to have an effect in the single generation needed to achieve Hardy–Weinberg equilibrium.
- **Drift**, or chance differences in the success of particular alleles from one generation to the next, will cause small deviations from the expected genotype frequencies, but this random error is allowed for in statistical tests of goodness of fit to the Hardy–Weinberg equilibrium. There is a possible source of confusion here. The term 'drift' properly applies to chance changes in allele frequencies between generations, which need not involve any deviation from Hardy–Weinberg equilibrium.

Testing the fit

A χ^2 test (see Topic C13) is used to compare the proportions of two or more samples falling into two or more categories. The smaller the value of χ^2, the better the fit. For testing fit to Hardy–Weinberg:

$$\chi^2 = \text{sum}\left[\frac{(\text{observed} - \text{expected})^2}{\text{expected}}\right]$$

There are two alleles, so there is one degree of freedom (number of categories minus 1). The fit of our sample of *M*, *MN* and *N* blood groups in white Americans (above) is tested in *Table* 2. The null hypothesis is that the sample frequencies are not different from those predicted by assuming Hardy–Weinberg equilibrium. The value of χ^2 obtained is 0.4236, very small, meaning that there is very little deviation between the observed and expected values. Using χ^2 tables we can see that we are very likely to get a deviation larger than this by chance, so we accept the null hypothesis, the frequencies fit the expected values. A value of χ^2 (with 1 degrees of freedom) greater than 7.879 would be required to indicate that there was only a 5% probability of getting the fit by chance, and thus a 95% probability that there was a genuine deviation from Hardy–Weinberg equilibrium. Note that a large deviation from the expected values gives a large value for χ^2, a small deviation (good fit) gives a small χ^2 value.

Estimating allele frequency

In many cases, (e.g. recessive alleles causing diseases in humans) it is easy to identify individuals homozygous for the recessive allele (who have the disease) but impossible to distinguish between heterozygotes and homozygous normal individuals. The frequency of the recessive allele can be estimated roughly by assuming Hardy–Weinberg equilibrium. The frequency of homozygous recessive individuals should then equal q^2, so q = square root (frequency homozygous recessives):

Table 2. Testing the fit of observed M, MN and N bloodgroups to Hardy–Weinberg using χ^2

	Genotype			
	MM	MN	NN	Total
Observed	357	485	158	1000
Expected	$1000p^2$	$1000 \times 2pq$	$1000q^2$	
	359.4	480.2	160.4	1000
Observed – expected	–2.4	4.8	–2.4	
(Observed – expected)^2	5.76	23.3	5.76	
(Observed – expected)^2/expected	0.016	0.0485	0.3591	

Frequency of *M* alleles = $p = 0.5995$; frequency of *N* alleles = $q = 0.4005$; $\chi^2_{(1\text{d.f.})} = 0.016 + 0.0485 + 0.3591 = 0.4236$.

$$q = \sqrt{q^2} = \sqrt{\frac{\text{homozygous recessives}}{\text{total population}}}$$

About 1/2500 Caucasian humans is born with cystic fibrosis. They are homozygous for nonfunctional alleles of a gene for a chloride transport protein involved in mucus secretion. Without this function, mucus is very viscous and accumulates causing damage to the pancreas, intestines, and most acutely the lungs.

Because the frequency of homozygotes = q^2 = 1/2500 = 0.0004, then:

$$q = \text{square root } (q^2) = 0.02 \quad (1/50)$$

This method is not very accurate for several reasons: (i) most populations are inbred. People marry those they meet, such as relatives (cousins), and also choose mates by neighborhood, race, religion, education, even height, increasing the frequency of homozygotes; (ii) the frequency of homozygous affected individuals is low, and there are large stochastic errors when counting such rare events. For example, if the true frequency was four affected homozygotes in a population of 100 000 people, any one study might find two more or less, that is two or six homozygotes, giving a twofold range of estimates of allele frequency; (iii) some genetic diseases have low penetrance, which means that not all homozygous individuals show the disease. This is because other conditions (genetic or environmental) are needed to trigger the disease, and so its frequency is underestimated.

We can calculate the frequency of heterozygotes (carriers) for cystic fibrosis as:

$$2pq = 2 \times (1 - 0.02) \times 0.02 = 2 \times 0.98 \times 0.02 = 0.392 \quad \text{(approximately 1/25)}$$

In other words, one Caucasian in 25 is a carrier of this lethal recessive allele. This is a remarkably high frequency for such an allele. The explanation seems to be that carriers of cystic fibrosis have an increased tolerance of cholera toxin, which causes death by loss of fluid in the gut. (They may also have increased tolerance to some *Salmonella* species.) Transgenic mice heterozygous for the cystic fibrosis allele show approximately half the rate of fluid loss of their homozygous normal relatives when exposed to cholera toxin. Mice homozygous for cystic fibrosis showed no fluid loss in response to the toxin. This is an example of a selective equilibrium due to heterozygote advantage, similar to sickle cell anemia and malaria (see Topic D4).

D4 GENETIC DIVERSITY

Key Notes

Introduction

There is a large amount of genetic variation in the amino acid sequence of proteins in most populations. Molecular techniques, especially DNA sequencing, reveal further variation in DNA which has no apparent effect. Synonymous changes (e.g. AAA to AAG; both code for lysine) are silent, and affect only mRNA secondary structure and the efficiency of protein synthesis. There is a continuing debate about the relative importance of mutation (introducing new alleles), drift (chance changes in the frequency of alleles which may be selectively neutral), and natural selection (either directional, changing alleles, or frequency-dependent selection maintaining an equilibrium) in maintaining this variation.

Source of variation

All new variation starts as a mutation, a change in the sequence of bases in the DNA. The mutated sequence may code for a protein which does not adequately fulfill its function, or has a reduced activity (rarely increased activity), or is phenotypically normal (a neutral mutation) or, rarely, may code for an advantageous variant. Null mutations have lost all function (i.e. a deletion of the gene) but loss of activity can follow a single amino acid change. Many, perhaps most, surviving mutations have an insignificant effect on function, so are neutral to selection, and susceptible only to drift.

Mutation vs. selection

Defective alleles are removed by selection, and an equilibrium frequency (q) will exist where rate of removal of defective alleles by selection equals rate of formation of new mutations. For fully recessive alleles, $q^2 = \mu/s$ (where q = frequency, s = selection against homozygotes and μ = mutation rate). For recessive lethal alleles, $s = 1$ and $q = \sqrt{\mu}$

Drift

Drift is the chance difference in transmission of alleles between generations, which causes fluctuations in allele frequency. This can cause differences in allele frequency between separate populations of one species. It has most effect in small populations, and on rare alleles, and its effects accumulate with time. Drift is important for establishing (or eliminating) new favorable mutations until they reach a frequency where selection becomes more important than drift. This is especially the case for recessive alleles which are hidden in heterozygotes. It is the main factor (with mutation) affecting the frequency of neutral alleles. The founder effect occurs when a small group is isolated and founds a new population that has a gene pool with allele frequencies different from those in the parent population. Bottle-necking is similar to the founder effect, but refers to a population or species reduced to a very small number of individuals by adverse circumstances.

Frequency-dependent selection

In some circumstances (e.g. a heterogeneous habitat) individuals which are different from each other can exploit the range of habitats and reduce competition. Common phenotypes which exceed their ideal habitat will face more competition than rare phenotypes which have surplus habitat. A range of genotypes will persist. Disease strains adapted to common genotypes will be less harmful to rare genotypes, again making rare genotypes fitter.

Balanced selection – heterozygote advantage

Selection which favors heterozygotes (heterozygote advantage, overdominance) will maintain two alleles and favor rare alleles which are more frequently found in heterozygotes than are common alleles. An example is sickle cell disease caused by a mutated allele of hemoglobin A. Normal homozygotes Hb^AHb^A suffer more from malaria than heterozygotes Hb^AHb^S who have a tolerance for malaria. Homozygous Hb^SHb^S individuals suffer from sickle cell disease. In the presence of malaria both the normal allele Hb^A and the sickle cell allele Hb^S are maintained in the population.

Related topics

Evolution by natural selection (D2)
Neo-Darwinian evolution: selection acting on alleles (D5)

Introduction

There are several alleles for most genes in most populations, in fact the limit to genetic variation seems to be the technical ability of researchers to discover it. DNA sequencing reveals differences which have no detectable phenotypic effect. This **synonymous variation** does not change the coded amino acid (e.g. codons AAA and AAG both code for lysine) so do not affect the polypeptide or protein product, and are called **silent mutations**. However, the alternative codons are often not used in equal numbers, which suggest that one codon may be translated more efficiently than another on the ribosomes. Sequence variation also exists at noncoding sites, which may or may not have regulatory roles. Only regions where the sequence is important will be subject to selection. There is thought to be too much variation in most populations for it all to be due to selection, because too many individual zygotes would have to die for sufficient selection to act to maintain the variation. The mechanisms which generate and maintain variability can be considered in turn to attempt to decide how much variation is due to selection, and how much is due to other forces like mutation and chance (drift).

Source of variation

All genetic variation originates from **mutations** (see Topic B5), which are changes in the sequence of bases in the DNA (or RNA of RNA viruses). These usually destroy the allele's ability to function, and so are usually detrimental to the fitness of the genotype. **Null mutations** are due to complete destruction of the gene's function, typically by deletion. Stop codons (nonsense mutations) or frameshifts early in the coding sequence, or point mutations changing essential amino acids (e.g. in enzyme active sites) have similar effects. All **loss-of-function** alleles can be considered together. They are recessive if a single copy of a functional allele is sufficient to produce a normal phenotype. They are usually deleterious (often lethal) when homozygous, and cause the typical human genetic diseases. Some null mutations may be advantageous (e.g. loss of the I^A allele prevents expression of A blood group antigen, which is not essential, and its loss confers some resistance to smallpox).

Point mutations which change a single amino acid in the encoded polypeptide may have no significant phenotypic effect. These may be very slightly deleterious, but not enough to matter. Such mutations are termed **neutral** because selection does not affect their frequency. An allele can be defined as neutral if its frequency is controlled by chance (drift, see below) more than by selection. Between these two extremes are mutations which vary in their deleterious effects. Very rarely, mutations will improve the fitness of the genotype. Such advantageous mutations will be selected for, and may replace the pre-existing alleles.

Mutation vs. selection

Dysfunctional alleles are continuously being created by mutation. Selection acts against them, but most are recessive, so selection only acts against homozygotes. An increase in frequency of mutant alleles causes an increase in homozygotes relative to heterozygotes, and increased selection against these homozygotes prevents further increase in the frequency of the mutant allele. At equilibrium the number of new dysfunctional mutations produced equals the number lost by selection. For simplicity we will assume that dysfunctional alleles are recessive, so selection only acts against homozygotes, which occur at a frequency of q^2. In a diploid population of N individuals there are $2N$ alleles. Assuming the frequency of mutants (q) is low, there will be $2pN\mu$ new mutations (where μ is the mutation rate per generation and p is the frequency of functional alleles), and $2Nsq^2$ mutant alleles will be lost by selection, where s is the coefficient of selection against homozygotes. ($s = 1$ for recessive lethal alleles.)

$$\textbf{At equilibrium, } 2pN\mu = 2Nsq^2$$

p is usually assumed to be 1, in which case, dividing through by $2N$ gives:

$$\mu = sq^2 \text{ and } q^2 = \frac{\mu}{s} \text{ and } q = \sqrt{\frac{\mu}{s}}$$

If there is no dominance, and heterozygotes are also subject to selection of $0.5s$, then $q = 2\mu/s$ (approximately).

Drift

The genotype of any gamete or zygote depends on which of the two homologous segments of chromosome it inherited by chance from its diploid parent(s). Each normal (Mendelian) allele has a probability of 1/2 of being transmitted to any one offspring, just as a tossed coin has an equal chance of coming down heads or tails. Two diploid parents will produce on average two offspring, so on average each allele will be transmitted once, each copy in the parental population will be represented once in the next generation. However, individual alleles will be more or less frequent, just as tossing many coins does not give exactly equal numbers of heads and tails. The gene pool will be changed in the next generation, and the whole process will be repeated, so the frequencies of particular alleles will drift (change) through time. The variance in frequency derives from the binomial distribution and can be calculated (but do not try to remember how). Variance S^2 of the frequency q of allele a is $S^2_q = p_0q_0/2N$ (where p_0 and q_0 are the initial frequencies of the two types of allele). Over a period of t generations this becomes:

$$S^2_q = p_0q_0\left\{1 - \left[1 - \frac{1}{2N}\right]^t\right\}$$

There are three important points to note:

(i) The proportional change depends on initial allele frequencies. Rare alleles are more susceptible to drift. If an allele only occurs in a single codfish, it might be included in half a million eggs, of which 1000 might survive (a huge increase) or it might be caught and eaten before it breeds, and the allele will be extinct. The more individuals an allele is found in, the nearer its transmission frequency will be to the average value of one.

(ii) The process is cumulative over time, but the frequency can go up or down in each generation (*Fig. 1*).

(iii) The change is inversely related to population size (for mathematicians, it is inversely proportional to the harmonic mean). Larger samples, like tossing more coins, more often give results near the expected mean value (*Fig. 1*).

Drift is important in three cases: (i) it is important for removing or promoting very rare alleles, especially new favorable mutations before they become established, because drift has a greater effect than selection on the transmission of rare alleles. It is important for increasing the frequency of new recessive mutations to a frequency where homozygotes occur in sufficient frequency for selection to become effective; (ii) drift is responsible for changing the frequency of neutral mutations, (caused by recurrent mutation) which by definition are affected more by drift than by selection. The probability of any new neutral mutation eventually becoming fixed in the population is $1/(2N_e)$ where N_e is the effective (breeding) population size. (Assuming they are neutral, each of the $2N_e$ copies which exist for each genetic locus in the population has an equal chance of becoming fixed.); (iii) drift in small populations can produce unrepresentative allele frequencies which would be very unlikely to occur in a large population. This is called **founder effect** (see Topic D7) when a small and unrepresentative group founds a new colony. A human example are the Amish, a religiously united group in the USA, established by a small number of immigrant families. It is also termed **bottle-necking** when a population is reduced to a small number (e.g. by disease or famine) who become parents for a large later population. Drift may cause small isolated populations to be very different from the species norm, and can be important in the production of new races

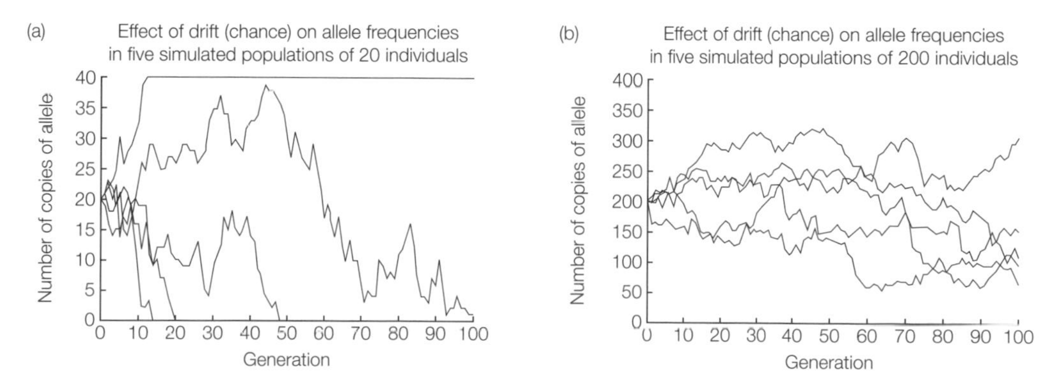

Fig. 1. Change in frequency of neutral alleles in five populations each of (a) 20 or (b) 200 diploid individuals produced by a computer simulation. The y axis shows the number of copies of a specific allele starting from a frequency of 0.5. Each trace shows the frequency (number of copies) of that allele in one population, changing by chance over successive generations. In four of the five populations of 20 individuals, the allele was fixed or lost (p = 1 or 0) within 50 generations.

or species. Inbreeding can be a special case of founder effect where the effective population is a small group of related individuals, possibly a single hermaphrodite individual.

Frequency dependent selection

An allele may have an advantage when it is rare, especially in a heterogeneous environment. If a species of finch lives on an island where there are several plant species producing seeds of different sizes, birds with larger beaks can exploit the larger seeds, those with smaller beaks exploit the small seeds, and there is less competition between them. If either large or small beak size becomes too frequent for the current seed crop, it will face more competition, and selection will decrease its frequency. For example, in *Geospiiza fortis*, a species of Darwin's finches on Daphne Major in the Galapagos islands, the mean beak size was found to be decreased after wet years when small seeds were numerous, and increased after drought years when only large-fruited plants were successful.

Disease is another selective agent which maintains variation in its host population. Strains of disease organism are adapted to particular genotypes of host, and the most successful disease strain is the one which can spread fastest through the most common host genotypes, reducing that genotype's fitness and frequency. This is a serious problem for agriculture, where large areas of a single variety of a single species are grown (monocultures). Merely mixing three varieties of barley seed, each variety susceptible to a different strain of fungal pathogen, can remove the need for fungicide treatment, because no single strain of fungus can spread through the entire crop. In a genetically variable population, no single disease strain spreads well, and this is the normal condition in natural populations.

Balanced selection – heterozygote advantage

If both homozygotes are less fit than the heterozygote, then both alleles will be maintained in the population (e.g. normal and **sickle cell disease** allele). An allele of beta hemoglobin, Hb^S, codes for a polypeptide with valine as the third amino acid, instead of the normal glutamic acid. When homozygous this allele causes sickle cell anemia, which is serious and often fatal, but the allele is common where malaria is prevalent because heterozygotes Hb^AHb^S have some tolerance to malaria. In parts of Africa the frequency of Hb^S reaches 0.2. At this frequency, the frequency of homozygous recessive genotypes is $0.2^2 = 0.04 = 4\%$ of the population, 4/5 of whom might die of sickle cell anemia. This is compensated by an increased susceptibility to malaria in normal homozygotes (Hb^AHb^A), to which heterozygotes Hb^AHb^S have some resistance. If the frequency of $Hb^S = q = 0.2$, and the fitness of sickle cell sufferers is 0.2 (four out of five do not reproduce), the selection due to malaria on the Hb^AHb^A genotype can be calculated. The equilibrium frequency of $Hb^A = q = s/(s + t)$, where s = selection coefficient on Hb^AHb^A and t = selection on Hb^SHb^S. The ratio of allele frequencies is the inverse of the ratio of selection coefficients:

Genotype	Hb^AHb^A	Hb^AHb^S	Hb^SHb^S
Fitness	0.8	1	0.2

The actual fitnesses depend on the local incidence of malaria and availability of health care.

D5 NEO-DARWINIAN EVOLUTION: SELECTION ACTING ON ALLELES

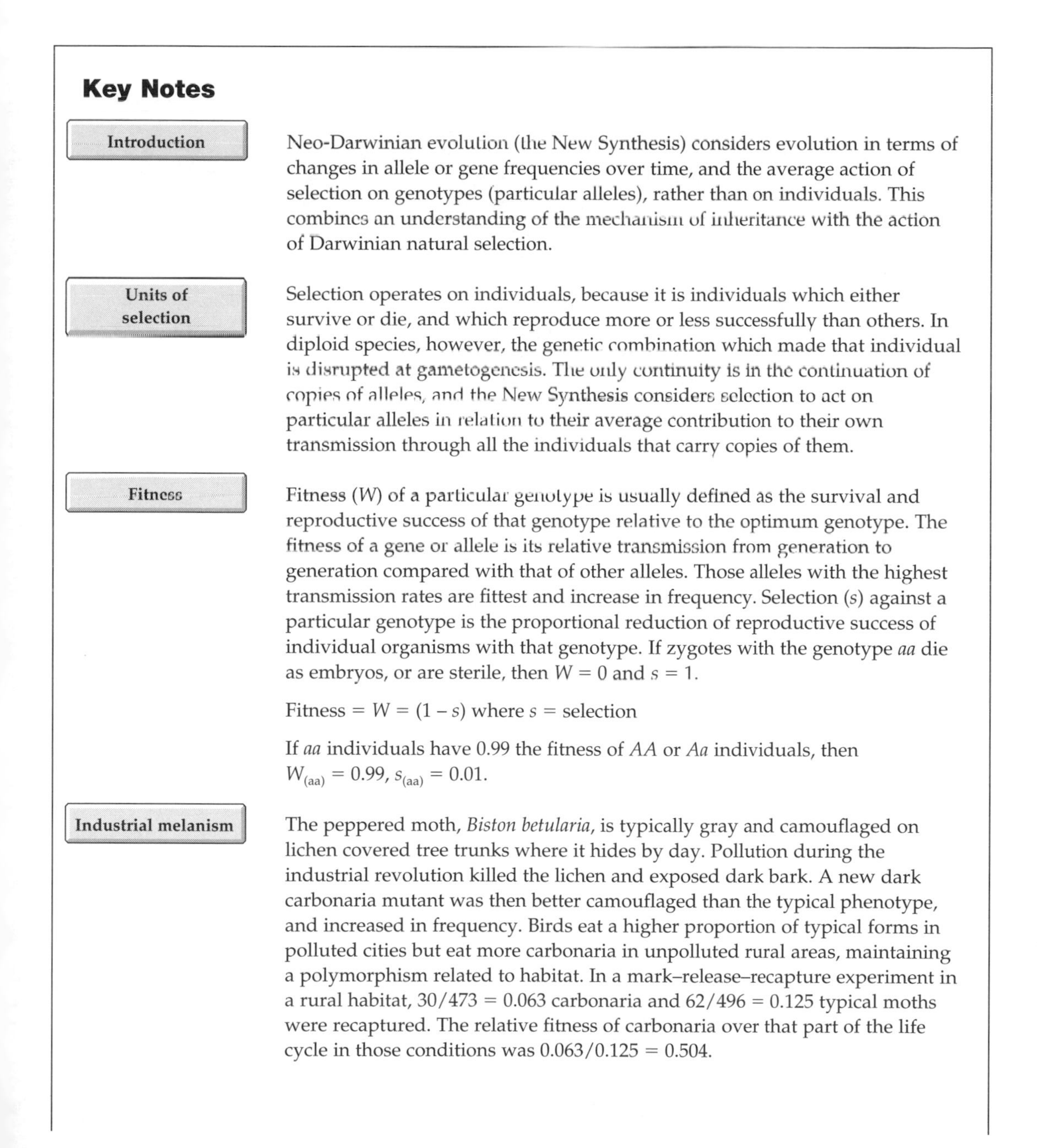

Key Notes

Introduction

Neo-Darwinian evolution (the New Synthesis) considers evolution in terms of changes in allele or gene frequencies over time, and the average action of selection on genotypes (particular alleles), rather than on individuals. This combines an understanding of the mechanism of inheritance with the action of Darwinian natural selection.

Units of selection

Selection operates on individuals, because it is individuals which either survive or die, and which reproduce more or less successfully than others. In diploid species, however, the genetic combination which made that individual is disrupted at gametogenesis. The only continuity is in the continuation of copies of alleles, and the New Synthesis considers selection to act on particular alleles in relation to their average contribution to their own transmission through all the individuals that carry copies of them.

Fitness

Fitness (W) of a particular genotype is usually defined as the survival and reproductive success of that genotype relative to the optimum genotype. The fitness of a gene or allele is its relative transmission from generation to generation compared with that of other alleles. Those alleles with the highest transmission rates are fittest and increase in frequency. Selection (s) against a particular genotype is the proportional reduction of reproductive success of individual organisms with that genotype. If zygotes with the genotype *aa* die as embryos, or are sterile, then $W = 0$ and $s = 1$.

Fitness $= W = (1 - s)$ where $s =$ selection

If *aa* individuals have 0.99 the fitness of *AA* or *Aa* individuals, then $W_{(aa)} = 0.99$, $s_{(aa)} = 0.01$.

Industrial melanism

The peppered moth, *Biston betularia*, is typically gray and camouflaged on lichen covered tree trunks where it hides by day. Pollution during the industrial revolution killed the lichen and exposed dark bark. A new dark carbonaria mutant was then better camouflaged than the typical phenotype, and increased in frequency. Birds eat a higher proportion of typical forms in polluted cities but eat more carbonaria in unpolluted rural areas, maintaining a polymorphism related to habitat. In a mark–release–recapture experiment in a rural habitat, 30/473 = 0.063 carbonaria and 62/496 = 0.125 typical moths were recaptured. The relative fitness of carbonaria over that part of the life cycle in those conditions was 0.063/0.125 = 0.504.

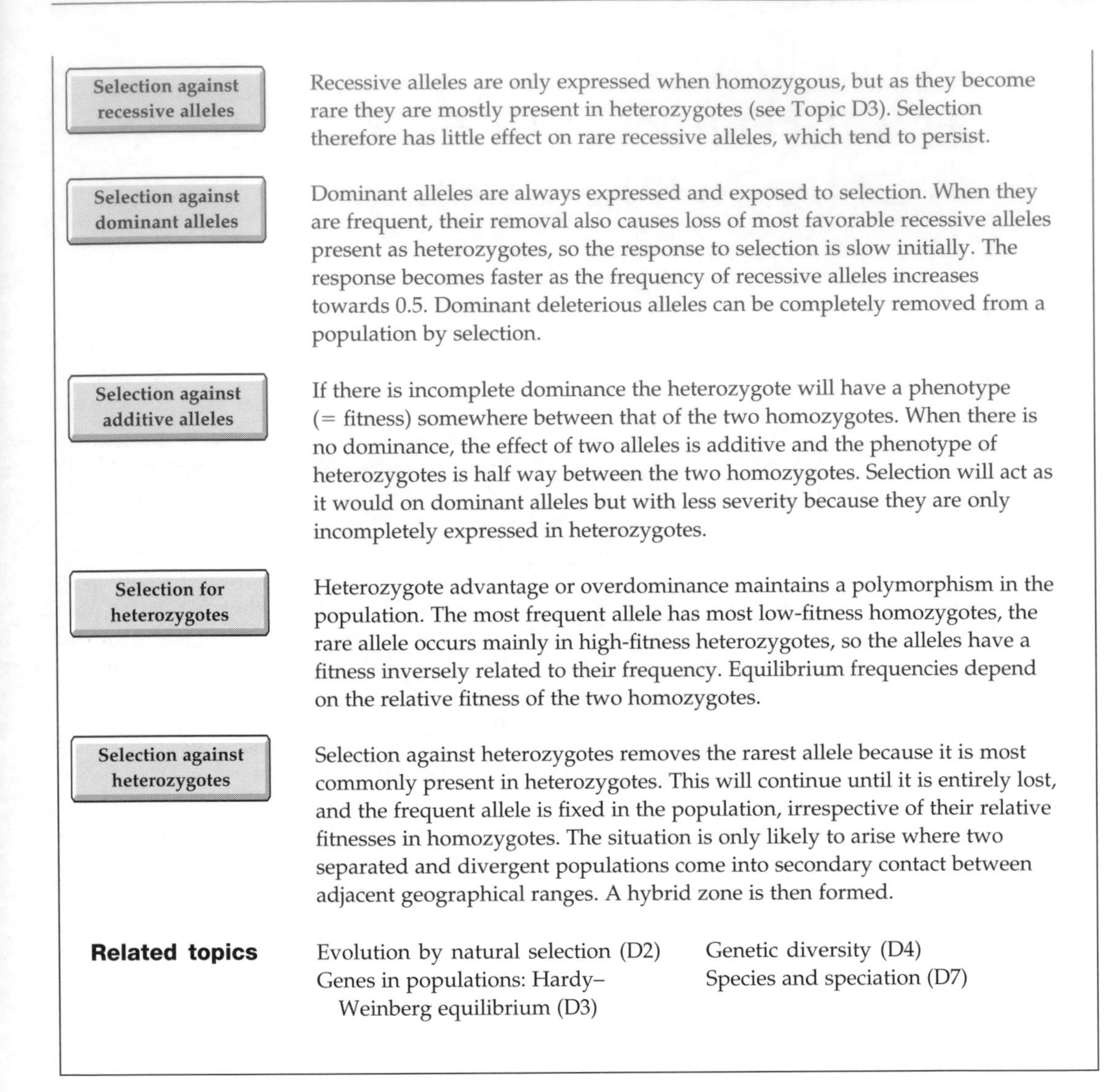

Selection against recessive alleles

Recessive alleles are only expressed when homozygous, but as they become rare they are mostly present in heterozygotes (see Topic D3). Selection therefore has little effect on rare recessive alleles, which tend to persist.

Selection against dominant alleles

Dominant alleles are always expressed and exposed to selection. When they are frequent, their removal also causes loss of most favorable recessive alleles present as heterozygotes, so the response to selection is slow initially. The response becomes faster as the frequency of recessive alleles increases towards 0.5. Dominant deleterious alleles can be completely removed from a population by selection.

Selection against additive alleles

If there is incomplete dominance the heterozygote will have a phenotype (= fitness) somewhere between that of the two homozygotes. When there is no dominance, the effect of two alleles is additive and the phenotype of heterozygotes is half way between the two homozygotes. Selection will act as it would on dominant alleles but with less severity because they are only incompletely expressed in heterozygotes.

Selection for heterozygotes

Heterozygote advantage or overdominance maintains a polymorphism in the population. The most frequent allele has most low-fitness homozygotes, the rare allele occurs mainly in high-fitness heterozygotes, so the alleles have a fitness inversely related to their frequency. Equilibrium frequencies depend on the relative fitness of the two homozygotes.

Selection against heterozygotes

Selection against heterozygotes removes the rarest allele because it is most commonly present in heterozygotes. This will continue until it is entirely lost, and the frequent allele is fixed in the population, irrespective of their relative fitnesses in homozygotes. The situation is only likely to arise where two separated and divergent populations come into secondary contact between adjacent geographical ranges. A hybrid zone is then formed.

Related topics

Evolution by natural selection (D2)
Genes in populations: Hardy–Weinberg equilibrium (D3)
Genetic diversity (D4)
Species and speciation (D7)

Introduction

When the principles of genetic inheritance and mutation were discovered it was realized that this filled in the detail about inheritance needed to complement Darwin's theory of evolution. Evolution can be considered as a change in the frequency of alleles and genes in a population over time, and selection can be considered as acting on particular alleles or genes through the average effect they have on all the individuals in a population which carry them. This approach was termed The New Synthesis, or Neo-Darwinism.

Units of selection

It is usually considered that the unit of selection is the individual organism, because it is individuals which either survive or die, and which reproduce more or less successfully than others. This is true within any one generation, but there is no continuity in individuals in a sexually reproducing species. Individuals grow old and die, and their genome is split up at gametogenesis.

However, copies of their alleles are passed on into new combinations. The only continuity is in the continuation of copies of alleles, and the New Synthesis considers selection to act on particular alleles in relation to their average contribution to all the individuals that carry copies of them.

Fitness

Fitness (*W*) has two components; viability (survival) and fecundity (reproductive success), and operates on individuals, each with a particular genotype. For simplicity, it is necessary to consider genes individually. The fitness of a genotype is the ratio of the average fitness (success in genetic transmission) of individuals with that genotype compared with the fitness of individuals which have the optimum genotype. For example consider an allele *A* which codes for a metabolic enzyme, and is dominant. Homozygous *AA* individuals and heterozygous *Aa* individuals have a fitness of one. If the enzyme is essential then all *aa* homozygotes will die, their fitness $W_{aa} = 0$, and the selection coefficient $s_{aa} = 1 - W_{aa} - 1$.

If the enzyme is not essential, but 1 in 100 *aa* individuals lacking it fail to reproduce successfully **because** they lack it, then the fitness of *aa* genotype individuals (W_{aa}) is 0.99, and the selection coefficient (s_{aa}) *is* 0.01. The rate at which the frequency of allele *a* changes in the population will be much faster with the higher selection coefficient (*Fig. 1*). Note that selection against a dominant allele is the same as selection for a recessive allele, and vice versa.

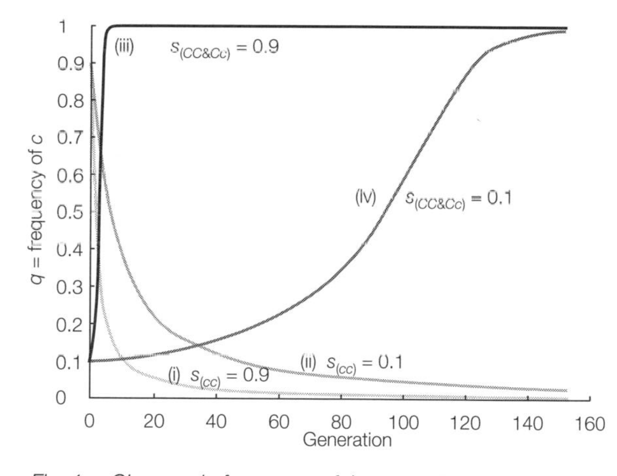

Fig. 1. Changes in frequency of the recessive allele under different types of selection (Table 1.) Each trace starts with the less fit allele at a frequency of 0.9. Data was calculated by repeatedly applying selection to genotypes at the Hardy–Weinberg equilibrium. The four conditions shown are selection against the recessive allele a at $s_{(cc)} = 0.9$ (i) and $s_{(cc)} = 0.1$ (ii); and selection against the dominant allele A at $s_{(CC\&cc)} = 0.9$ (iii) and $s_{(CC\&cc)} = 0.1$ (iv). $s_{(ccaa)} = 0.9$ means that 90% of individuals with the aa genotype are removed at each generation.

Industrial melanism

B. betularia, the peppered moth, provides a classic example of natural selection. They are nocturnal, and the 'typical' form is gray and speckled, which is camouflaged very well on lichen-covered bark on trees where they rest during the daytime. In 1849 the first dark melanic specimen was collected in Manchester, England. Collectors are very eager to obtain rarities, so we can assume they were very rare before this. The melanic form is known as carbonaria and is caused by a dominant allele of a single gene (*C*) so typical moths are homozygous *cc*. Carbonaria turned from a collectors rarity to the common form and by 1895 about 98% of that population of moths was dark. Pollution, principally acid rain from the industrial revolution, had killed the lichens on the trees, exposing the dark bark, which may have been darkened further by soot. The gray typical moths were now not camouflaged as well as the dark carbonaria. Direct observations and mark–release–recapture experiments supported the hypothesis that predators, in this case birds, were catching a higher proportion of typical than carbonaria in polluted industrial regions. When both types were released in unpolluted rural woods, the situation was reversed and carbonaria were eaten proportionately faster than typical.

In one experiment in Birmingham, UK, 123 out of 447 (0.275) released carbonaria were recaptured, and 18 out of 137 (0.131) typical moths. The carbonaria are given a fitness of 1 (= 0.275/0.275), while the typical form had a fitness of 0.131/0.275 = 0.476. (This assumes equal migration and equal efficiency of retrapping of both forms). A similar experiment in rural Dorset, UK recaptured 30 out of 473 carbonaria (0.063) and 62 out of 496 typical (0.125). Here typical forms are designated a fitness of 1 (= 0.125/0.125) and carbonaria a fitness of 0.063/0.125 = 0.504. The relative fitness reversed between the polluted and unpolluted habitats.

These figures are not conclusive because the moths might already have laid their eggs the night before, in which case any postreproductive mortality is irrelevant. In another experiment, Bishop calculated a fitness of typical, relative to carbonaria, of about 0.8 in Liverpool and 1.8 in rural Wales, 50 km away, based on egg laying rates and survival of adults. This still ignores selection at other stages of the life cycle, and there is evidence that there are also affects on the survival of the caterpillar stage. Genotype frequencies at all stages of the life cycle must be determined for more than one generation before really accurate numerical conclusions can be drawn. The effects of selection on allele frequency over a long period of time can be calculated if the selection coefficients are known (*Fig. 1*).

Selection against recessive alleles

Recessive alleles are only exposed to selection in homozygous *cc* genotypes. As the frequency (q) of *c* decreases, q^2 becomes very small (Hardy–Weinberg equilibrium, see Topic D3) and most *c* alleles are in dark heterozygotes ($2pq$) where they are not expressed and so cannot be acted on by selection. For this reason, selection is very ineffective when acting on rare recessive alleles, whether they are of high or low fitness in homozygotes (*Table 1a*, and *Fig. 1b* and *f*). The typical allele will tend to persist at low frequency even in highly polluted areas.

Selection against dominant alleles

The carbonaria allele is dominant, and therefore subject to selection in both homozygous (*CC*) and heterozygous (*Cc*) individuals. Selection against carbonaria in an unpolluted wood will tend to remove all the *C* alleles because they are all expressed in the phenotype, even when rare (*Table 1b* and *Fig. 1c* and *d*).

Notice that selection against dominant alleles does not lead to an immediate replacement by rare recessive alleles because these are mostly in the heterozygotes being selected against because of their one dominant allele. It is only when the recessive alleles reach a frequency (q) around 0.3, and the frequency of homozygotes (q^2) reaches about 0.09 that they start to increase in frequency rapidly (*Fig. 1*). Even then, more than two-thirds of the recessive alleles are in heterozygotes ($2pq = 0.42$).

Table 1. Change in allele frequency per generation produced by selection against different genotypes

Type of selection	Genotype fitness			
	CC	*Cc*	*cc*	dq
a Complete dominance selection against *c* (in *cc*)	1	1	$(1 - s)$	$-sq^2\dfrac{(1-q)}{1-sq^2}$
b Complete dominance selection against *C* (in *CC* and *Cc*)	$1 - s$	$1 - s$	1	$\dfrac{sq^2(1-q)}{1-s(1-q^2)}$
c Incomplete dominance selection against *c* (in *cc* and *Cc*)	1	$1 - hs$	$1 - s$	$-\dfrac{hsq(1-q)}{1-sq(2hp+q)}$

q is the frequency of the recessive allele, dq the change in its frequency. s is the fraction of that genotype lost to selection, and h is the extent of dominance, between 0 in *a* and 1 in *b*. These values were obtained by multiplying the frequencies of zygotes of each genotype by the appropriate selection factor to obtain the new allele frequencies, and subtracting algebraically to find the difference. The derivation of this can be found in specialized population genetics textbooks.

Selection against additive alleles

In cases where there is no dominance the effects of the two alleles adds up, so the heterozygotes have a fitness halfway between those of the homozygotes. If there is incomplete dominance, the relative selection against heterozygotes (denoted by h) will be somewhere between that of the homozygotes. In these cases the rate of change of allele frequency due to selection will be intermediate between the effects of selection against wholly dominant or wholly recessive alleles (*Table 1c*).

Selection for heterozygotes

Overdominance, or heterozygote advantage, occurs when the heterozygote is fitter than either homozygote. A classic example are carriers (heterozygotes) of sickle cell anemia trait (see Topic D4), who are more tolerant of malaria than normal homozygotes, and who do not get the sickle cell disease of homozygous recessive individuals. The rarest allele is found most frequently in heterozygotes and so is favored. This gives a frequency-dependent advantage to the rare allele and maintains a polymorphism. Both alleles persist in the population at an equilibrium frequency determined by the relative fitness of the two homozygous genotypes. Selection for heterozygosity may be an important factor in maintaining genetic variation in a population.

Selection against heterozygotes

Heterozygote (or hybrid) disadvantage occurs when heterozygotes are less fit than homozygotes. The rare allele is most commonly found in heterozygotes, so suffers the greatest rate of loss. It is lost faster as it becomes rarer and there are fewer homozygotes. The rare allele will therefore become extinct once a trend is established. This extinction depends on frequency rather than on the fitness of the two homozygous genotypes. This situation is only likely to arise when two adjacent populations are isolated and become homozygous for different alleles, and then come into secondary contact at the borders of their ranges. If some members of the two homozygous groups do not interbreed (i.e. they have a premating isolation mechanism) then they may become two species (see Topic D7) but this is outcome is uncertain.

D6 Chromosome change in evolution

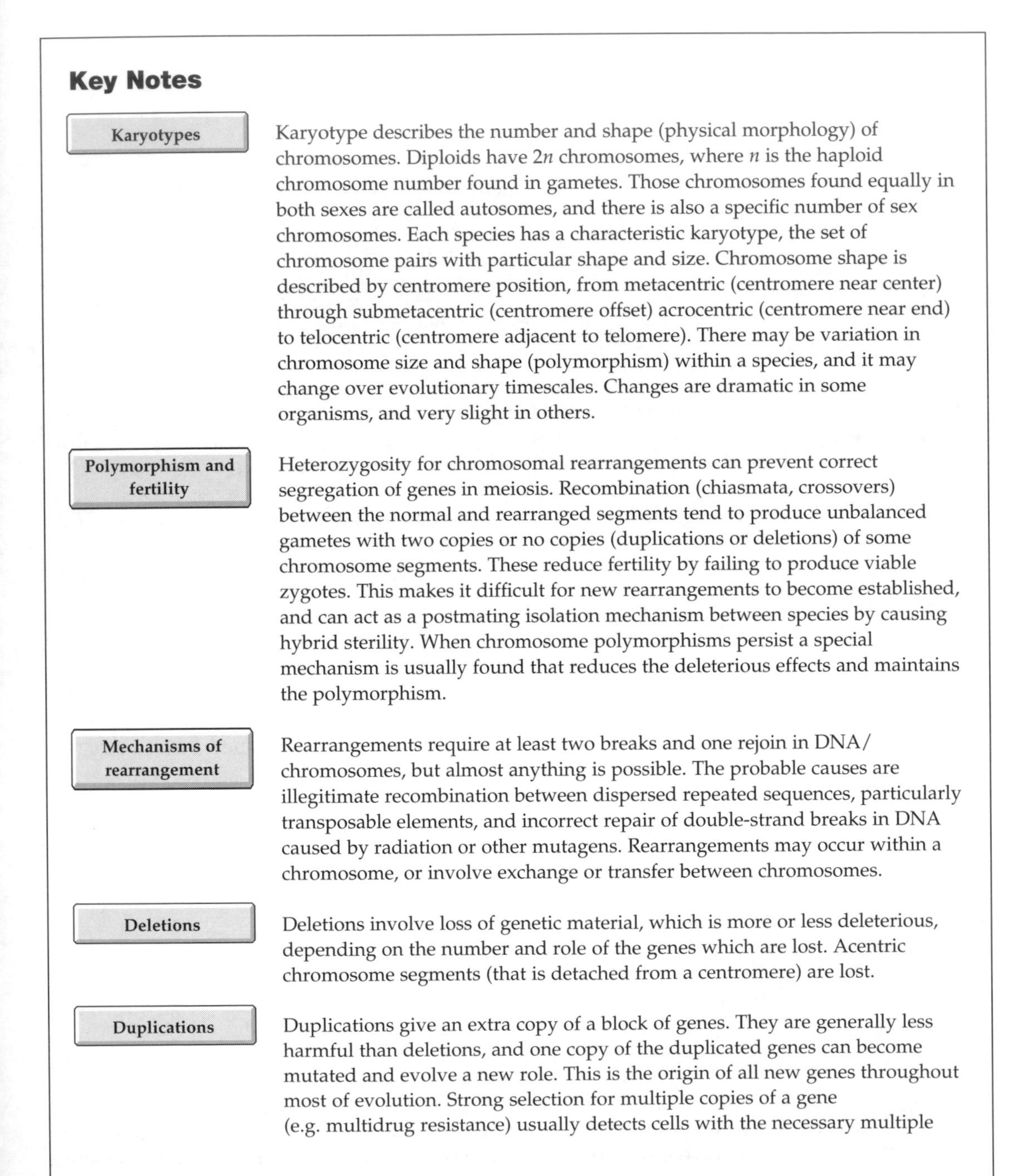

Key Notes

Karyotypes

Karyotype describes the number and shape (physical morphology) of chromosomes. Diploids have $2n$ chromosomes, where n is the haploid chromosome number found in gametes. Those chromosomes found equally in both sexes are called autosomes, and there is also a specific number of sex chromosomes. Each species has a characteristic karyotype, the set of chromosome pairs with particular shape and size. Chromosome shape is described by centromere position, from metacentric (centromere near center) through submetacentric (centromere offset) acrocentric (centromere near end) to telocentric (centromere adjacent to telomere). There may be variation in chromosome size and shape (polymorphism) within a species, and it may change over evolutionary timescales. Changes are dramatic in some organisms, and very slight in others.

Polymorphism and fertility

Heterozygosity for chromosomal rearrangements can prevent correct segregation of genes in meiosis. Recombination (chiasmata, crossovers) between the normal and rearranged segments tend to produce unbalanced gametes with two copies or no copies (duplications or deletions) of some chromosome segments. These reduce fertility by failing to produce viable zygotes. This makes it difficult for new rearrangements to become established, and can act as a postmating isolation mechanism between species by causing hybrid sterility. When chromosome polymorphisms persist a special mechanism is usually found that reduces the deleterious effects and maintains the polymorphism.

Mechanisms of rearrangement

Rearrangements require at least two breaks and one rejoin in DNA/ chromosomes, but almost anything is possible. The probable causes are illegitimate recombination between dispersed repeated sequences, particularly transposable elements, and incorrect repair of double-strand breaks in DNA caused by radiation or other mutagens. Rearrangements may occur within a chromosome, or involve exchange or transfer between chromosomes.

Deletions

Deletions involve loss of genetic material, which is more or less deleterious, depending on the number and role of the genes which are lost. Acentric chromosome segments (that is detached from a centromere) are lost.

Duplications

Duplications give an extra copy of a block of genes. They are generally less harmful than deletions, and one copy of the duplicated genes can become mutated and evolve a new role. This is the origin of all new genes throughout most of evolution. Strong selection for multiple copies of a gene (e.g. multidrug resistance) usually detects cells with the necessary multiple duplications (gene amplification), often visible as a heterogeneous staining region on the chromosome.

Centromeric fusions and fissions

Two telocentric or acrocentric chromosomes (with negligible short arms) may fuse at the centromere to produce a single metacentric chromosome (centromere near the middle) or submetacentric (two arms clearly unequal). This is called a centric fusion or Robertsonian translocation. The reverse is centric fission, that is splitting of the centromere in a metacentric to give two telocentrics. In this way chromosome number can change. Correct meiotic segregation in a heterozygous cell requires that the two telocentrics segregate to one pole, and the metacentric (with both arms) goes to the other pole.

Translocations

Translocations involve exchange of distal regions of nonhomologous (genetically different) chromosomes. It is necessary that either both rearranged or both original chromosomes occur in the same gamete to ensure genetic balance. In meiosis, normal chromosomes and chromosomes with translocations pair up alternately to form a ring or chain of several chromosomes. If two adjacent chromosomes segregate to the same pole, genetically unbalanced gametes will be produced, but if alternate chromosomes go together the gametes will contain a complete balanced genome. Some species are heterozygous for many translocations, and form long chains or rings of chromosomes in meiosis.

Inversions

Inversions turn part of the chromosome around, reversing its polarity. A pericentric inversion includes the centromere, a paracentric inversion is contained in one arm of the chromosome, and does not move the centromere. A chromosome with an inversion can pair with a normal chromosome in meiosis in a heterozygote if they form an inversion loop. Meiotic recombination within an inversion loop duplicates one end of each chromatid and deletes the other end in a reciprocal manner. These recombined chromatids are therefore not viable, and do not reach the next generation. Inversions therefore act as recombination suppressors, preventing recombination between the inverted segment and its normally oriented counterpart. Some species have polymorphic chromosomal inversions and have mechanisms to prevent loss of fertility (see below).

Paracentric inversions

These do not include the centromere. Meiotic recombination in a paracentric inversion loop generates a dicentric chromatid (two centromeres) and an acentric (no centromere) fragment. Most dipterans (true two-winged flies) can tolerate this because they have no recombination in males so sperm all contain complete chromosomes, and female dipterans have mechanisms to ensure that only unrecombined chromatids reach the egg cell, so zygotes are viable.

Pericentric inversions

These invert the centromeric region, and the effects of meiotic recombination in the loop duplicating and deleting the ends of the chromatid cannot be avoided because all chromatids have one centromere. Many zygotes will have unbalanced genomes. Pericentric inversions only persist as polymorphisms in species where they do not recombine at meiosis.

Changes in sex chromosomes

Y chromosomes have few genes and are lost in some species (XX females, XO males). Fusion of an X chromosome with an autosome causes a copy of the free autosome to become a neo-Y chromosome, only found in males which have another copy attached to their single X. Fusion of a Y chromosome with an autosome causes the unattached autosome to become a neo-X chromosome, one free copy being in males (to balance the copy attached to the Y), two copies in females as before. When two or more different X or Y chromosomes exist they are numbered X_1, X_2, Y_1, Y_2, Y_3 etc.

Evolutionary effects

Polymorphisms for paracentric inversions are common in dipterans where they may contain sets of alleles coadapted to control sex, or adapted to particular ecological environments. The suppression of recombination protects the cluster of genes and facilitates their coevolution. Chromosomal change is common in evolution, and can cause hybrid sterility between closely related species, forming a postmating isolation mechanism.

Related topics

Chromosomes (B1)
Meiosis and gametogenesis (C3)
Species and speciation (D7)

Karyotypes

The term **karyotype** describes the number, size and centromere position/arm length ratio of the chromosomes in a cell (see Topic B1). **Autosomes** are the chromosomes found in all individuals, irrespective of sex. **Metacentric** chromosomes have a centromere near the center, and roughly equal arm lengths. **Submetacentric chromosomes** have noticeably unequal arm lengths. **Acrocentric chromosomes** have one long arm and one very short arm. **Telocentric chromosomes** have their centromere adjacent to one **telomere**. Telomeres are specialized sequences and associated proteins which protect the end of the DNA from degradation, and interact with the nuclear envelope and the cytoskeleton during meiosis.

The number of chromosomes in a single set, as found in a gamete, is called the **haploid number** and is denoted by ***n***, so **diploids** (which are zygotes formed by fusion of two haploid gametes) have **2*n*** chromosomes. Put another way, diploids have n pairs of chromosomes. In some species the two sexes have different numbers of sex chromosomes, so males and females can have different diploid numbers. The **sex chromosomes** may be noted separately (e.g. in a species where $n = 8$, females have two X chromosomes, males have one X, and there is no Y chromosome, then in males $2n = 16$XO, and females are 16XX). The karyotype is usually a constant feature within a species, but variant types of chromosome carrying homologous information may exist in a species. These are known as **chromosomal polymorphisms** (many shapes).

Karyotypes may be very stable through evolution. Most dragonflies have $n = 13$; snakes and birds both have about 18 large chromosomes and several microchromosomes. These data suggest karyotype conservation over hundreds of millions of years. On the other hand, closely related species may be very different suggesting rapid change. The Indian muntjac *Muntiacus muntjac vaginalis* has $2n = 7$ for males and $2n = 6$ for females, whereas the Chinese muntjac *Muntiacus muntjac reevesi* has $2n = 46$. Different populations (or races) of the European mole cricket, *Gryllotalpa gryllotalpa* have diploid numbers from 12 to 23. This variation may relate to speciation (see Topic D6).

Polymorphism and fertility

Sexual reproduction requires meiosis to segregate genetic information in the DNA of chromosomes so that each gamete gets one copy of each gene (and each chromosome), and all zygotes gets two copies of each (except the sex chromosomes of course). If the cell is heterozygous for a chromosomal rearrangement then the homologous genes are not in the same structural locations, and segregation may be distorted, producing genetically unbalanced gametes with missing or duplicated information. The problem is complicated further if there is a chiasma (also called meiotic recombination or crossover; see Topics C3 and B7) in the heteromorphic (different shaped) chromosomes. This is only a problem in heterozygotes, and therefore has its worst effects on rare chromosome types (morphs), making it difficult for new chromosomal rearrangements to become established in a population, because rare genetic elements are usually heterozygous (see Topic D3). Where chromosomal polymorphisms are found in populations, they are either transient states during replacement of one chromosome type by another, or there is likely to be some special genetic mechanism operating to achieve balanced gametes or to minimize fertility losses. Chromosomal differences between races can act as an effective postmating isolation mechanism by causing hybrid sterility (see Topic D7).

Mechanisms of rearrangement

Rearrangement of chromosomes requires at least two breaks (four ends) in DNA (chromosomes), followed by efficient rejoining and/or healing. These break/rejoins may occur by recombination between DNA sequences repeated and dispersed throughout the genome. Transposable elements (see Topic B4) dispersed throughout the genome may promote such rearrangements directly by their transposition mechanism. Incorrect repair of ends of double-strand breaks caused by ionizing radiation (or other mutagens) is another source of rearrangement which can be demonstrated in cultured cells (see Topic C10). All linear chromosomes must have a telomere, special repeated sequences with associated proteins which protect the end of the chromosome from degradation (see Topic B1). Chromosome breaks which do not rejoin to a normal chromosome end or acquire a telomere themselves remain unstable and reactive and promote further cycles of chromosome breakage and possible rearrangement. **Terminal** rearrangements involve the end of a chromosome, **interstitial** rearrangements are internal. The rearrangement may only affect one chromosome, or may involve transfer or exchange between two nonhomlogous (genetically different) chromosomes.

Deletions

Terminal deletions involve a single break, then the fragment with a centromere obtaining a new telomere. The **acentric fragment** (no centromere) will be lost at cell division. Interstitial deletions involve the loss of a region within the chromosome, followed by rejoining of the ends of the flanking pieces. This may occur by recombination between repeated sequences in similar orientation in the same chromosome to delete the intervening region as a closed circle (*Fig. 1a–c*). Alternatively, misaligned recombination between repeats on homologous chromosomes can create a duplication–deletion pair (*Fig. 1d* and *e*, see duplications below). Deletions are also produced by meiotic recombination in chromosome regions heterozygous for other rearrangements (e.g. translocations or inversions, see below).

Heterozygous zygotes carrying a deletion will be partially haploid, unless the region is duplicated elsewhere. In the fruitfly *Drosophila melanogaster* there are two loci which are dominant lethals if present in a single copy (haplolethals),

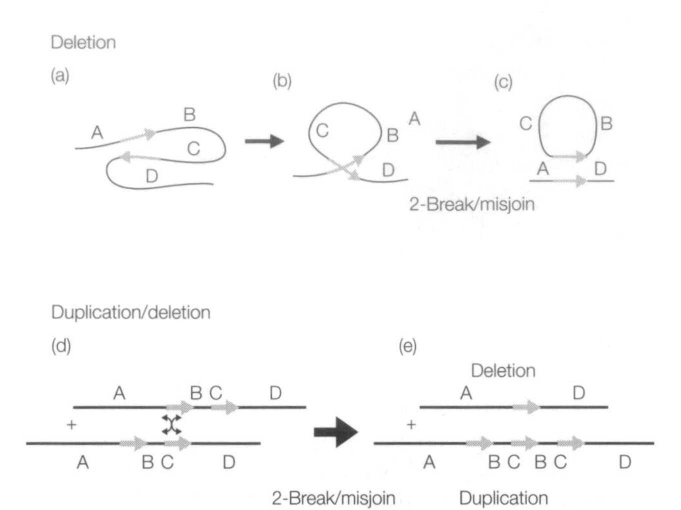

Fig. 1. Deletion of region BC as a closed circle (a–c) or duplication/deletion of region BC (d–e) by misaligned breakage – rejoining or recombination, possibly between repeated sequences or transposable elements (gray arrows) on the same chromosome (a–c, deletion) or on homologous chromosomes or sister chromatids (d–e).

11 are deleterious and 41 give small flies (haplo-minutes) if haploid. Single copies of genes may often be inadequate for full fitness, and combined loss of about 3% of the genome is lethal, probably because of the combined effects of loss of one copy of many genes. Deletions are probably less harmful in polyploid species. Heterozygous deletions, especially if large, may inhibit meiotic pairing/synapsis, and so cause nondisjunction and lower fertility.

Duplications

Since cellular life arose, and possibly before, duplications have been the main source of new genetic material, followed by divergence and separate evolution of the two copies. Sequencing the entire genome of the yeast *Saccharomyces cerevisiae* has revealed 53 groups of genes duplicated onto nonhomologous chromosomes. In some cases where numerous copies of a gene are required (e.g. ribosomal RNA genes) the tandem copies evolve in concert in one species, all with the same sequence. Misaligned recombination between dispersed repeated sequences on homologous chromosomes is the most likely origin of tandem duplications on one chromosome (*Fig. 1d* and *e*). Occasionally a small translocation will survive in its new location as a duplication. When selection acts for a large increase in copy number (**amplification**) of a particular gene, they usually appear, often as microscopically visible **heterogeneous staining regions** containing many tandem repeats. Examples are selection by insecticide on aphids to amplify the gene for an esterase which breaks down the insecticide, thus becoming insecticide resistant, and by chemotherapy on cancer cells to amplify genes for a protein which pumps toxic drugs out of the cell. Duplications of the centromere do not survive because the two centromeres may orientate to opposite poles at division, breaking the chromosome. Duplications are generally less harmful than deletions.

Centromeric fusions and fissions

Two telocentric chromosomes, or acrocentric chromosomes with no significant genetic material on the short arm, may fuse at the centromeres to give one metacentric or submetacentric chromosome, reducing the number of chromo-

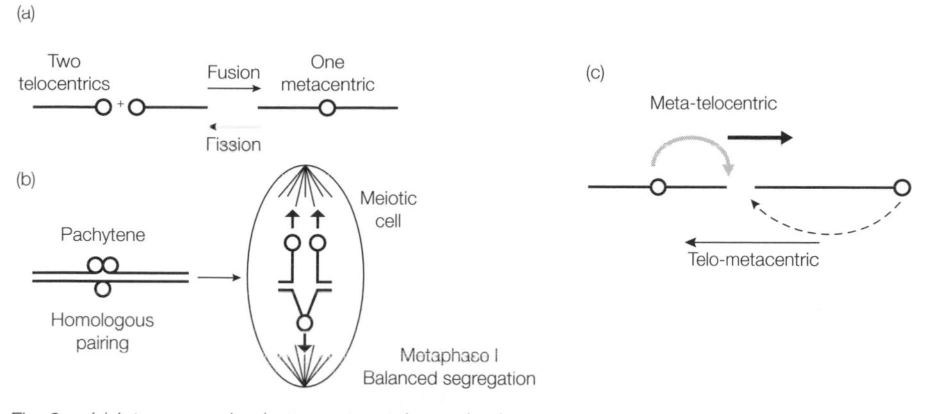

Fig. 2. (a) Interconversion between two telocentric chromosomes and a single metacentric chromosome by centromeric fusion and fission respectively (circles represent centromeres). (b) Meiosis in a heterozygote produces balanced gametes if the two telocentric chromosomes go to the pole opposite the metacentric. (c) Movement of the centromere by pericentric inversion can change the chromosome's shape between telocentric and metacentric.

somes (*Fig. 2a*). This is also called a **Robertsonian fusion** or **Robertsonian** translocation. The short arms are lost. The reverse of fusion is **centric fission** where the centromere of a two-armed metacentric chromosome splits into two and heals, producing two one-armed telocentric chromosomes, increasing the chromosome number (*Fig. 2a*). When heterozygous, the two telocentric chromosomes may pair with the metacentric homolog to make a meiotic trivalent (three chromosomes paired up). If the two telocentrics orientate to the same pole opposite the metacentric (*Fig. 2b*) segregation is normal and the gametes will be balanced, containing either the two telocentric or the single metacentric chromosome. If a telocentric enters the same gamete as the metacentric chromosome (**missegregation or nondisjunction**), that gamete carries two copies of the genes on the telocentric, and will produce a trisomic (one chromosome too many) zygote. The corresponding gametes will have no copies of that telocentric chromosome, and will produce monosomic zygotes (one chromosome missing). In mice and humans, heterozygous Robertsonian fusions are associated with much more sterility than expected from observed nondisjunction, but the reason for this is unknown. Inversion (see below) with a breakpoint near a centromere can convert between metacentric and acrocentric chromosome shapes (*Fig. 2c*). Further centromeric fusion can follow, reducing the chromosome number still further.

Translocations

A translocation is the transfer of material between two nonhomologous chromosomes. This usually involves a reciprocal exchange of the ends of chromosome arms (*Fig. 3a*). Heterozygotes will only produce genetically balanced gametes when both original chromosomes or both rearranged chromosomes pass into the same gamete. Recombination complicates this segregation, but there is one arrangement, **alternate disjunction**, which ensures balanced gametes. This requires metacentric chromosomes, and no chiasmata (i.e. no recombination) between the translocation breakpoint and the centromere. This is favored if the translocation breakpoints were near the centromere, and chiasmata are near the telomeres (*Fig. 3b*). (These conditions may arise by centric fusion as easily as by translocation.) Translocation heterozygotes can form **multivalent rings or**

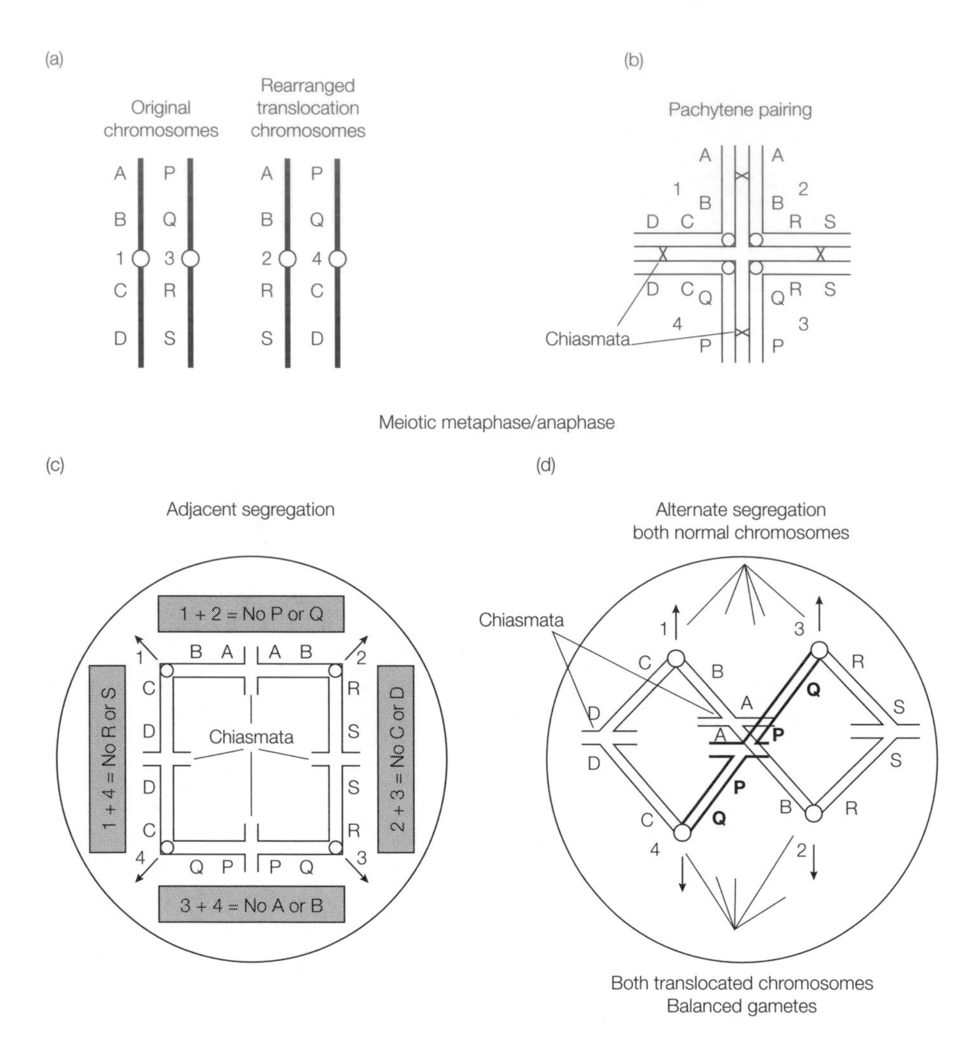

Fig. 3. (a) A translocation involves transfer or exchange between two nonhomologous chromosomes. (b) Two normal and two rearranged chromosomes provide a diploid genome and can pair to form a quadrivalent ring or chain in meiosis. The centromeres are numbered in their sequence in this ring. (c) If any two adjacent chromosomes segregate to the same pole they produce unbalanced gametes with a duplication and a deletion. (d) Segregation of alternate chromosomes to the same pole gives gametes with a copy of all genetic segments ABCD and PQRS.

chains in meiosis. These are closed rings or open chains composed of several chromosomes joined by chiasmata. The translocated chromosomes alternate with their normal homologues, and alternate centromeres must align towards opposite poles for correct segregation (*Fig. 3d*) hence the name alternate disjunction. **Adjacent segregation** (*Fig. 3c*) where any two chromosomes which are side-by-side (adjacent) go to the same pole after meiosis produces gametes with duplications and deletions.

In the evening primroses, genus *Oenothera*, many or all of the chromosomes are involved in translocations, forming multivalents in meiosis. In one species, *O. lamarkia*, six out of the seven pairs of chromosomes are always heterozygous for translocations. All of the rearranged chromosomes go to one pole, the original type to the other (original and recombined are arbitrary), so each set of chromosomes always segregates together, linking most of their genes, which cannot be reassorted by recombination. The two sets of chromosomes may carry alleles for quite different phenotypes, one adapted to a desert habitat, the other to a humid habitat. Each set of chromosomes in *O. lamarkia* also has gametic lethal alleles, so one set dies in egg cells and only survives in pollen, the other only survives in egg cells, and all zygotes (and so all individual plants) are fully heterozygous. In primitive termites, (e.g. genus *Kalotermes*), the X and Y chromosomes are translocated into a chain with most of the autosomes, effectively sex-linking a large part of the genome. This makes genetic relatedness brother-to-brother and sister-to-sister high, greater than between parent and offspring, and encourages the evolution of sociality in termites through kin selection (see Topic D2).

Inversions

An inversion occurs when part of a chromosome is turned around, possibly by recombination across an omega-loop (named after its shape, see *Fig. 4*). Heterozygous inversions can pair in meiosis to form an **inversion loop**. Recombination in an inversion loop produces a duplication of one end of the chromosome and loss of the other, so recombined chromatids do not survive. The exact outcome depends on whether the centromere is inside (**pericentric**) or outside (**paracentric**) the inversion. As with other heterozygosities which may reduce fertility, rare inversions are likely to be selected against. Some species avoid the adverse affects by not pairing the inverted segment. Inversions are **recombination suppressors** because only unrecombined parental chromosome arrangements of the inverted region can be transmitted, allowing genes linked in the inversion to evolve together, differently from homologous genes on the standard chromosome.

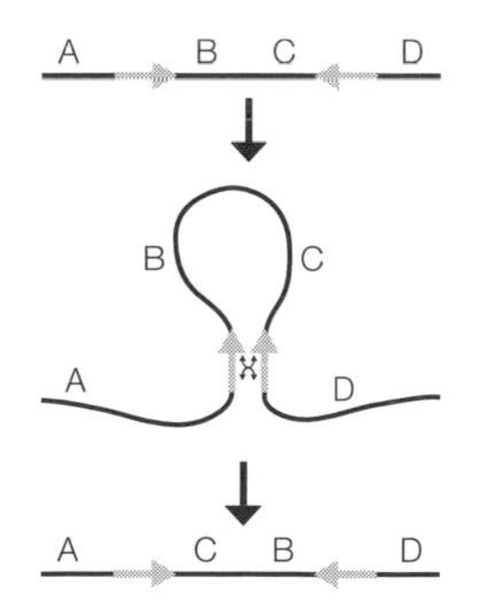

Fig. 4. Production of an inversion by recombination between inverted repeats (gray arrows) in an omega loop configuration

Paracentric inversions

Paracentric inversions have break points on the same side of the centromere (*Fig. 5a*). Meiotic recombination (a chiasma) in an inversion loop (at pachytene, *Fig. 5b*) produces one recombined chromatid with two centromeres (**dicentric**) and another chromatid with no centromeres (**acentric**) (*Fig. 5c–e*). The acentric fragment lags at the first anaphase and is lost, and the dicentric bridge breaks to give incomplete chromosomes. Dipterans are often polymorphic for paracentric inversions, which are important in their ecology and evolution. In females the bridge holds the dicentric into the center of the spindle (*Fig. 5d*) so

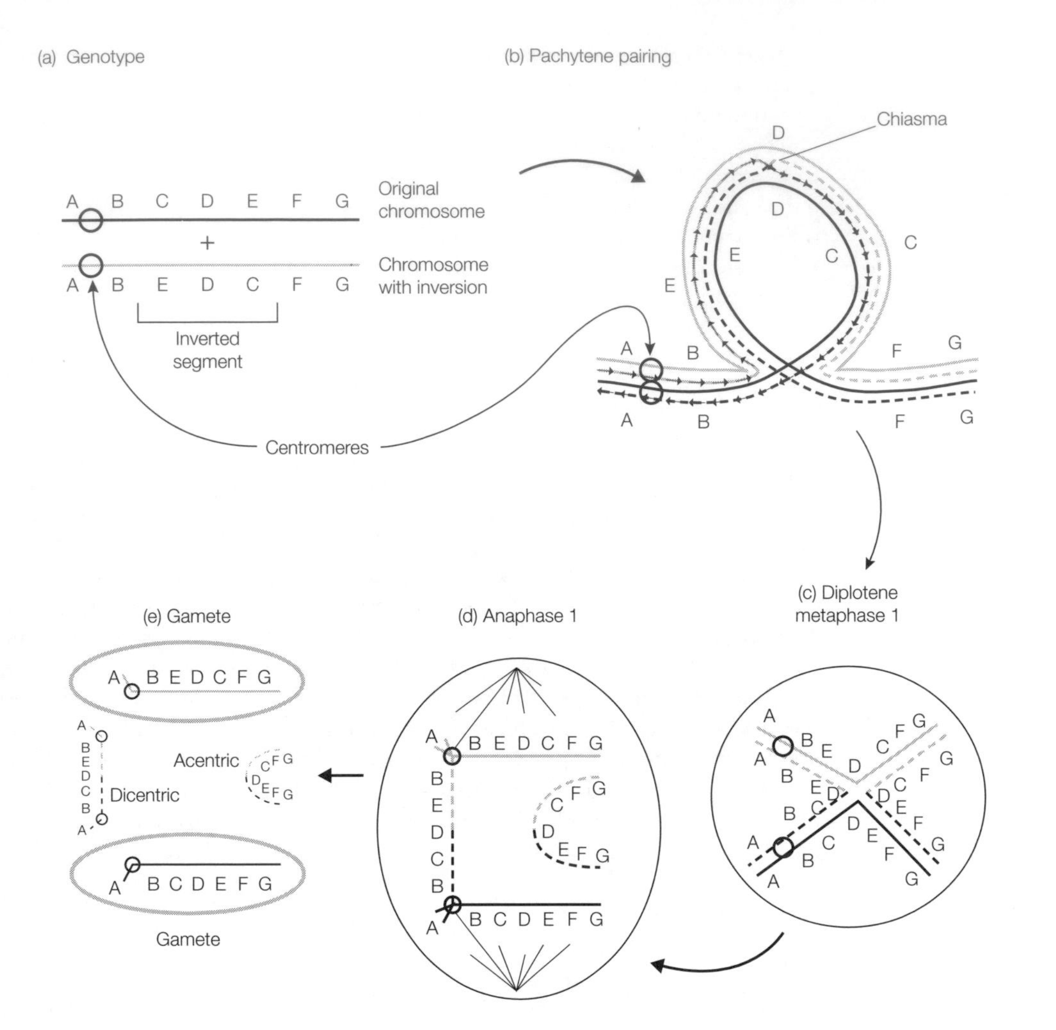

Fig. 5. Effects of recombination in a heterozygous paracentric inversion in a female Drosophila. *(a) The polymorphic chromosome pair: the gray chromosome has the inversion. (b) At pachytene the inverted region pairs to form an inversion loop. There are two chromatids in each chromosome. The chiasma creates a dicentric chromatid ABEDCBA, traced by small arrows, and an acentric fragment GFEDCFG. (c) At diplotene the homologous pairing ceases and a chiasma is visible in the inverted region. (d) At anaphase the dicentric chromatid forms a bridge between the poles and breaks, and the acentric fragment is lost. (e) One of the unrecombined chromatids at the outside of the cell enters the gamete nucleus with a full complement of genes.*

that the genetically correct unrecombined chromatids are at the outer poles. In female meiosis it is one of these outer daughter nuclei which becomes the egg nucleus (gamete; *Fig. 5e*), the other three become polar bodies. This mechanism ensures that the egg is viable with unrecombined chromosomes, all nonviable products are excluded in polar bodies, and fertility of the females is not impaired. Male dipterans generally have no recombination, so they do not generate acentric/dicentric chromatids, their sperm are normal, and so they also escape the ill effects of heterozygous paracentric inversions.

Pericentric inversions

Pericentric inversions span the centromere (*Fig. 6a*). A crossover within the inversion loop swaps the ends of one chromatid on each chromosome (*Fig. 6 b* and *6c*), duplicating one end and deleting the other (as with paracentric inversions). With pericentric inversions, however, all chromatids have a single centromere (*Fig. 6d* and *Fig. 5d*), so they can all segregate normally at meiosis and pass into gametes which can be genetically unbalanced in either sex, and so produce no viable zygotes. Polymorphisms for pericentric inversions are not found in *Drosophila* but are found in other species, for example where the inverted region pairs nonhomologously in meiosis, preventing crossing-over (e.g. in the grasshopper *Keyacris scurra*).

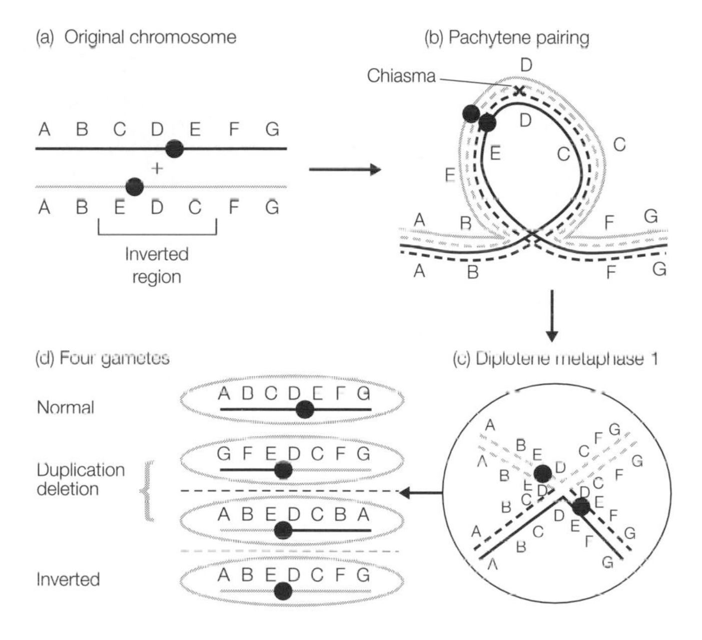

Fig. 6. Effects of recombination in a heterozygous pericentric inversion (a). This is similar to the paracentric inversion (Fig. 5) but the centromeres are now within the inversion loop at pachytene (b). A crossover in the inversion gives duplications and deletions (c) but all chromatids now have a single centromere, so the duplication and deletion chromatids can be incorporated into gametes (d).

Changes in sex chromosomes

The chromosome only found in the heterogametic sex (e.g. the Y chromosome in humans and *Drosophila*) has few genes on it, and in some species it is lost, giving XO males and XX females and a reduction in the number of sex chromosomes. Fusions or translocations between sex chromosomes and autosomes can also increase the number of sex chromosomes by converting autosomes into neo-sex chromosomes. These are forced to continue to exist in the free state to balance segregation of the homologous copy attached to a sex chromosome and to maintain diploidy. For example, when an X chromosome fuses with an autosome, that autosome becomes part of the X. This is balanced in females with two X chromosomes plus attached autosome, but in males, one copy of the autosome will be attached to the X, the second will be free, only found in males, and behave just like a Y chromosome (*Fig. 7a*). It is known as a **neo-Y chromosome**, and is found after an X–autosome fusion. If a Y already exists they will be numberer Y_1 and Y_2. If a Y

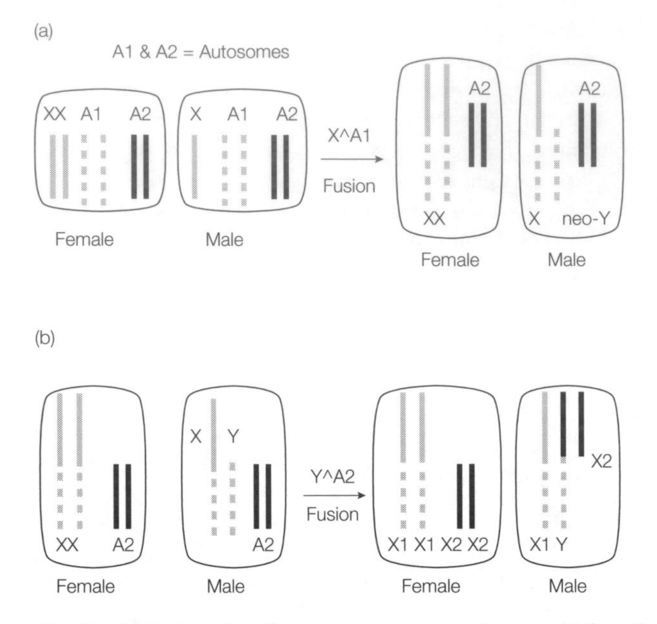

Fig. 7. (a) Fusion of an X chromosome to an autosome A1 (creating A1^X) causes the free copy of the autosome to behave as a neo-Y chromosome only found in males, as a second copy to the one attached to the X. (b) Fusion of a Y chromosome to an autosome A2 (creating A2^Y) causes the free copy of A2 to behave like an X-chromosome (called X_2), with two copies in females, but only one in males to balance the single copy attached to the Y.

chromosome fuses to an autosome (*Fig. 7b*), the male will still have one free autosome, and the female will have two free autosomes. These are called **neo-X chromosomes** and form as a result of a Y-autosome fusion. In this way, multiple X and Y chromosomes can evolve from autosomes. Multiple sex chromosomes are numbered X_1, X_2, X_3 and Y_1, Y_2 and so on.

Evolutionary effects

Dipterans (two-winged flies) have no recombination in males, an adaptation which avoids fertility loss from paracentric inversions. The females also avoid passing recombined chromatids with duplications and deletions into eggs, so heterozygous paracentric inversions can persist in dipterans, where they may be common. Suppression of recombination between the inverted and normally orientated chromosome can allow coevolution of the genes locked together in the inversion. They may become adapted to a different habitat (e.g. hotter or colder, higher or lower altitude) than the alleles on the other chromosome. Particular paracentric inversions in *Drosophila* species are associated with sex determination, taking the role of sex chromosomes. Many populations have stable polymorphisms for inversions, which are said to be 'floating' (as opposed to 'fixed' when only one chromosome type is found in a population or species). There are about 2000 species of *Drosophila*, and their evolution is estimated to have seen the evolution of some 350 million paracentric inversions, of which 20 000–60 000 have become fixed, and 18 000–28 000 are still floating. Different types of chromosomal rearrangements are common in different evolutionary lines. In primate lineages, great apes including humans have mostly pericentric inversions, lemurs mainly have centric fusions, and *Cercopithidae* have many fissions.

D7 Species and speciation

Key Notes

Species

The genetic definition of a species is a set of organisms actually or potentially interbreeding, and sharing a common gene pool. Most species are clearly different (domestic cats and dogs) but two closely related species may appear identical to humans (they are hidden or cryptic), and members of a single species may be quite different. Taxonomy is the scheme or science used to classify species. Species which share a more recent common ancestor should be classified closer together than distantly related species. A morphological species may be defined on the basis of appearance or shape as a group of organisms sharing a unique set of morphological characters (this is the only technique that can be applied to fossils). An ecological species can be defined as a group of (interbreeding) organisms utilizing a unique resource–habitat combination (a niche). These definitions are complementary, and should not conflict.

Species isolation mechanisms

Isolation mechanisms prevent species interbreeding or exchanging genetic material (gene flow). This is the most important requirement for speciation, because populations can only diverge into species if they stop exchanging genes. Gene flow dilutes the differences. Premating isolation mechanisms are most efficient because they avoid wasting resources. These are: (i) behavioral isolation (mate choice, sexual, ethological); (ii) spatial (mating in different places, ecological or habitat isolation); (iii) temporal isolation (mating at different times); (iv) mechanical isolation (mating physically impossible); (v) pollinator isolation (some plants); or (vi) gametic isolation (restricted attraction and/or fusion between gametes).

Postmating isolation mechanisms are inefficient because mating produces unfit or inviable offspring, wasting resources and reducing successful reproduction. They are not compatible with extended cohabitation (sympatry). They are: (i) hybrid inviability; (ii) hybrid sterility; and (iii) hybrid breakdown (where second generation hybrids and backcross progeny are unfit).

Speciation

Speciation is the process by which one species splits into two or more. Study of this splitting is called cladistics, and the tree of descent of related species is called a phylogeny. (The science is phylogenics.) Speciation requires that populations are isolated so that they do not interbreed to exchange genetic information, and their genomes can then diverge. This isolation can happen either because they mate (or live) in different places (geographic isolation or allopatry), or they mate (or live) in different habitats in one geographic area (habitat isolation in sympatry). The physical separation required depends on the motility of the organism, and may be small for sedentary species. Divergence occurs by accumulated genetic differences which may prevent successful interbreeding when these populations establish secondary contact. It will be faster if the populations experience different selection pressure. When they develop premating isolation mechanisms they are separate species.

Secondary contact

The test of speciation occurs when two divergent populations meet and mix following range expansion because of habitat change. The outcome depends on the extent of development of pre- and postmating isolation and ecological (niche) divergence. If hybrids are fit the two populations merge into a hybrid swarm. If there is some hybrid unfitness for genetic or ecological reasons then there will be selection to reinforce premating isolation. If, however, limited interbreeding and low postmating isolation transmits some genes between them (gene flow) then divergence may be reduced or prevented and a permanent hybrid zone can be established between parapatric (side-by-side) populations. If there is little or no gene flow because hybrids are either inviable or sterile (strong postmating isolation), divergence can continue until the populations become isolated species. If premating isolation exists or arises after contact between the populations, then individuals of the two populations can intermingle as different species, but if they still utilize the same ecological niche, the more efficient species will replace the other.

Phylogenic patterns

There is a complete gradation between races (no isolation), subspecies (postmating isolation), semi-species (some premating isolation) to sibling species which look alike but do not interbreed. A single species may change through time. The succession appears in the fossil record as a series of chronospecies. Phylogenic algorithms always produce a dichotomous pattern of species splitting, but in reality climate change may divide a widespread species into many isolated refugia, each population a potential new species. Expansion of a few species after an environmental catastrophe may also have an enormous adaptive radiation with populations diverging as they enter new habitats free from competition from adapted species (e.g. mammalian radiation after the extinction of the dinosaurs).

Related topic Evolution (D9)

Species

The **genetic definition of a species** is a set of actually or potentially interbreeding organisms. Members of a species share a common **gene pool** with all other members of that species. Members of different species do not exchange genetic material. There are clear differences in most cases (e.g. between cats and dogs) however in many species there are geographic races which look different but which can interbreed if they come into contact. This makes it difficult to define species, and causes conservation problems, e.g. Mallard drakes from Europe mate with Canadian Black Duck females, and Ruddy Duck from America mate with European White Headed Duck, in both cases destroying the native species by hybridization. Conversely, similar organisms which do not appear so noticeably different to humans, (e.g. 'small brown birds', small mammals or insects), may be classified as single species. One named species may contain many **hidden (cryptic) species** which we simply do not notice. The commonest British bat, *Pipistrellus pipistrellus*, has recently been discovered to include two species, first differentiated by the use of electronic bat detectors which indicated two types of ultrasonic call.

Taxonomy is a scheme of **classification** for defining all species and grouping related species together. This should ideally group together species which share a common ancestral species. The first step is usually based on a **morphological species** concept. All members of a species share a set of physical characteristics which can distinguish them from members of other species. This is the only criterion which can be applied to fossils. It has problems with very variable species, in severe cases even males and females may be described as different species. The reverse problem arises with very similar species such as sheep and goats whose skeletons cannot be distinguished unambiguously. There is also an **ecological species** concept. Each species uses a particular set of resources, a **niche**, from their environment, and they do this most efficiently in a particular set of conditions, or **habitat**. Two species cannot share the same niche in the same geographic location, because the more efficient will succeed, the other will become extinct. These three concepts of species are complementary, and generally in agreement except for variable or closely related species.

Species isolation mechanisms

Members of different species are prevented from interbreeding or exchanging genetic material by **isolation mechanisms**. These must evolve in the process of speciation. The list of isolation mechanisms below is based on an original list by Dobzhansky.

Premating isolation mechanisms

Premating isolation mechanisms prevent interbreeding and are efficient because resources and effort in reproduction are not wasted producing inviable or sterile offspring.

(i) **Behavioral (sexual, ethological) isolation**. Individuals choose members of their own species and reject other species as mates. In species with external fertilization this behavior is expressed by the gametes.

(ii) **Spatial (geographic, ecological or habitat) isolation**. Individuals of different species do not meet when mating because they mate in different locations. The separation must be considered on the scale of the motility of the organism, and includes different continents, ecosystems, food plants, or different dryness, acidity or temperature preferences in the habitat. The effects are identical whether the cause is genetically determined selection of habitat or passive geographical separation of isolated populations on different islands, lakes, mountains, etc. Sedentary insects on one tree may be effectively isolated from a population on a similar tree 100 m away.

(iii) **Temporal isolation**. Different species mate or flower at different times of the day or year.

(iv) **Mechanical isolation**. The genitalia or flower parts prevent copulation or pollen transfer respectively.

(v) **Isolation by different pollinators**. Related plants species may attract different pollinators.

(vi) **Gametic isolation**. In externally fertilized species gametes show selective attraction and/or acceptance. Sperm may only swim towards eggs of their own species. In plants, styles may prevent entry of incompatible gametophytes. In internally fertilized species the male gametes may not survive in the sex ducts of the other species.

Postmating isolation mechanisms

Postmating isolation mechanisms reduce the viability and/or fertility of hybrids produced by matings between members of two species. This is very inefficient because reproductive effort is wasted producing useless offspring. This prevents the two species coexisting in the same habitat (prevents sympatry) except as migrants, because the rarer species predominantly meets the other (wrong) species when looking for a mate, and so produces mostly low fitness heterozygous hybrid offspring. The rarer species therefore faces greatest selection (see Topic D3).

(i) **Hybrid inviability**. Hybrid zygotes are inviable or have reduced viability. This could be because the products of genes from the two parental species do not interact together correctly.
(ii) **Hybrid sterility**. One or both sexes of the F1 hybrids are sterile or have reduced fertility. This can be caused by sequence changes from random mutations. When *Saccharomyces* DNA sequences diverge by more than approximately 0.6% (which probably takes about 300 000 years isolation) mismatch repair systems start to prevent meiotic synapsis because they do not identify chromosomes as being homologous. If pairing fails, the chromosomes do not segregate properly into gametes and aneuploidy is caused. More rapid isolation can occur following fixation of chromosomal rearrangements in one of the isolated populations. These can also cause sterility in heterozygous hybrids because of failure of chromosomes to segregate properly in meiosis (see Topics D6 and C3).
(iii) **Hybrid breakdown**. F_2 or backcross hybrids are inviable, sterile, or have reduced viability or fertility. This occurs because F1 individuals with one complete set of chromosomes from each parent are viable, but after meiotic assortment the gametes do not carry a full genome. This will occur if there are small translocations, so some genes are no longer on homologous chromosomes (see Topic D6).

Speciation

Speciation is the process of forming two or more species from a single species. Study of this splitting arrangement is termed **cladistics**. The 'family tree' of descent of related species is a **phylogeny**, and study of this subject is called **phylogenics**. Separate species can only form if: (i) gene flow between populations of one species is restricted to allow them to diverge, and then (ii) a component of the divergence produces a complete barrier to interbreeding, called a mating isolation mechanism. The initial barrier can be either spatial or temporal. Changing climate may isolate populations on mountain peaks or in remnant lakes, or rare migrants may colonize new areas such as the Galapagos and Hawaiian islands (*Fig. 1*). Geographic isolation is termed **allopatry** and gives rise to **allopatric speciation**.

Disruptive selection (see Topic D2) in a polymorphic species may select for two species to form if the intermediate hybrid is less fit than its parents. If the two types continue to interbreed, however, it is difficult to see how they can diverge sufficiently for two species to arise.

Sometimes selection may produce ecotypes living in adjacent habitats in the same geographic area. This is termed **sympatry**. There are some suggested cases of sympatric populations evolving into separate species (e.g. fish in the same lake and insects living on different species of food plants in the same habitat). This is called **sympatric speciation**, however it is increasingly being discovered that the patches of habitats involved may be so distant compared with the motility or dispersal of the organisms when mating that they are effectively micro-geographically isolated. These isolated populations are incipient species.

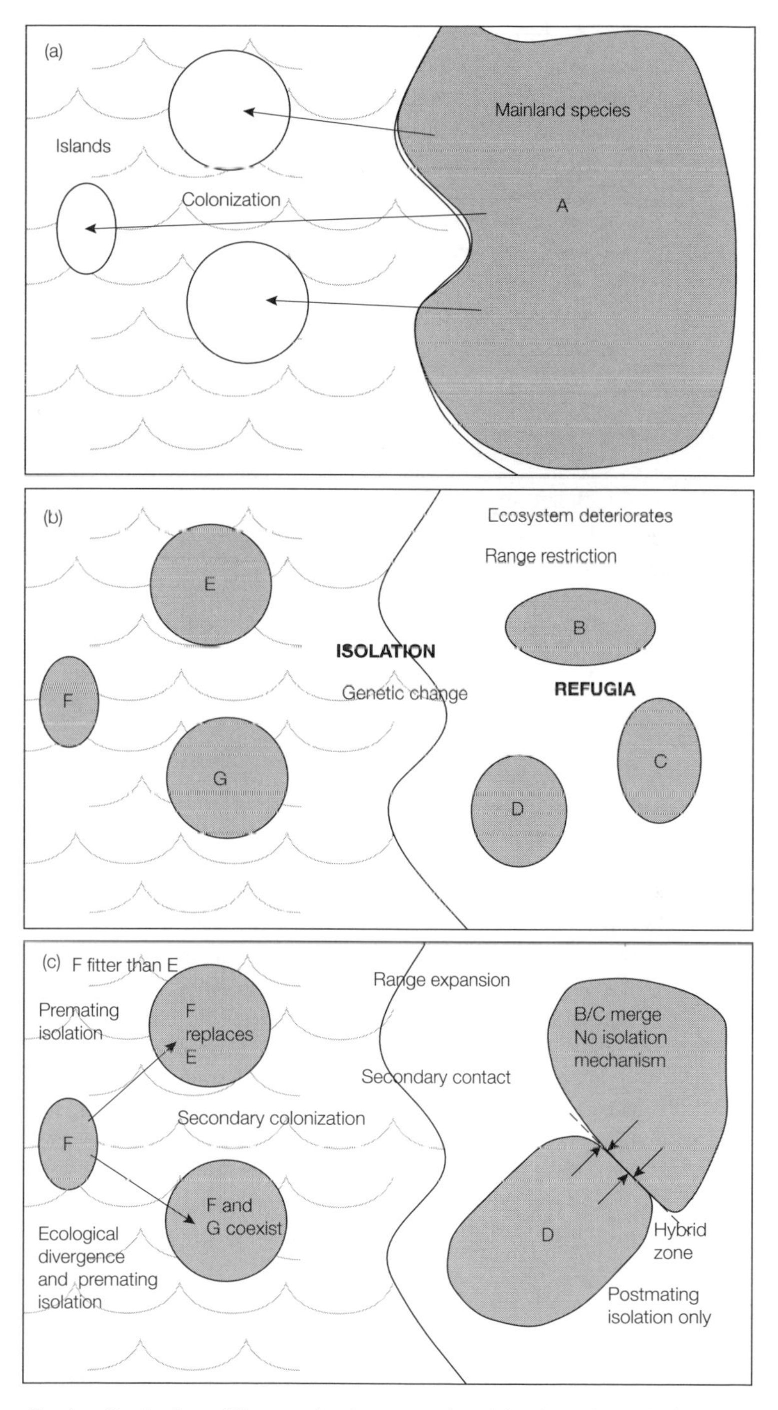

Fig. 1. Production of five species from one. A mainland species colonizes three offshore islands (a). Climatic change separates the mainland population into three refugia, and the six isolated populations diverge (b). On future expansion and secondary contact (c) the outcome depends on mating isolation and ecological divergence. B and C have no divergence and merge. B/C have post mating isolation from D, so B/C and D meet at a hybrid zone. F has premating isolation from E and G. It uses the same niche as E and is more efficient, so F replaces E. Ecological differences have evolved between F and G so they can coexist.

Random genetic change by mutation and drift over a long period (probably exceeding about 300 000 years) will produce enough divergence in the DNA for mismatch detection in meiosis in hybrids to prevent recombination and chiasmata. The resulting nondisjunction will produce inviable gametes and cause hybrid infertility. This makes speciation inevitable if populations are separated for long enough. Divergent selection may produce more rapid changes increasing the effect. If two populations evolve adaptations to different environments then interbreeding may produce poorly adapted hybrids. There is then selection for premating isolation mechanisms to prevent interbreeding. This can only happen if the postmating isolation mechanisms are strong enough for selection to increase the premating isolation mechanisms faster than gene flow through hybrids dilutes the necessary differences. Selection for premating isolation also occurs when there is any postmating isolation for whatever reason (usually genetic differences).

Secondary contact

When conditions change, the isolated and diverged populations may meet again; this is termed **secondary contact** at a **contact zone**. The outcome then depends upon the degree to which postmating isolation, premating isolation and ecological (niche) divergence has arisen while they were isolated. If they interbreed freely and hybrids have normal fitness, the populations will merge, producing a **hybrid swarm** (*Fig. 1c*, populations B and C). If interbreeding produces unfit hybrids, selection will act to **reinforce** premating isolation mechanisms, because any individual which correctly mates with its own type will gain an advantage. Those individuals mating incorrectly will produce unfit hybrid offspring. The locally rarer type has a disadvantage because it is more likely to mate mistakenly with the (wrong) majority type and produce low fitness offspring.

If postmating isolation is weak, and genetic exchange (gene flow) through hybrids occurs, the genetic differences will be diluted rather than reinforced at the contact zone, and divergence cannot occur. Premating isolation will not develop, but the two populations will live **parapatrically**, side-by-side. They will meet in a **hybrid zone** in a region of low population density (i.e. poor habitat), supported by immigration from both populations (*Fig. 1c*, D vs. B/C). Such hybrid zones will be stable for thousands of years if the environment is stable. Genetic loci which do not produce hybrid unfitness will be exchanged between populations, but genes which are unfit in hybrids cannot cross the barrier. If hybrid inviability or sterility is effectively complete there will be no gene flow, and divergence can continue to reinforce premating isolation. When premating isolation is effective, interbreeding stops, and populations which are ecologically different can coexist in one geographic region (*Fig. 1c*, F and G), however if they still use the same niche then the more efficient species will displace the other (*Fig. 1c*, F and E).

Phylogenic patterns

In nature there is a complete gradation between **geographic races** with no reproductive isolation, through **subspecies** with partial hybrid sterility and **semispecies** with some premating isolation, to **sibling species** which appear to be very similar, but which do not interbreed. Note that quite different species may appear identical to humans, while races which interbreed freely may appear to be very different.

Speciation need not be a dichotomous event, one species splitting to two then four, etc. All the algorithms used in phylogenics will artificially enforce a dichotomous series of events, but this may give a very misleading impression. It is likely that unspecialized species with a wide geographic and habitat distribution will survive in many isolated **refugia** during adverse conditions. For example, the ice ages caused ice caps on mountain ranges and separated the populations in adjacent valleys on either side. Each isolated population then started to diverge from the others, becoming an incipient species ready to spread when the climate improved. When a catastrophe destroys many species (e.g. a comet strike killing off the dinosaurs), descendants of surviving species can colonize many new habitats during an **adaptive radiation** when conditions improve. The populations isolated in each newly occupied habitat may simultaneously diverge into many new species. This produces clusters of new sibling species in a very short geological time and confuses attempts to create phylogenies.

As a single species lineage changes through time, it will produce a series of **chronospecies** in the fossil record, but identification of these is arbitrary, and often influenced by the patchy availability of fossils.

D8 POLYPLOIDY

Key Notes

Introduction

Polyploidy is the state of having more than two complete sets of chromosomes. The base haploid number is x, diploids have a diploid number $(2n) = 2x$, hexaploids have $2n = 6x$ and a haploid number $n = 3x$. Polyploidy can cause problems in meiotic segregation if the multiple copies of chromosomes do not divide evenly. Odd-number polyploids are sterile because they cannot divide their chromosomes evenly into two cells at meiosis (it is impossible to divide three or five exactly by two). Polyploidy is rare in animals, but 47% of flowering plants (angiosperms) are polyploid. Vegetative reproduction helps plants through the initial sterile period, and they are less severely affected than animals by chromosomal imbalance. Polyploidy is usually lethal in animals but may have happened in evolutionary history. There are many parthenogenic polyploid animal species whose reproductive cycle avoids sterility. The sudden change in chromosome number and consequent hybrid sterility caused by polyploidy can act as postmating isolation mechanisms, causing instant speciation.

Autopolyploids

Autopolyploidy is caused when the chromosome number doubles in an individual (e.g. mitotic separation fails). They are usually infertile because there are more than two copies of each chromosome per cell and they may not pair and segregate evenly in meiosis.

Allopolyploids

Allopolyploids are formed from hybrids between closely related species. The different sets of chromosomes are sufficiently different in DNA sequence not to pair between sets in meiosis. The chromosome number doubles accidentally by failed mitosis so that each cell has two complete sets of chromosomes from each parent, and these pair like with like, restoring fertility.

Introgression

Crosses between species produce a hybrid with a higher level of polyploidy than the parent species because of the doubling required to restore fertility. Genomes introgress from low ploidy species into higher ploidy species, but chromosomes from high polyploids cannot return to diploids.

Polyploid complexes

A cluster of related species (e.g. grasses and magnolias) can hybridize in many combinations producing many polyploid species. Further hybridization leads to more polyploids with higher ploidy, and the eventual loss of diploid species. Eventually a few highly polyploid species remain with no apparent related diploid species.

Characteristics of polyploids

Polyploids tend to be large with large cells. This makes them attractive as food, with a soft texture and high yield. They also produce large flowers. The sterility of triploids is used to produce seedless fruit. The multiple genomes in polyploids tend to stabilize the genotype and phenotype because segregation of extreme genotypes is rare.

Related topics

Chromosomes (B1)
Chromosome changes in evolution (D6)
Species and speciation (D7)

Introduction

Polyploidy is the state of having more than two complete sets of chromosomes. (It can be confused with aneuploidy which is an incorrect number of particular chromosomes.) Species with three sets are called triploid, four sets tetraploid, six sets hexaploid and so on. To avoid confusion, the base (original haploid) number may be called x, and the current haploid chromosome number n, so a hexaploid which behaves as a diploid has a diploid number $2n = 6x$, haploid number $n = 3x$.

Polyploid cells have difficulty in meiosis unless they can behave as diploids, so that each gamete gets a balanced set of chromosomes. This requires chromosomes to pair in twos at meiosis. Only even-number polyploids are fertile, because odd numbers of chromosomes cannot be divided in two at meiotic reduction division. Polyploidy is usually lethal to animals. The few polyploid animal species are hermaphrodite (e.g. some earthworms and planarians) or parthenogenic (e.g. some beetles, moths, crustaceans, fish and salamanders). Parthenogenic animals usually double their chromosome number by replication without division just before meiosis, then pair identical sister chromosomes, and separate them in meiosis to restore the original karyotype. This avoids problems of segregation.

Plant genomes are much more tolerant of changes in chromosome number and 47% of all flowering plants are polyploid. Polyploidy is important as a speciation mechanism in plants because it can prevent interbreeding in a single step. A new polyploid can only produce fertile offspring if its gametes fuse with a gamete of the same ploidy. For example, a new tetraploid produces $2x$ gametes. If these fuse with a haploid gamete from a diploid plant the offspring will be sterile triploids. Fertility is restored by a chance failure of mitotic division that creates a hexaploid cell. Many plants are able to reproduce vegetatively which may allow them to survive the sterile phase until fortuitous doubling occurs. They also have the advantage of not having a differentiated germ line. Any dividing cell whose descendent clone could become a flowering shoot and produce gametophytes could undergo chromosome doubling; its seeds would produce fertile polyploids.

There are claims that polyploidy (polyploidization) has happened in the vertebrate lineage because many genes have been duplicated and diverged within the genome. Others claim that this cannot be distinguished from numerous small duplications that are known to have occurred (over 50 are apparent in the small genome of the yeast *S. cerevisiae*). Small duplications avoid any deleterious effects of sudden polyploidy. The question will only be answered persuasively when extensive genome sequencing reveals the size and number of ancient duplications.

Autopolyploids

Autopolyploids have double the normal chromosome number, but all the chromosomes come from the same species, often the same individual. This can arise by a failure of mitosis or from a diploid gamete produced by a failure of the second meiotic division. Autopolyploids are usually sterile because the three or more homologous chromosomes will not form bivalents (pairs) at meiosis but rather multivalents (three or more synapsed chromosomes). This does not lead to even reduction at division, and gametes will not contain complete sets of chromosomes. Experimentally produced autopolyploids are always less fit than their diploid parents.

Allopolyploids

Allopolyploids have a hybrid origin. They are produced by fusion of gametes from related species. This is common in plants. The two sets of chromosomes in the diploid hybrid may be sufficiently different from each other that they cannot pair correctly at meiosis. This prevents them from consistently orientating towards different poles at the first meiotic division and segregation may be random. Gametes then contain too many copies of some chromosomes, and not enough copies of others, resulting in gamete failure or the formation of inviable zygotes. An accident of mitosis failing to separate the daughter mitotic nuclei can double the chromosome number in that cell, forming an allotetraploid. Each chromosome can now pair with its own duplicate, so the allotetraploid behaves like a normal diploid, but the gametes are now $2x$ 'diploid'. This can be seen in the reconstructed evolution of wheat (*Fig. 1*). Two species, *Triticum monococcum* (*AA* genomes) and *Aegilops speltiodes* (*BB* genomes) each with seven pairs of chromosomes ($x = 7$) formed a hybrid (*AB*). This doubled its chromosome number to 14 pairs ($2n = 28$) becoming tetraploid *Triticum duococcum* (*AABB* genomes). Now the two *A* sets could pair together and the two *B* sets could pair together, producing balanced *AB* gametes. This specific homologous pairing *A–A* and *B–B* (rather than *A–B* pairing, which is termed homoeologous) requires activity of a particular genetic locus (called *Ph* for pairing homology) on the long arm of chromosome 5B. This activity is only found in cultivated wheats, it has never been found in its wild ancestor. Later, an *AB* gamete (14 chromosomes) from *T. duococcum* ($2n = 28$) fused with a haploid (*D* genome) gamete from *Aegilops squarrosa* ($2n = 14$) to give a triploid hybrid (*ABD*) which doubled up its chromosomes to give the fertile hexaploid *AABBDD* ($2n = 42$) *Triticum aestivum*. This produced bread wheat, one of the most important human food plants. It contains two complete genomes from each of its three diploid ancestors.

Introgression

Introgression is the movement of genes from one race or species into another. It is possible for hybrids to arise between species with different levels of ploidy, but the outcome is usually an overall increase in ploidy because the chromosome number doubles to restore fertility. This causes introgression because genes from the low ploidy species enter, or introgress into, the higher ploidy species, but the process cannot simply reverse. Genes from high ploidy species cannot return to low ploidy species.

Polyploid complexes

The ploidy in a group of hybridizing species steadily increases because each new hybrid doubles its chromosome number to restore fertility. Such a group is called a polyploid complex. The *Magnoliaceae* provide a good example (*Fig. 2*). Diploid species exist with haploid numbers from seven to 10 chromosomes (7–10 pairs), and extinct species are thought to have had five and six pairs. The tetraploid

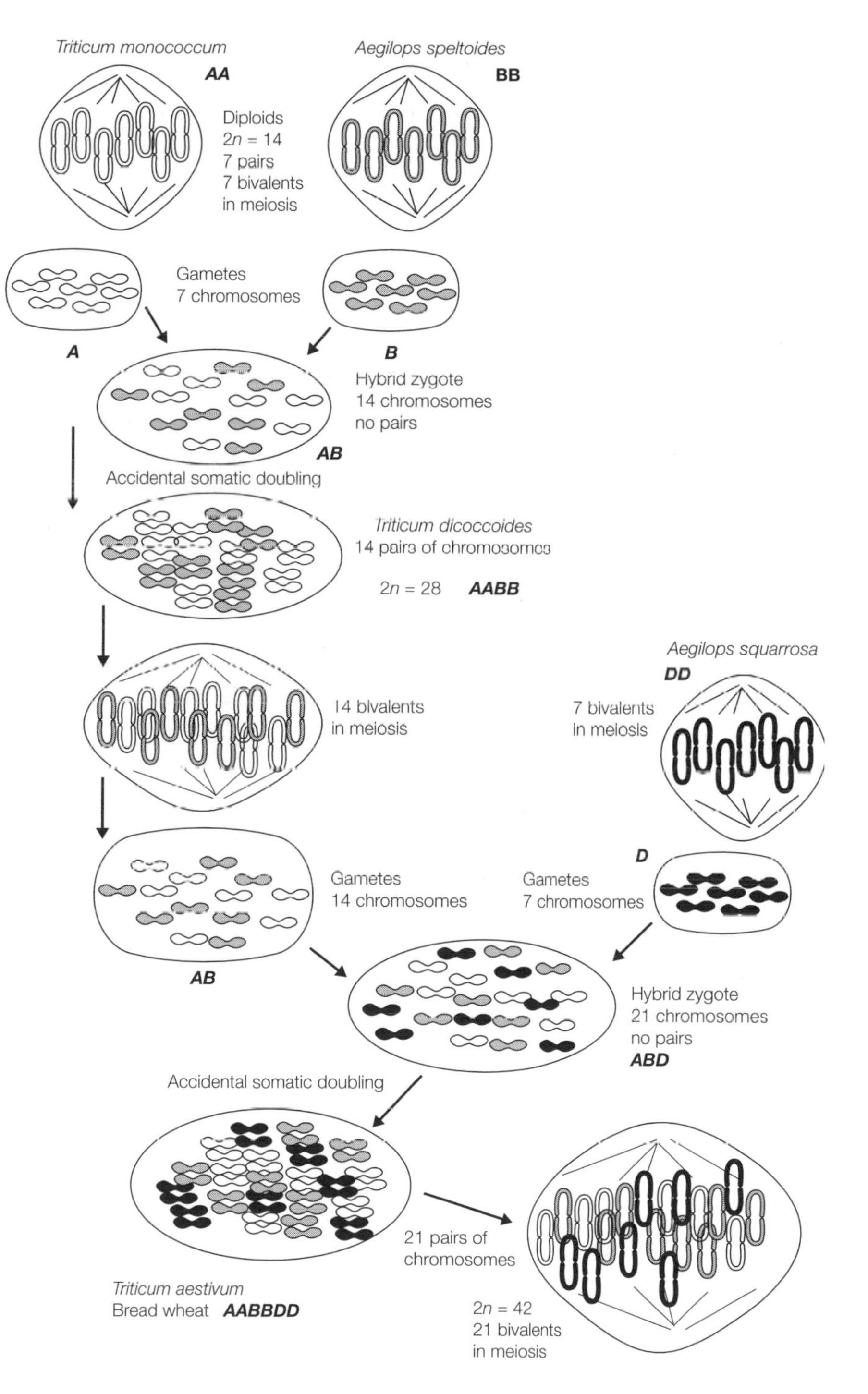

Fig. 1. The evolution of hexaploid bread wheat T. aestivum *by repeated hybridization and polyploidization.*

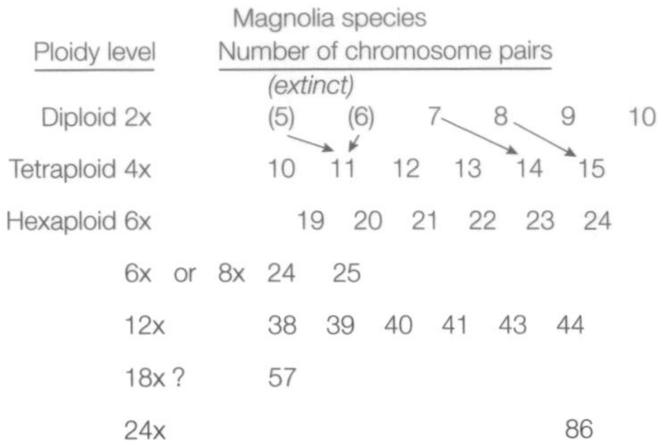

Ploidy level	Magnolia species Number of chromosome pairs
	(extinct)
Diploid 2x	(5) (6) 7 8 9 10
Tetraploid 4x	10 11 12 13 14 15
Hexaploid 6x	19 20 21 22 23 24
6x or 8x	24 25
12x	38 39 40 41 43 44
18x ?	57
24x	86

Fig. 2. The magnolia polyploid complex showing chromosome numbers. Numbers in brackets are for species that have not been found, presumed extinct. Arrows suggest two possible parental combinations to produce 11 and 15 chromosome pairs in hybrid offspring.

species have 10 to 15 pairs of chromosomes. There are several ways these could have formed. The 12-pair species could combine two six-pair genomes or a five-pair and a seven-pair genome. The hexaploid species have between 19 and 24 pairs of chromosomes. There are also presumed octaploids with 24 and 25 pairs. Note that 24 pairs could arise as tetraploids: $4 \times 6 = 24$, or by different triploid routes: $3 \times 8 = 24$, or $(7 + 8 = 15) + 9 = 24$. The 12-ploid (duodecaploid) species have 38, 39, 40, 41, 43 or 44 pairs of chromosomes, and there are two species with 57 pairs (perhaps 18-ploid) and 86 chromosome pairs (24-ploid).

Polyploid complexes start with many diploid species and a few polyploids. The polyploid species spread and hybridization is common. As the complex matures after 500 000 to 10 000 000 years the diploid species become rarer, eventually only surviving in isolated locations free from hybridization, or as extreme phenotypes. The hybrid complex is then said to decline. The original diploids become extinct or differentiated beyond recognition, and eventually only a few highly polyploid species remain with no obvious relatives.

Characteristics of polyploids

Polyploids tend to have larger cells. This makes them larger and slower maturing. Most garden flowers and most crop plants are polyploid. The larger cells improve the texture of fruit and vegetables (e.g. strawberries, apples, potatoes). They have more cell contents and less cell wall. The larger size also gives higher yields (e.g. wheat, cotton). The sterility caused by triploidy is useful to produce seed-free fruit that is easier to eat (e.g. bananas) or better tasting (e.g. less bitter cucumbers). Triploid F1 can be produced by crossing a tetraploid and a diploid. This is useful for sugar beet where seeds are not wanted, but is impossible for inbreeding species or species where the seed is the crop (e.g. wheat). Triploid bananas are propagated vegetatively, but the sterility causes a problem when trying to breed improved varieties.

Polyploidy **stabilizes the genome** because extreme genotypes segregate out much more rarely. A heterozygous tetraploid *AAaa* will produce gametes in the ratio 1/6 *AA* : 4/6 *Aa* : 1/6 *aa*. The next generation genotype ratio will be: 1/36 *AAAA* : 8/36 *AAAa* : 18/36 *AAaa* : 8/36 *Aaaa* : 1/36 *aaaa*.

Only 1/18 of the progeny are extreme genotype, compared with 1/2 for a diploid monohybrid cross. If a new hybrid polyploid happens to arise in an environment to which its genotype is well adapted its progeny will tend to maintain the adaptation. Polyploids, however, may be slow to evolve because they cannot respond rapidly to selection.

D9 EVOLUTION

Key Notes

Evolution by divergence

Evolution occurs by changes in allele frequencies over time. Species form when populations diverge genetically to the extent that they cannot interbreed. They then continue to evolve and diverge independently. Degree of divergence for unselected characters is a measure of time since separation. Shared characteristics which evolved in a common ancestor are said to be homologous (e.g. mammals back legs). Structures that have evolved similar functions separately are analogous (e.g. bat's wings and bird's wings). Evolving to be more similar from different starting points is called convergent evolution (e.g. whales and fish). Related species share homologous structures from their common ancestors, but do not share characters arising after separation.

Populations

Differences between populations within a species can show patterns of migration and colonization, and indicate degrees of divergence preceding speciation. Divergence may be selected. Dark skin in humans from equatorial regions may reduce damage from sunlight, and pale skin aids photosynthesis of vitamin D by sunlight in northern lands. There are large differences in blood group frequencies in different human populations, which may be due to selection by disease or drift and founder effect in small colonizing tribes.

Ring species

A ring species has an extended continuous range around an obstacle, but the populations at the extreme ends of the range are sufficiently diverged to be different species where they meet. An example is a boreal gull whose global range overlaps in Europe as the herring gull *Laurus argentatus* and the lesser black backed gull *L. fuscus*. This shows that distance alone is an effective isolating mechanism.

Molecular clocks

The rate at which amino acid changing mutations accumulate in genes for specific proteins tends to be constant over time. Similarly, changes in ribosomal RNA sequence, synonymous changes in coding sequences (silent mutations) and changes in noncoding sequences all have particular rates. These can all be used as molecular clocks, faster diverging sequences for more recent events. The degree of divergence between two species reflects the duration in time of their independent evolution. There is debate about whether the clock rate is faster at times of rapid evolution (adaptive radiations) and slower during stasis.

Phylogenics

Arranging species in order of increasing divergence gives a phylogenic tree which represents evolutionary history. Species within a group are compared with a distantly related outgroup to provide an ancestral root to the tree. Animal mitochondrial DNA evolves rapidly and does not recombine, so is excellent for relatively recent divergences. Human mitochondrial DNA suggests that all humans are descended by the female line from one woman who lived in Africa 140 000–290 000 years ago. Conserved proteins (e.g. cytochrome *c*) and ribosomal RNA are useful for studying the whole period of life on earth. Phylogenies reveal discrepancies in evolutionary rates in particular lineages. In addition, the algorithms used to construct them artificially force dichotomous branching, which is only necessarily true for mitochondrial DNA and Y chromosomes which cannot recombine.

Human evolution

All current human populations are the same species. Humans are clearly mammals, and are closely related to chimpanzees and gorillas. Hominids separated from the apes about 4 million years ago. Bipedalism was an early development, freeing hands to use tools and weapons for hunting and defence, and seems to have been a key evolutionary step. The first humans almost certainly evolved in Africa, where stone tools were in use by *Homo habilis* for cutting up big game 2.5 million years ago. *Homo erectus* appeared 1.5 million years ago in Africa with a larger brain and more sophisticated tools. *Homo erectus* colonized Eurasia around 700 000 years ago, to be replaced by modern humans, *Homo sapiens*, again from Africa, about 120 000 years ago.

Related topics

Natural selection (D2)
Neo-Darwinian evolution: selection acting on alleles (D5)
Species and speciation (D7)

Evolution by divergence

Evolution is a change in allele frequencies over time, with different changes in different populations causing them to become separate species. We can compare populations to discover how much variation is expected between members of one species (within species) and compare species to see how different they have become. The extent of the differences between populations or species indicates how long they have been diverging, which gives the length of time since they became separated. On the other hand, similarities in some characteristics between populations or species suggest that they share a common ancestor in whom those characteristics first evolved. The more recently two species diverged, the more they should have in common. Such characteristics in descendants are then said to be **homologous** (e.g. back legs of mammals). This is different from **analogous (analogy)**, where species evolve similar structures independently (bat's and bird's wings), a process called **convergent evolution**. There is still room for confusion. For example, all mammals, reptiles and birds have homologous forelimbs descended from the front legs of an early amphibian ancestor, and particular fins on a fish before that. However, while bats' wings and birds' wings are homologous as forelimbs, they are analogous as wings, because they evolved as wings independently. The last common ancestor of bats and birds did not have wings.

Populations

Comparisons between the populations of a species shows the degree of divergence within species before speciation. Similarities and differences can reveal migration patterns, changes in selection pressure across the species range, and chance divergence due to drift. The problem is to identify which cause produces which effect. Humans are selected for dark skin in equatorial climates (presumably to avoid sunburn), and for pale skin in less equatorial regions, supposedly so that sunlight can penetrate the skin and synthesize vitamin D, thus avoiding

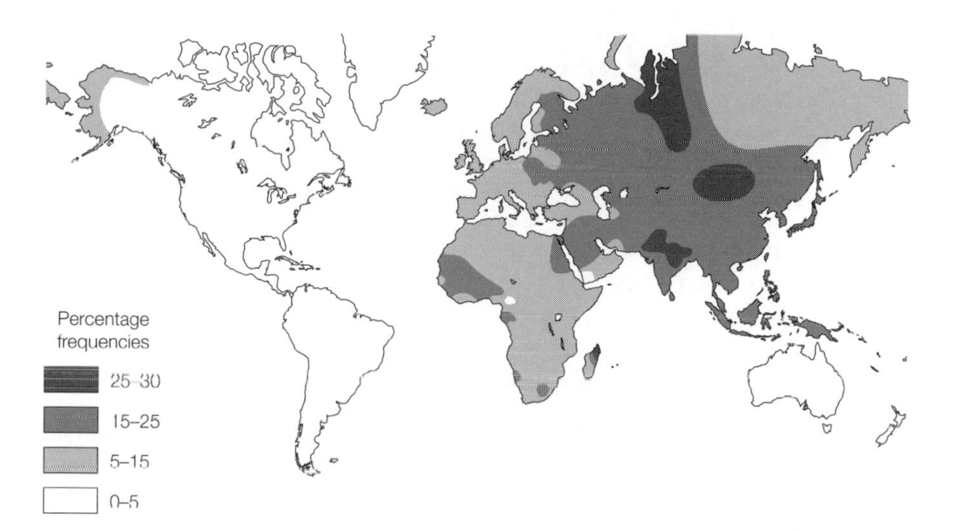

Fig. 1. *Global distribution of I^B (B) allele of human ABO blood group.*

rickets. Some support for this comes from observing that although all native Americans entered the continent from North East Asia and Alaska, they have distinctly darker skin color in the dry sunny regions (e.g. California/Nevada and the Andes). Another human example is sickle-cell anemia, selected for by the tolerance of heterozygotes for malaria. Both malaria and the allele for sickle cell hemoglobin are most frequent in equatorial West Africa.

Blood group distributions are more difficult to explain. The I^B Allele (of the ABO blood groups) shows a very uneven global distribution (*Fig. 1*). I^B did not exist in native Americans apparently having failed to enter America or Australia from Asia. This may have been due to founder effect, the relatively small band of colonizers losing I^B by chance drift. Alternatively these populations may have become isolated from Asia when the I^B allele was very rare. The patchy distribution of I^B in the rest of the world may be due to drift or to selection by disease. The A and B antigens are saccharide groups on the surface of red blood cells and many bacteria and some viruses also have these. A large range of diseases appear to have a more severe effect on individuals with particular blood groups. For example, I^A confers susceptibility to smallpox in unvaccinated populations, and may have been selected against. This initial selection or drift may have been followed by a population explosion in some tribes, and migration or invasion, sometimes on a large scale, may have carried the alleles across continents. The Mongol hordes who invaded Eastern Europe and the Middle East in the twelfth and thirteenth centuries have been proposed to explain the frequency of I^B in these areas, but cannot explain the high frequency in West Africa. Such hypotheses are easy to produce but almost impossible to test.

Ring species

When a species covers a sufficient range, the populations at the extremes may diverge sufficiently to be different species. This is apparent when the range circles an uninhabitable obstacle, and the ends overlap forming a ring species. An example is a boreal (northern) gull which surrounds the Arctic. In Europe the ends of the range overlap as the herring gull *Laurus argentatus* and the lesser black backed gull *L. fuscus*. Thus distance alone restricts gene flow across large population ranges sufficiently for a widespread species to be many incipient species if the center of the range is removed. The tendency of many bird species to nest and mate near their own birthplace may restrict geneflow despite large seasonal migrations.

Molecular clocks

Mutations occur at a relatively constant rate determined mainly by mistakes during DNA replication. A proportion of these are not deleterious (they may be neutral), and survive, increasing in frequency by drift or selection. This proportion depends upon how tightly the particular protein (or RNA) product is conserved by selection (see Topic C4). As the differences between populations or species accumulate, they give a guide to the degree of divergence between them. Some proteins (e.g. fibrinopeptide) change relatively rapidly and are useful for studying closely related species, while others change slowly (e.g. cytochrome *c* and ribosomal RNA) and can be used to follow divergence from the earliest living organisms (*Fig. 2*). Because the rate of change in any particular sequence is approximately constant it can function as a **molecular clock**.

Clocks must be calibrated, and this relies on fossil evidence to say when the taxonomic lines being compared became separated. In hemoglobin the rate of amino acid change is about 1.2 changes per amino acid site per 10^9 years (or 1.2×10^{-9} site^{-1} year^{-1}). The value for fibrinopeptide is higher at 8.3×10^{-9} (less conserved) and for histone H4 is only 0.01×10^{-9} (highly conserved). Changes in protein sequence seem to occur at the same rate in mice and whales, per year rather than per generation. Rates of change in noncoding regions and synonymous changes in coding sites of DNA are higher, and appear to be (per generation), faster in mice than in whales. This suggests a considerable degree of selection on proteins to change at a constant time-dependent rate as the environment changes. If correct, it supports the hypothesis that selection, rather than drift of neutral mutations, is the cause of most amino acid changes. There is some doubt about the constancy of molecular clocks. They may run faster at times of rapid evolution such as the adaptive radiation of mammals following

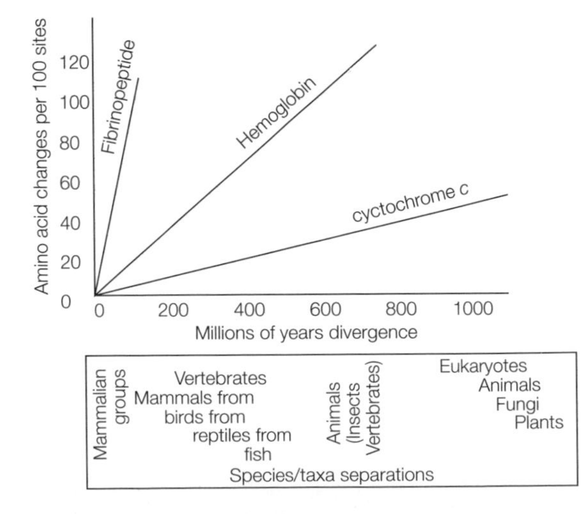

Fig. 2. Relative rates of change of amino acids in fibrinopeptide, hemoglobin and cytochrome.

the extinction of the dinosaurs, and many other animals, at the end of the Cretaceous period. They may also slow during periods of stasis. The molecular clock suggests that the main groups of mammals diverged substantially earlier than 65 million years ago, but the first fossils of these groups are unequivocally younger than this.

Phylogenics

Phylogenies are produced by arranging similar species close together, and drawing lines between species and species clusters. The line-lengths represent the number of genetic sequence changes that have occurred between them, and hence represent evolutionary time since separation. Phylogenics is a science in itself. The product is a tree-, bush-, or star-shaped diagram (*Fig. 3*). One problem is to find the point of origin, or root. This can be done by using an **outgroup** species which is equally distantly related to them all. For example, by using a fungus or plant to root the cytochrome *c* tree for animals (*Fig. 3*). Faster mutating sequences (e.g. mitochondrial DNA) are used to construct phylogenies of more closely related species. Fast mutating sequences cannot be used to go back far in evolutionary time because the mutations effectively randomize the sequence. Reverse mutations restore the original sequence at some sites, and comparisons become error prone, or impossible.

Animal mitochondrial DNA (mDNA) evolves much faster than nuclear DNA, presumably because it is a small genome under less selection for accuracy, so DNA synthesis is more error prone. This makes it useful for examining relatively recent events. It also has the advantage of only being inherited from

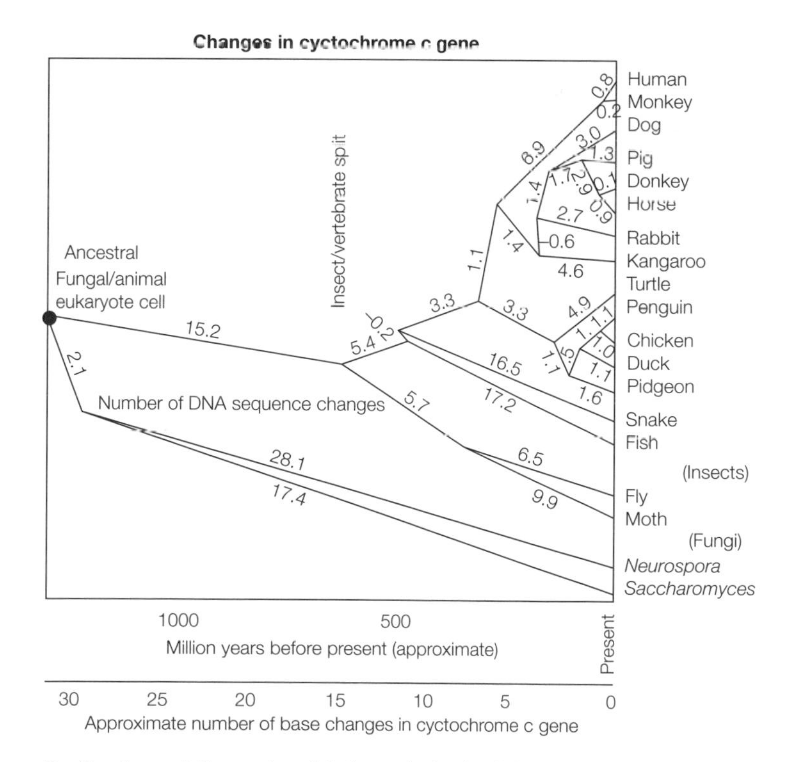

Fig. 3. An evolutionary tree (phylogeny) of animals based on nucleotide changes in the cytochrome c gene. Numbers show average number of nucleotide changes along each branch. Fungi are used to root the tree.

females (uniparentally) so it never recombines, and each lineage carries its own history. Analysis of human mDNA suggests that it all comes from a single ancestral sequence (one woman, hence 'Eve') who lived around 140 000–290 000 years ago. Note that it is a mathematical certainty that any particular mutation or unrecombined segment of DNA can be traced back to a single origin if we can follow it far enough, so 'Eve' is not controversial. All the other lineages became extinct, probably by chance rather than by selection. There is some controversy over Eves' place of residence. The most favored suggestion is that she lived in Africa. The two pieces of evidence for this are: (i) most branches of the phylogeny are represented there; and (ii) the most divergent types of mDNA are found there. This suggests that parallel female lineages have been living in Africa longer than anywhere else. Most European groups are relatively closely related, with little divergence. The data does not clearly separate individuals from Europe, Africa and Asia and is probably not adequate to construct the tree absolutely accurately. There is too much divergence for this one character to be conclusive on this timescale.

Cytochrome *c* is a conserved protein that can be used to compare all eukaryotes (*Fig. 3*). The cytochrome *c* based phylogeny confirms our view that mammals are one related group and birds are another, as was supposed from comparative morphology and common sense. The whole evolutionary tree of animal evolution is very well supported. Perhaps the most surprising observation is that the two most divergent species by far are two species of fungus, both ascomycetes, that separated twice as long ago as the ancestors of mammals separated from the ancestors of insects.

Some **problems with phylogenies** can be seen in *Fig. 3*. If mutation rates were the same in all lineages then all ends of branches would be level because each would have diverged by the same amount. The numbers on the branches for the cytochrome *c* tree would also add up to the same value. It appears that *Neurospora* has a mutation rate 50% faster than that of *Saccharomyces*. Amino acid changes must happen in whole numbers, but the averaging algorithms (numerical techniques) used to construct the tree produce fractional changes. For example there is one amino acid difference in cytochrome *c* between humans and rhesus monkey, and it occurred in the human lineage, not the monkey, which still has the sequence of our common ancestor. This is shown as 0.8 human 0.2 monkey in the tree. The algorithms always give a dichotomous branching patter. This is correct for mDNA, but not for nuclear genes which can recombine, and it is misleading for species that exist as populations and can split simultaneously into many isolated species.

Human evolution

All current human races are the same species. There are larger differences between individuals within any racial group than there are between the averages of each race. Humans are obviously mammals, and careful anatomical and biochemical analysis shows humans to be very closely related to the apes. Current estimates are that apes (hominoids) diverged from monkeys 25 million years ago. Gibbons and orangutans diverged from other apes 10 and 8 million years ago respectively. About 4 million years ago an ancestral ape population divided into three lineages which would separately give rise to bipedal humans, chimpanzees and the massive vegetarian gorillas. There is no convincing evidence that one lineage branched off first.

Raymond Dart found a 2.8 million year old hominid fossil in South Africa in 1924. He gave the genus the name *Australopithecus*, meaning southern ape.

The oldest hominid fossils have been found in Laetolil, Tanzania, and Hadar, Ethiopia. The most famous collector is Mary Leakey, and the most famous fossil is 'Lucy', a 40% complete female skeleton from Hadar. These fossils are 3.6–3.8 million years old. The skeletons have strong muscle attachments, long arms and large canine teeth similar to apes but they were bipedal, walking upright on two legs like modern humans. Males were much larger than females. Another line of hominid may have diverged from the human line but become extinct, because a large form, *A. robustus*, appeared, together with a small 'gracile' form (possibly the females). Both seem to have been chewers. The larger robust form in particular had huge jaw muscles, molar and premolar teeth. This suggests a tough vegetarian diet and some similarity in ecology to modern gorillas. Robust australopithecines coexisted with early human ancestors around 1.5–2 million years ago.

Development of bipedalism was completed very early, freeing hands to manipulate tools and weapons. A reduction in jaw size and increased intelligence followed together, but brain size continued to develop. The skills involved in tool manufacture must be taught and learned, and probably provided selection for intelligence and language. Cooperation was also needed to allow humans to compete with big cats as top predator, rather than being their prey, and would also benefit from language. The oldest stone tools are about 2.5 million years old, and were found associated with bones of *Homo habilis*. These were replaced around 1.5 million years ago by the larger-brained species *Homo erectus* who produced improved tools. *Homo erectus* migrated from Africa and had colonized the accessible world by about 700 000 years ago. The African population continued to evolve, and gave rise to modern humans, *Homo sapiens*, between 100 000 and 200 000 years ago. Some *Homo sapiens* migrated from Africa, replacing *Homo erectus*. There may have been more human groups, possibly subspecies.

Neanderthal people were rather robust 'Eurasians' with heavy brow ridges and larger brains than modern humans. They are classified as *Homo sapiens* and lived from 120 000–35 000 years ago with all the trappings of modern hunter–gatherer family life. They cared for their sick and buried their dead with artifacts, suggesting belief in an afterlife. Neanderthals may have been a stocky human race adapted to the cold of ice age Eurasia. They were replaced by anatomically modern humans (Cro-magnon man) who were present in the middle east 40 000 years ago and spread through Europe by 35 000 years ago. This is approximately the time that all of Asia and Australia were first colonized by anatomically modern humans. It is not clear where Neanderthals fit in the hominid phylogeny. In Africa, the species *H. habilis*, *H. erectus* and *H. sapiens* were probably a continuum of arbitrary chronospecies. These repeatedly expanded out of Africa and replaced the previous colonizers of the Eurasian land mass.

E1 NUCLEIC ACID HYBRIDIZATION

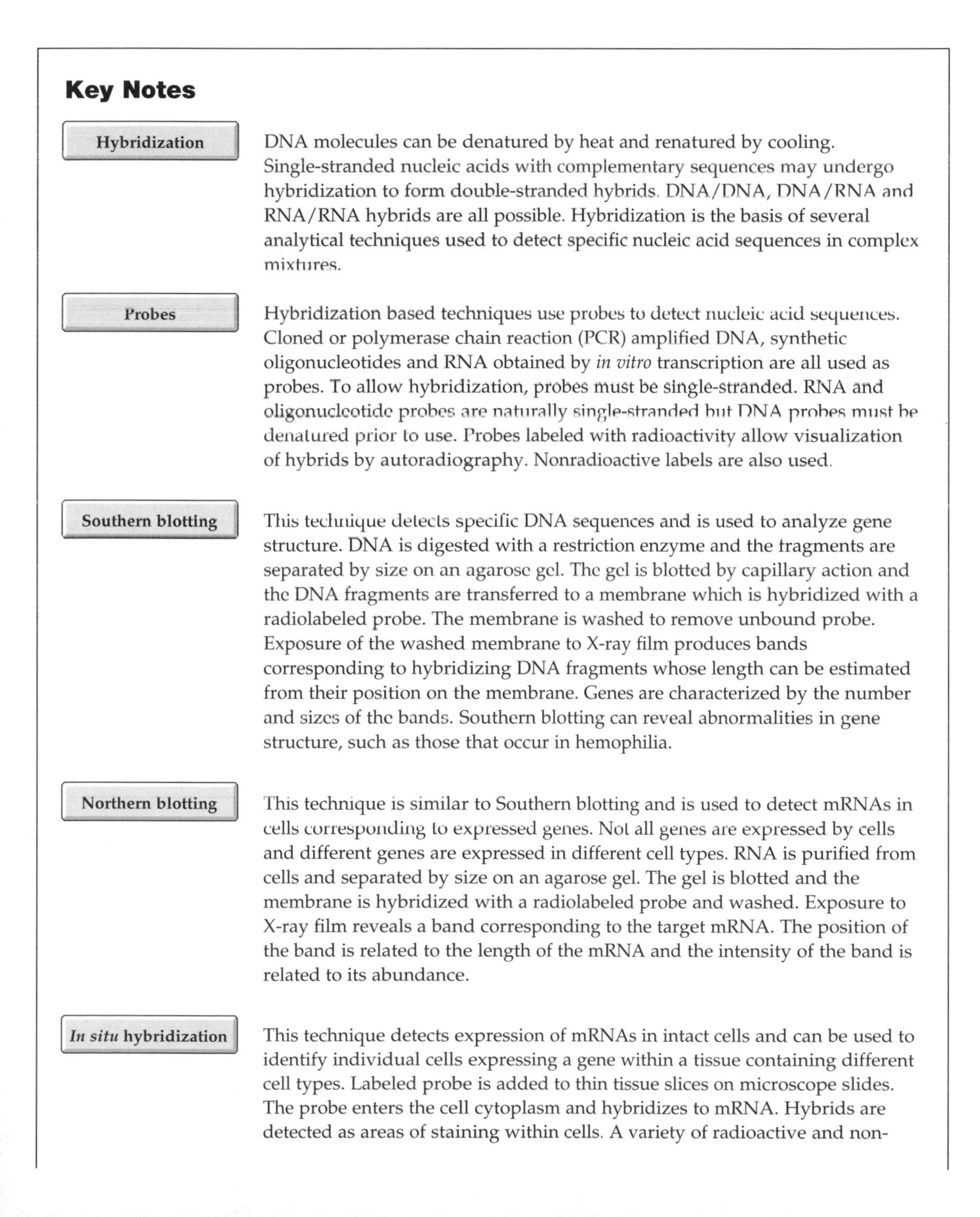

Key Notes

Hybridization

DNA molecules can be denatured by heat and renatured by cooling. Single-stranded nucleic acids with complementary sequences may undergo hybridization to form double-stranded hybrids. DNA/DNA, DNA/RNA and RNA/RNA hybrids are all possible. Hybridization is the basis of several analytical techniques used to detect specific nucleic acid sequences in complex mixtures.

Probes

Hybridization based techniques use probes to detect nucleic acid sequences. Cloned or polymerase chain reaction (PCR) amplified DNA, synthetic oligonucleotides and RNA obtained by *in vitro* transcription are all used as probes. To allow hybridization, probes must be single-stranded. RNA and oligonucleotide probes are naturally single-stranded but DNA probes must be denatured prior to use. Probes labeled with radioactivity allow visualization of hybrids by autoradiography. Nonradioactive labels are also used.

Southern blotting

This technique detects specific DNA sequences and is used to analyze gene structure. DNA is digested with a restriction enzyme and the fragments are separated by size on an agarose gel. The gel is blotted by capillary action and the DNA fragments are transferred to a membrane which is hybridized with a radiolabeled probe. The membrane is washed to remove unbound probe. Exposure of the washed membrane to X-ray film produces bands corresponding to hybridizing DNA fragments whose length can be estimated from their position on the membrane. Genes are characterized by the number and sizes of the bands. Southern blotting can reveal abnormalities in gene structure, such as those that occur in hemophilia.

Northern blotting

This technique is similar to Southern blotting and is used to detect mRNAs in cells corresponding to expressed genes. Not all genes are expressed by cells and different genes are expressed in different cell types. RNA is purified from cells and separated by size on an agarose gel. The gel is blotted and the membrane is hybridized with a radiolabeled probe and washed. Exposure to X-ray film reveals a band corresponding to the target mRNA. The position of the band is related to the length of the mRNA and the intensity of the band is related to its abundance.

***In situ* hybridization**

This technique detects expression of mRNAs in intact cells and can be used to identify individual cells expressing a gene within a tissue containing different cell types. Labeled probe is added to thin tissue slices on microscope slides. The probe enters the cell cytoplasm and hybridizes to mRNA. Hybrids are detected as areas of staining within cells. A variety of radioactive and nonradioactive agents are used to label probes. Fluorescent *in situ* hybridization (FISH) is a variation of the technique used to determine the chromosomal locus of a gene.

Related topics

DNA structure (A1)
DNA cloning (E2)
The human genome project (F5)

Hybridization

When double-stranded DNA is heated the hydrogen bonds that stabilize the double helix are disrupted. The helix becomes unstable and the two strands of the DNA molecule separate. This process is called **denaturation**. If the temperature is then reduced the double helix reforms and the original double-stranded DNA molecule is recovered. This process is called **renaturation** or **reannealing**. In fact, any two single-stranded nucleic acid molecules are capable of forming a double-stranded molecule as long as their base sequences are mostly complementary. The double-stranded molecule is called a **hybrid** and its formation is called **hybridization**. Hybrids can form between two strands of DNA, between DNA and RNA and between two strands of RNA. This feature of nucleic acids is used in a series of analytical techniques in which specific DNA or RNA sequences in complex mixtures are detected by using a nucleic acid of complementary sequence.

Probes

The detection of nucleic acid sequences using techniques based on hybridization requires the use of probes. A probe is simply a DNA or RNA molecule that can be used to detect nucleic acids of complementary sequence. Probes must be pure and free from other nucleic acids with different sequences. Typically, probes are cloned DNA sequences or DNA obtained by polymerase chain reaction (PCR) amplification. Synthetic oligonucleotides and RNA obtained by *in vitro* transcription of cloned DNA sequences are also used as probes. To permit the detection of target nucleic acids by complementary base pairing, probes must be single-stranded. Oligonucleotides and RNA probes are naturally single-stranded, however DNA molecules are normally double-stranded and must be made single-stranded before hybridization with target sequences can occur. This is usually achieved by heating the probe to cause denaturation. Rapid cooling will lower the temperature to a point where renaturation occurs only very slowly keeping the probe single-stranded. The detection of target DNA molecules requires that the probe is labeled with an agent that allows the hybrids to be visualized. Probes are most commonly labeled with radioactive isotopes such as $\mathbf{P^{32}}$, $\mathbf{S^{35}}$, $\mathbf{C^{14}}$ or $\mathbf{H^{3}}$. Hybrids emit radiation and can be visualized by exposure to X-ray film. This is called **autoradiography**. The risk to health associated with the use of radioactivity has led to the development of nonradioactive labeling systems. A commonly used system involves labeling probes with the steroid **digoxigenin** (DIG) which can be detected by specific antibodies labeled with a dye or linked to an enzyme that catalyzes the formation of a colored product.

Southern blotting This was the first analytical technique based on nucleic acid hybridization to be developed and is named after its inventor, Ed Southern of the University of Oxford. Southern blotting is used to detect DNA molecules and is typically used to analyze the structure of genes (*Fig. 1*). For this application, the DNA for analysis is purified from cells and is obtained as large fragments of chromosomal DNA, typically 20 000 bp or more in length. The first step in Southern

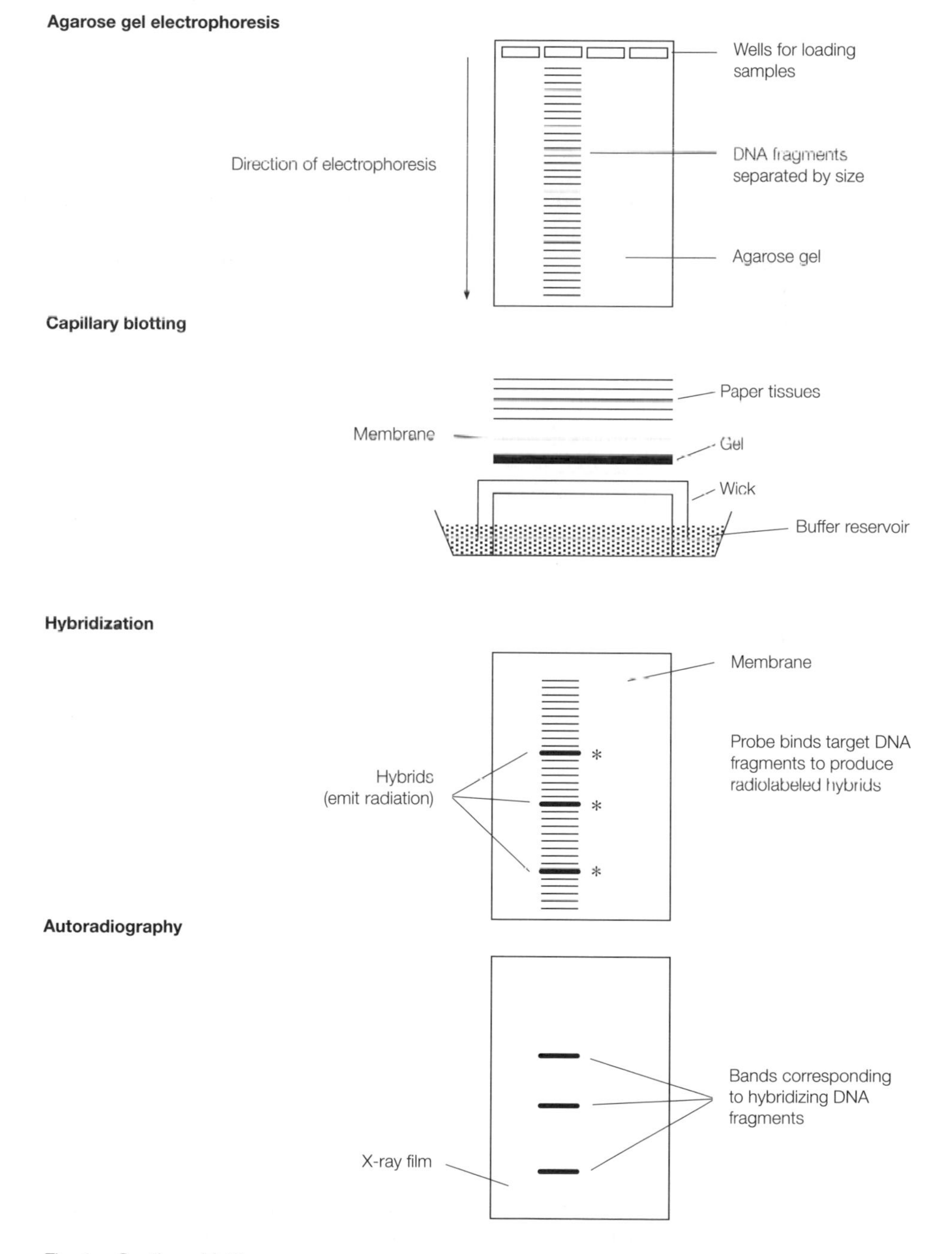

Fig. 1. Southern blotting.

blotting involves digestion of the DNA with a **restriction enzyme**. These enzymes are derived from bacteria and cut DNA molecules specifically at each point where a DNA sequence of 4–8 bases recognized by the enzyme occurs (see Topic E2). Digestion of the chromosomal DNA produces thousands of DNA fragments ranging in size from just a few bases to several thousand bases. The DNA fragments are then subjected to electrophoresis in an agarose gel. This separates them according to size with the large ones near the top of the gel and the small fragments near the bottom. After electrophoresis, the gel is soaked in an alkaline solution that denatures the fragments making them suitable for hybridization. The next stage involves transferring the denatured fragments out of the gel and onto a membrane made from nylon (or sometimes nitrocellulose) where they become accessible for analysis using a probe. Transfer is achieved by **blotting**. The gel is placed on a platform above a dish containing a buffer. A wick made from filter paper runs from the platform into the buffer reservoir. A sheet of membrane is placed on top of the gel, covering it, and a stack of absorbent paper tissues is placed on top of the membrane. The gel is blotted for several hours during which time the buffer in the reservoir is drawn, by capillary action, from the wick through the gel and the membrane into the stack of paper towels. The flow of buffer causes the DNA fragments to pass out of the gel and onto the membrane above, where they become attached. After several hours, a replica of the pattern of fragments in the gel forms on the membrane. At this point the blot is dismantled and the membrane containing the fragments is removed. The membrane is treated to firmly attach the DNA fragments either by baking it at 80°C or by exposing it to ultraviolet radiation.

The next stage involves incubating the membrane with a labeled probe. This is known as **hybridization** and is carried out at a temperature and in a buffer that favor the formation of hybrids between the probe and fragments of DNA bound to the membrane whose sequence is complementary to the probe. The membrane is then washed with a buffer to remove probe that is bound nonspecifically so that only labeled probe bound to target sequences remains. If the probe is labeled with radioactivity, the membrane is placed in the dark against a sheet of X-ray film. After several hours, development of the film reveals one or more dark bands which correspond to the position of fragments bound by the probe. The length of the fragments can be calculated from their position relative to marker DNA molecules of known length. In this way the structure of individual genes can be characterized in terms of the number of hybridizing fragments and their sizes. When a probe for a new gene is isolated one of the first experiments carried out is to use Southern blotting to analyze the structure of the gene in the chromosomal DNA. Southern blotting can also be used to reveal abnormalities in the structure of genes. For example, about 20% of individuals with the inherited bleeding disorder, hemophilia, show alterations in the structure of the gene for the blood clotting protein, Factor VIII. These alterations can be used to identify relatives who are carriers of hemophilia and may need genetic counseling (see Topic F1).

Northern blotting

This technique is similar to Southern blotting but is used to analyze RNA rather than DNA. The DNA in every human cell contains copies of all of the genes present in the human genome. However, only about 15% are active in any particular cell. The remaining 85% are inactive. These are not transcribed and do not lead to the synthesis of protein. Furthermore, different genes are active in different cell types. For example, the genes that are active in a muscle cell are very different from those active in red blood cells. The active genes reflect the very different protein compositions of these cells. Muscle cells contain proteins such as actin and myosin that are required for contraction and the genes for these proteins are very active. In contrast, in red blood cells the major protein present is hemoglobin. Consequently, the globin genes which encode the polypeptides that make up hemoglobin are very active. Northern blotting is used to identify which genes are active in different cell types. The method used for Northern blotting is similar to Southern blotting. The main difference is related to the starting material. For Northern blotting, RNA is isolated from cells. This contains messenger RNAs (mRNAs) derived from all the active genes in that cell type. The mRNAs are different sizes, depending on the size of the protein they encode. mRNAs that encode proteins present at high concentrations in a cell are also more abundant than those encoding proteins present at low concentrations. The RNA is separated by agarose gel electrophoresis. Each mRNA migrates to a position on the gel determined by its size, with the larger transcripts near the top and the smaller transcripts near the bottom. As was the case for Southern blotting, the gel is blotted and the membrane is hybridized with a probe specific for the mRNA being investigated. The membrane is then washed and exposed to X-ray film. Development of the film reveals usually a single band corresponding to the mRNA recognized by the probe. The position of the band relative to the top of the membrane can be used to estimate the length of the mRNA and the intensity or darkness of the band is a measure of how much of that mRNA was present in the original cells. Thus mRNAs of genes that are expressed at low levels appear as faint bands and very abundant mRNAs appear as dark bands. Northern blotting can, therefore, provide useful information about which genes are switched on in a cell, the level at which they are expressed and the sizes of the different mRNAs.

In situ hybridization

This technique is different from other hybridization methods in that it is used to detect nucleic acid sequences present in intact cells. The main use of *in situ* hybridization is to identify expression of specific mRNAs in individual cells present in tissues containing a number of different cell types. For example, *in situ* hybridization can be used to show that insulin mRNA is produced only by the β cells of the pancreas. Thin slices of tissue, known as sections, are cut using an instrument called a microtome and are placed on a microscope slide. Probe is added to the cells on the microscope slide and is taken up into the cytoplasm where it binds to target mRNA forming hybrids. These can be detected by the label on the probe and appear as areas of staining within cells which can be seen when the section is viewed under a microscope. Probes may be labeled in a number of ways. Radioactive isotopes, fluorescent molecules, enzymes that catalyze the formation of colored products and the steroid digoxigenin are all used.

A variation on the technique called **fluorescent *in situ* hybridization (FISH)** involves hybridizing probe to chromosomes in cells as a way of identifying the position of a gene on a chromosome. Cells undergoing metaphase contain chromosomes in a noncondensed form which have characteristic shapes. By hybridizing probe to these cells it is possible to determine the chromosomal locus of the hybridizing sequence (see Topic F5).

E2 DNA CLONING

Key Notes

DNA cloning

This technique allows individual DNA sequences in complex mixtures to be isolated and copied permitting detailed analysis and manipulation. The DNA to be cloned is recombined with vector DNA and introduced into host cells where it is copied. Recombinant vector is purified from cultures of host cells and the cloned DNA can be recovered for analysis.

Restriction enzymes

Recombining DNA molecules for cloning depends on the use bacterial restriction enzymes which cut DNA molecules at palindromic sequences and produce sticky ends. Two DNA molecules cut with the same restriction enzyme can be joined by complementary base-pairing between the sticky ends and covalently linked using DNA ligase.

Plasmids

These are small circular DNA molecules found in bacteria that are frequently used as cloning vectors. Plasmids are easily purified and confer antibiotic resistance to host bacteria allowing easy identification of recombinants. Linearized plasmid is recombined with foreign DNA and the recombinant plasmid is transformed into bacteria. Colonies containing recombinant vector are isolated from agar plates containing antibiotic and vector containing the cloned DNA is purified from bacterial cultures. Early plasmids such as pBR322 contain twin antibiotic resistance genes which allow identification of recombinants. Later plasmids such as pUC identify recombinants by blue/white selection based on disruption of the *lac Z* gene by insertion of the foreign DNA. Additional plasmid modifications include multiple cloning sites, phage promoter sequences for *in vitro* transcription and the ability to express cloned sequences as protein.

Lambda (λ) phage

Bacteriophage λ which infects *E. coli* has been adapted as a cloning vector. The central portion of the phage DNA is deleted and can be replaced with foreign DNA. Recombinant phage DNA is packaged *in vitro* into capsids which infect *E. coli* producing plaques on agar plates. λ vectors are used to construct genomic libraries. These are collections of recombinant phage containing cloned sequences representative of an entire genome. Libraries can be screened for sequences of interest by hybridizing plaque lifts with a probe. cDNA libraries are constructed from mRNA and contain clones representing expressed sequences only. Expression libraries are cDNA libraries that allow screening with antibodies.

Cosmids

These cloning vectors resemble plasmids but contain λ phage *cos* sequences which allow them to be packaged into λ capsids. Packaged cosmids infect *E. coli* and are replicated. Cosmids do not contain λ genes and so produce bacterial colonies instead of plaques. Cosmids can accommodate large inserts up to about 44 kbp.

Yeast artificial chromosomes (YACs)

These vectors use eukaryotic host cells and replicate in the same way as host cell chromosomes. They contain features required for chromosome replication including an origin of replication, a centromere and telomeres. YACs can accommodate inserts of several hundred kbp and can be used to construct chromosome maps.

Plant cloning vectors

Most plant cloning vectors are based on the Ti plasmid from *Agrobacterium tumefaciens* which causes crown gall disease in plants. Part of the Ti plasmid can integrate into the host cell chromosome and is used to carry useful genes into the plant genome.

Applications of DNA cloning

Gene cloning has made important contributions to many areas of research in biology. These include: identification of genes involved in disease processes; construction of genome maps; production of recombinant proteins; and the creation of genetically modified organisms.

Related topics

Chromosomes (B1)
Prokaryotic genomes (B3)
Bacteriophages (B8)
Genetic engineering and biotechnology (F7)

DNA cloning

The human genome is estimated to contain 50–100 000 genes. In DNA isolated from human cells, individual gene sequences are present in only very small amounts. DNA cloning is a powerful technique that allows specific DNA sequences to be separated from other sequences and copied so that they can be obtained in large amounts permitting detailed analysis or manipulation. An important use of DNA cloning is to isolate new genes allowing them to be investigated and characterized.

All DNA cloning experiments are based on the construction of **recombinant DNA** molecules. This involves joining different DNA molecules together. The DNA molecule to be cloned (often a fragment of human DNA containing a gene of interest) is inserted into another, usually circular, DNA molecule called a **vector**. The recombinant vector is introduced into a **host cell**, usually the bacterium *E. coli*, where it produces multiple copies of itself. When the host cell divides copies of the recombinant vector are passed on to daughter cells. Large amounts of the vector are produced which can be purified from cultures of the host cells and used for analysis of the foreign DNA insert.

Restriction enzymes

The ability to join different DNA molecules together for cloning is dependent on the use of enzymes from bacteria called restriction endonucleases. These enzymes cut DNA molecules at specific sequences, usually of 4–8 bases. The sequences recognized are **palindromes**. This means that the sequence is the same reading 5′→3′ on both strands. Each enzyme has a specific target sequence. For example the enzyme ***Eco*RI**, which is obtained from the bacterium *E. coli*, will cut any DNA molecule that contains the sequence GAATTC. Other restriction enzymes recognize different sequences. The cut made by restriction enzymes is usually staggered such that the two strands of the double helix are cut a few bases apart. This creates single-stranded overhangs called **sticky ends** at either end of the cut DNA molecule. Some restriction enzymes cut the DNA

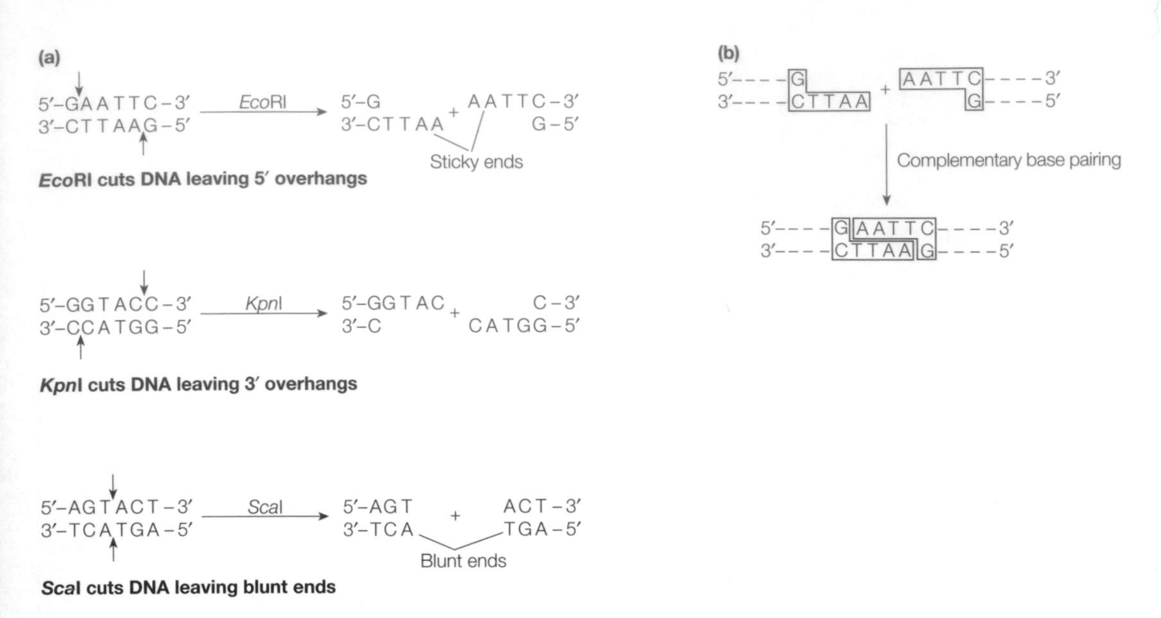

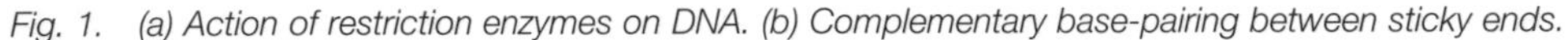
Fig. 1. (a) Action of restriction enzymes on DNA. (b) Complementary base-pairing between sticky ends.

leaving the 5′ end overhanging and others leave a 3′ overhang. A few restriction enzymes cut both strands of the double helix at the same position creating what is known as a **blunt end** (*Fig. 1a*). Two DNA molecules cut with the same restriction enzyme can be joined together by complementary base-pairing between the sticky ends (*Fig. 1b*). Although the molecules are joined by the sticky ends, they are not covalently linked. However, an enzyme called **DNA ligase**, which is found in all cells, can be used to catalyze the formation of a phosphodiester bond between the two DNA molecules, thus permanently recombining them. A number of systems exist for cloning DNA molecules based on the use of different types of vector. Each has its own individual characteristics and uses.

Plasmids

These are small, circular DNA molecules present in bacteria. They occur in addition to the main bacterial chromosome and can replicate autonomously (see Topic B3). Plasmids are the commonest type of cloning vector used. The cloning procedure involves a number of steps (*Fig. 2*): first, the plasmid is digested with a restriction enzyme that cuts it at a single site converting it from a circular molecule into a linear molecule with sticky ends. The foreign DNA to be cloned is also digested with the restriction enzyme to produce the same sticky ends. When the plasmid and the foreign DNA are mixed, molecules of plasmid become joined to molecules of foreign DNA via their common sticky ends and circular recombinant plasmids are obtained. DNA ligase is then used to covalently join the two. The recombinant plasmid is introduced into host bacteria (usually *E. coli*) by a process called **transformation**. The transformed bacteria are spread on agar plates and bacterial colonies comprised of cells that have taken up the recombinant plasmid are grown. Individual colonies are isolated and cultured in liquid medium. Large amounts of plasmid can then be purified from the cultures and the cloned DNA recovered for analysis.

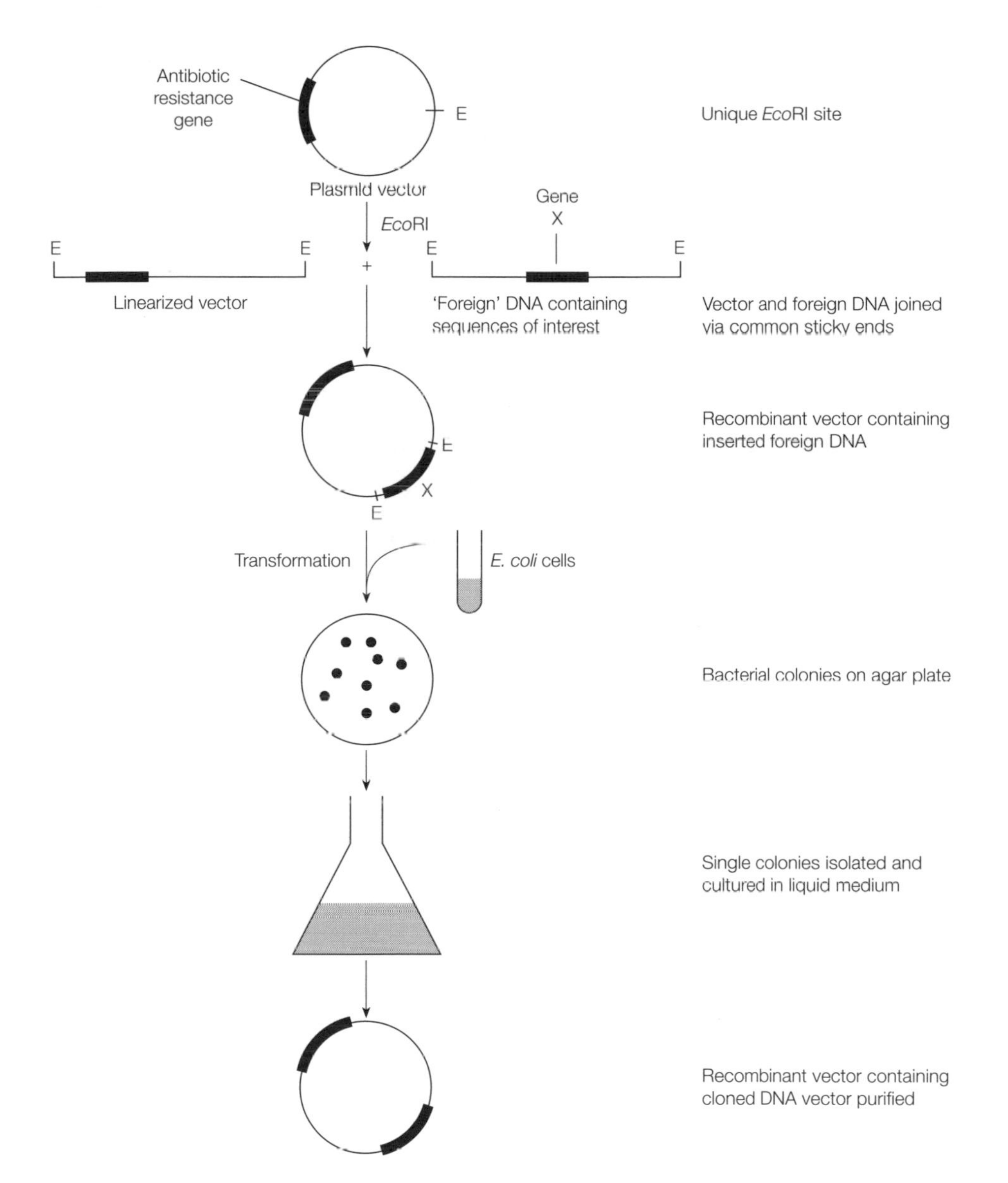

Fig. 2. Cloning with plasmids.

Several features of plasmids make them especially suitable as cloning vectors. Their small size (usually about 3 kbp) makes them easy to purify from bacterial cultures allowing the cloned DNA to be recovered easily. In addition, plasmids often contain genes encoding proteins that make the bacteria resistant to antibiotics such as ampicillin and tetracycline. By growing colonies on agar plates containing antibiotic, it is possible to isolate bacteria that have taken up the plasmid during transformation because only these will be resistant to antibiotic and will be able to grow.

Plasmid cloning vectors were initially based on naturally occurring plasmids. These have gradually been replaced by improved vectors whose DNA sequences have been altered to include features useful for cloning.

One of the earliest plasmid vectors to be developed was **pBR322**. This plasmid contains two genes that confer resistance to the antibiotics **ampicillin** and **tetracycline**. During cloning, foreign DNA is inserted into the tetracycline gene, thereby inactivating it. Transformed bacteria containing recombinant plasmid could therefore be identified by being resistant to ampicillin but not to tetracycline. Bacteria which had taken up plasmid that did not contain foreign DNA but had simply been religated to itself could be identified by being resistant to both antibiotics.

The pBR322 plasmid was followed by the **pUC** series of vectors which allowed identification of colonies containing recombinant plasmid by a method called **blue/white selection** (*Fig. 3*). This method relies on the presence of a gene called *lac* Z which encodes the enzyme β-galactosidase and is located on the plasmid at the point where the foreign DNA is inserted. Bacteria that contain the intact plasmid synthesize β-galactosidase which acts on a synthetic substrate called **X-gal** (5-bromo-4-chloro-3-indolyl-β-D-galactopyranoside) to produce a colored product. When colonies are grown on agar plates containing X-gal they take on a blue color. However, when foreign DNA is inserted into the pUC plasmid, the *lac* Z gene is disrupted and β-galactosidase is no longer produced. As a result, colonies containing recombinant plasmid remain white when grown on X-gal and are easily distinguished from colonies containing religated vector which are blue.

Another useful feature of pUC vectors is that the sequence of part of the *lac* Z gene is modified to create a series of clustered restriction enzyme sites. This is called the **multiple cloning site (MCS)**. Its purpose is to create extra flexibility during the cloning procedure by allowing the foreign DNA to be inserted at any one of several restriction sites (*Fig. 4*).

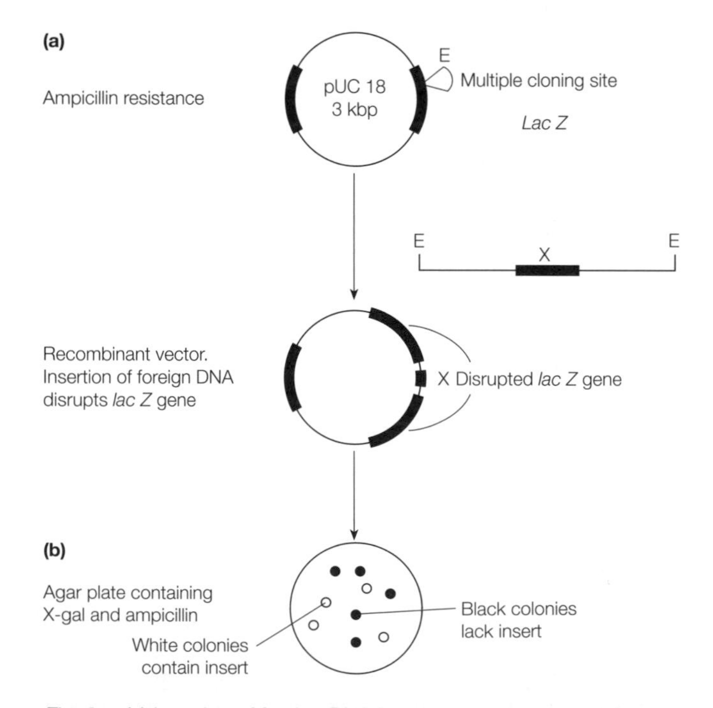

Fig. 3. (a) Insertion of foreign DNA inactivates lacZ *gene. (b) Recombinant colonies appear white on agar plates containing X-gal.*

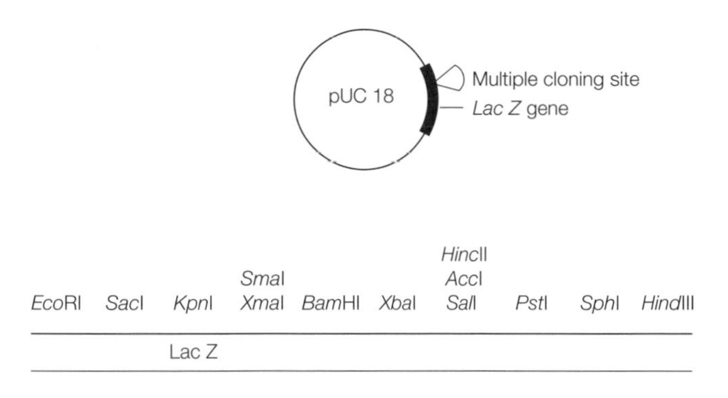

Fig. 4. Multiple cloning site of the pUC18 vector.

Other useful plasmid modifications include the presence of promoter sequences from bacteriophages inserted on either side of the MCS which allow *in vitro* transcription of the inserted foreign DNA by RNA polymerase. This feature is useful for producing RNA probes from cloned sequences (see Topic E1). Some plasmids are modified to allow cloned sequences to be translated into protein. These are known as **expression vectors**.

Lambda (λ) phage

Bacteriophages (phages) are viruses that infect bacteria. They consist of a nucleic acid genome inside a protective protein coat called a capsid. The bacteriophage λ which infects *E. coli* has a linear double-stranded DNA genome (see Topic B8). The phage attaches itself to the surface of a bacterium and injects its DNA into the cell. Inside the cell, the λ DNA is copied and capsid proteins are synthesized. The DNA is packaged into capsids and new phage particles are produced that are released by lysis of the infected cells.

Lambda phage has been adapted for use as a cloning vector. The central portion of the λ DNA, which is not essential for infection, is deleted leaving 5′ and 3′ fragments known as **arms** (*Fig. 5a*). The deleted region can be replaced by foreign DNA to produce recombinant phage DNA (*Fig. 5b*). This is inserted into phage capsids *in vitro* by a process called **packaging** which involves mixing the recombinant phage DNA with a **packaging extract** containing phage capsid proteins and processing enzymes. Recombinant phage particles are produced which are highly efficient at infecting *E. coli*. Infected cells are spread on an agar plate and produce a continuous sheet of bacteria called a **lawn** which contains small clear areas about the size of a pin head. These correspond to areas of lysis produced by infection with phage and are known as plaques. Individual **plaques** can be isolated and used to generate large amounts of cloned DNA by infection of fresh cultures of *E. coli*.

The main advantage of λ as a cloning vector is that the size of the fragments that can be cloned is much larger than for plasmids. Lambda vectors can accommodate fragments up to 25 kb as compared to less than 10 kb for plasmids. The ability to clone larger fragments has led to the development of the main use of λ vectors which is the construction of **DNA libraries**. A library is a collection of recombinant phage which together contain clones representative of all the DNA sequences present in the genome of an organism. Because λ can accommodate relatively large fragments of DNA, many fewer clones are required to represent an entire genome than would be required if plasmids

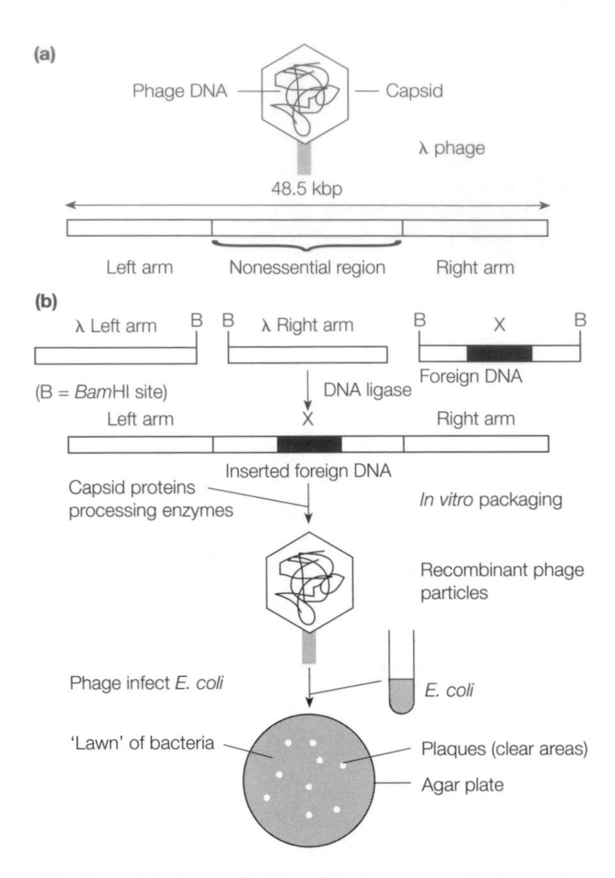

Fig. 5. (a) λ phage. (b) Use of λ phage as a cloning vector.

were used. **Genomic libraries** are constructed from DNA purified from cells which has been broken randomly into fragments of around 20 kbp either by digestion with a restriction enzyme or due to physical shearing by pipeting or sonication. The fragments are ligated to the λ arms and are cloned at random. Many thousands of plaques are produced, each of which contains a different cloned sequence. Plaques corresponding to a cloned sequence of interest can be identified by screening the library with a probe using a procedure called a **plaque lift**. This involves taking an agar plate containing plaques and laying a sheet of special nylon membrane on top of it. Some of each plaque adheres to the membrane and a replica of the pattern of plaques on the plate forms. The membrane is then treated with alkali to denature the DNA in the plaques and is hybridized to a DNA probe labeled with radioactivity. After washing away unbound probe, exposure of the membrane to X-ray film produces a series of black dots corresponding to the position of plaques containing the desired sequence. These can then be isolated from the agar plate and used to obtain the cloned DNA.

Libraries can also be produced using RNA. The enzyme reverse transcriptase is used to convert the RNA into complementary DNA (cDNA) which can then be cloned in the same way as for genomic libraries. Libraries made this way are called **cDNA libraries** and contain clones that are representative of the genes that are active in the cells used to isolate the RNA. Thus, a cDNA library from blood cells will have many different clones from one derived from lung cells or kidney cells because the active genes in each cell type will be different. cDNA libraries have the advantage that the cloned sequences do not contain introns. This greatly simplifies the characterization of cloned genes because the complete coding sequence of the gene may be present in a single clone. **Expression libraries** are a type of cDNA library in which the cloned sequences are translated into protein by the host bacteria. This allows the library to be screened using antibodies specific for the protein encoded by the cloned sequence.

Cosmids

This type of vector combines features found in plasmids and λ phage. Cosmids contain all the normal features found in plasmids, including a MCS and genes conferring antibiotic resistance, but also include sequences found in λ called ***cos* sequences**. These occur at either end of the λ DNA molecule and are responsible for its insertion into the phage capsid. The presence of *cos* sites on cosmids allows them to be packaged into phage capsids. Cloning with cosmids combines features associated with the use of both λ and plasmids as cloning vectors

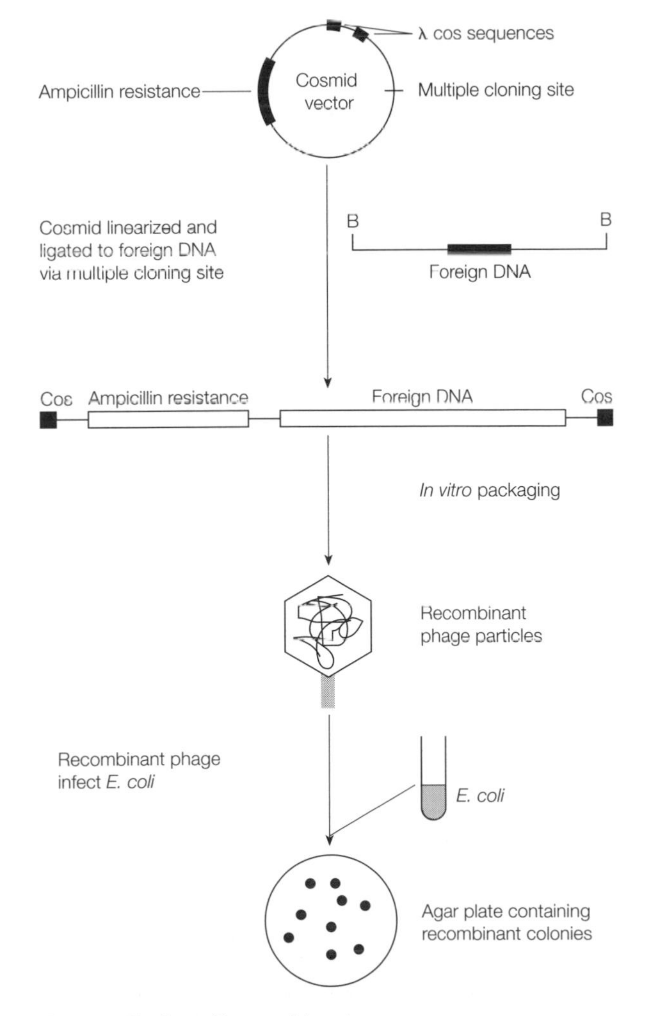

Fig. 6. Cloning with cosmid vectors.

(*Fig. 6*). Cosmid DNA is cleaved with a restriction enzyme and ligated to foreign DNA. The recombinant cosmid is then packaged into λ capsids and used to infect *E. coli*. Cosmids do not contain any λ genes and so do not form plaques after infection. Instead, infected cells are grown on agar containing antibiotic and resistant colonies containing recombinant cosmids are obtained which can be propagated in the same way as plasmids. Cosmids have the advantage of being able to accommodate very large inserts. Because cosmids are small, typically 8 kbp or less, and the λ capsid can accommodate up to 52 kbp, inserts of up to 44 kbp can be cloned.

Yeast artificial chromosomes (YACs)

These vectors were developed recently and represent a new approach to gene cloning which uses eukaryotic host cells with a vector that is replicated in the same way as a host cell chromosome. YACs contain all the essential features of a chromosome required for its propagation in a yeast cell including an **origin of replication**, a **centromere** to ensure segregation into daughter cells and **telomeres** to stabilize the ends of the chromosome (*Fig. 7*). Very large DNA molecules up to several hundred kbp can be cloned using YACs. This is significant because individual clones are large enough to encompass an entire mammalian gene. YACs are also used to construct maps of parts of the human genome by identifying clones containing adjacent regions of the genome. Two related types of vector are bacterial artificial chromosomes (BACs) and P1 artificial chromosomes (PACs) which have uses similar to those of YACs.

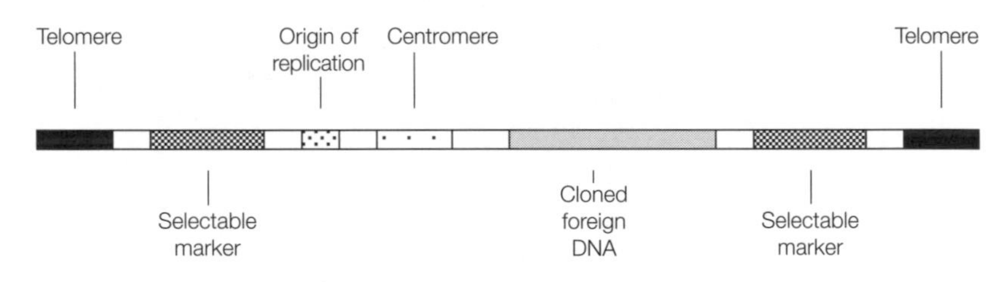

Fig. 7. Structure of a yeast artificial chromosome (YAC).

Plant cloning vectors

Most vectors for cloning with plants as the host organism are based on the **Ti plasmid** which occurs in a soil bacterium called *Agrobacterium tumefaciens*. The bacterium invades plant tissue and causes a cancerous growth called a **crown gall**. During infection part of the Ti plasmid called the **T-DNA** integrates into the plant chromosomal DNA. Cloning vectors based on the Ti plasmid use the ability of T-DNA to integrate to carry useful genes into the plant genome; such genes may confer useful features to the plant such as resistance to disease (see Topic F7).

Applications of DNA cloning

Gene cloning is a powerful technique that has made important contributions to a variety of areas in biological research. These include:

- **Identification of genes with involvement in disease processes**. Cloning has led to the identification of defective genes that cause inherited diseases such as hemophilia and muscular dystrophy and oncogenes and tumor supressor genes that play an important role in tumor development. Characterization of these genes has greatly improved our understanding of disease processes.

- **Genome mapping.** The identification and characterization of clones corresponding to adjacent regions on chromosomes is being used to construct maps which will allow the relative positions of genes to be determined.
- **Recombinant proteins.** Gene cloning is used to produce large amounts of medically useful proteins such as insulin which is used in the treatment of diabetes and Factor VIII which is used to treat hemophilia. Genes encoding useful proteins are cloned in expression vectors and introduced into a suitable host organism such as yeast which expresses the cloned sequence in large amounts.
- **Genetically modified organisms.** Gene cloning can be used to transfer foreign genes into an organism creating transgenic plants and animals with modified characteristics.

E3 Polymerase chain reaction

Key Notes

Principle

Polymerase chain reaction (PCR) allows specific DNA sequences to be copied or amplified over a million fold in a simple enzyme reaction. DNA, corresponding to genes or fragments of genes, can be amplified from samples of chromosomal DNA containing thousands of genes. Amplified DNA is used for the analysis or manipulation of genes.

Components

Each PCR contains four important components: (i) template DNA containing the target DNA sequence to be amplified; (ii) oligonucleotide primers – short single-stranded DNA molecules that bind by complementary base-pairing to opposite strands of the template DNA at either end of the sequence to be amplified; (iii) DNA polymerase. This enzyme copies the target sequence and is thermostable. Several polymerases are used. *Taq* DNA polymerase is the most common; (iv) dNTPs, the substrates for the polymerase. Four are present corresponding to the bases in DNA.

How the PCR works

Target DNA is amplified by 20–40 cycles of DNA synthesis. Each cycle has three stages carried out at different temperatures: (i) denaturation – the reaction is heated to above 90°C to separate the strands of the double helix; (ii) annealing – the reaction is cooled to 40–60°C to allow the primers to bind to the single-stranded template DNA; (iii) extension – the reaction is heated to 72°C where the polymerase is most active and the target DNA sequence is copied. Each molecule of target DNA acts as a template for the synthesis of new DNA in the next cycle. This leads to a rapid increase in the amount of target DNA with successive cycles.

Applications

PCR is used extensively to study genes. It has many applications throughout biology and medicine and has made important contributions to the study of inherited diseases and in cancer research. It also has practical applications in other areas including forensic science and biotechnology.

Related topics

DNA structure (A1)
DNA replication (A9)
Genetic diseases (F1)
Genes and cancer (F2)
Genetics in forensic science (F6)
Genetic engineering and biotechnology (F7)

Principle

The polymerase chain reaction (PCR) is a powerful and widely used technique that has greatly advanced our ability to analyze genes. Chromosomal DNA present in cells contains thousands of genes. This makes it difficult to isolate and analyze any individual gene. PCR allows specific DNA sequences, usually corresponding to genes or parts of genes, to be copied from chromosomal DNA in a simple enzyme reaction. The only requirement is that some of the DNA sequence at either end of the region to be copied is known. DNA corresponding to the sequence of interest is copied or **amplified** by PCR more than a million fold and becomes the predominant DNA molecule in the reaction. Sufficient DNA is obtained for detailed analysis or manipulation of the amplified gene.

Components

DNA is amplified by PCR in an enzyme reaction which undergoes multiple incubations at different temperatures. Each PCR has four key components:

- **Template DNA.** This contains the DNA sequence to be amplified. The template DNA is usually a complex mixture of many different sequences, as is found in chromosomal DNA, but any DNA molecule that contains the target sequence can be used. RNA can also be used for PCR by first making a DNA copy using the enzyme reverse transcriptase.
- **Oligonucleotide primers.** Each PCR requires a pair of oligonucleotide primers. These are short single-stranded DNA molecules (typically 20 bases) obtained by chemical synthesis. Primer sequences are chosen so that they bind by complementary base-pairing to opposite DNA strands on either side of the sequence to be amplified.
- **DNA polymerase.** A number of DNA polymerases are used for PCR. All are thermostable and can withstand the high temperatures (up to 100°C) required. The most commonly used enzyme is *Taq* DNA polymerase from *Thermus aquaticus*, a bacterium present in hot springs. The role of the DNA polymerase in PCR is to copy DNA molecules. The enzyme binds to single-stranded DNA and synthesizes a new strand complementary to the original strand. DNA polymerases require a short region of double-stranded DNA to get started. In PCR, this is provided by the oligonucleotide primers which create short double-stranded regions by binding on either side of the DNA sequence to be amplified. In this way the primers direct the DNA polymerase to copy only the target DNA sequence.
- **Deoxynucleotide triphosphates (dNTPs).** These molecules correspond to the four bases present in DNA (adenine, guanine, thymine and cytosine) and are substrates for the DNA polymerase. Each PCR requires four dNTPs (dATP, dGTP, dTTP, dCTP) which are used by the DNA polymerase as **building blocks** to synthesize new DNA.

How the PCR works

PCR allows the amplification of target DNA sequences through repeated cycles of DNA synthesis (*Fig. 1*). Each molecule of target DNA synthesized acts as a template for the synthesis of new target molecules in the next cycle. As a result, the amount of target DNA increases with each cycle until it becomes the dominant DNA molecule in the reaction. During the early cycles, DNA synthesis increases exponentially but in later cycles, as the amount of target DNA to be copied increases and the reaction components are used up, the increase becomes linear and then reaches a plateau.

Each cycle of DNA synthesis involves three stages (**denaturation**, **primer annealing, elongation**) which take place at different temperatures and together result in the synthesis of target DNA.

Denaturation

The reaction is heated to greater than 90°C. At this temperature the double helix is destabilized and the DNA molecules separate into single strands capable of being copied by the DNA polymerase.

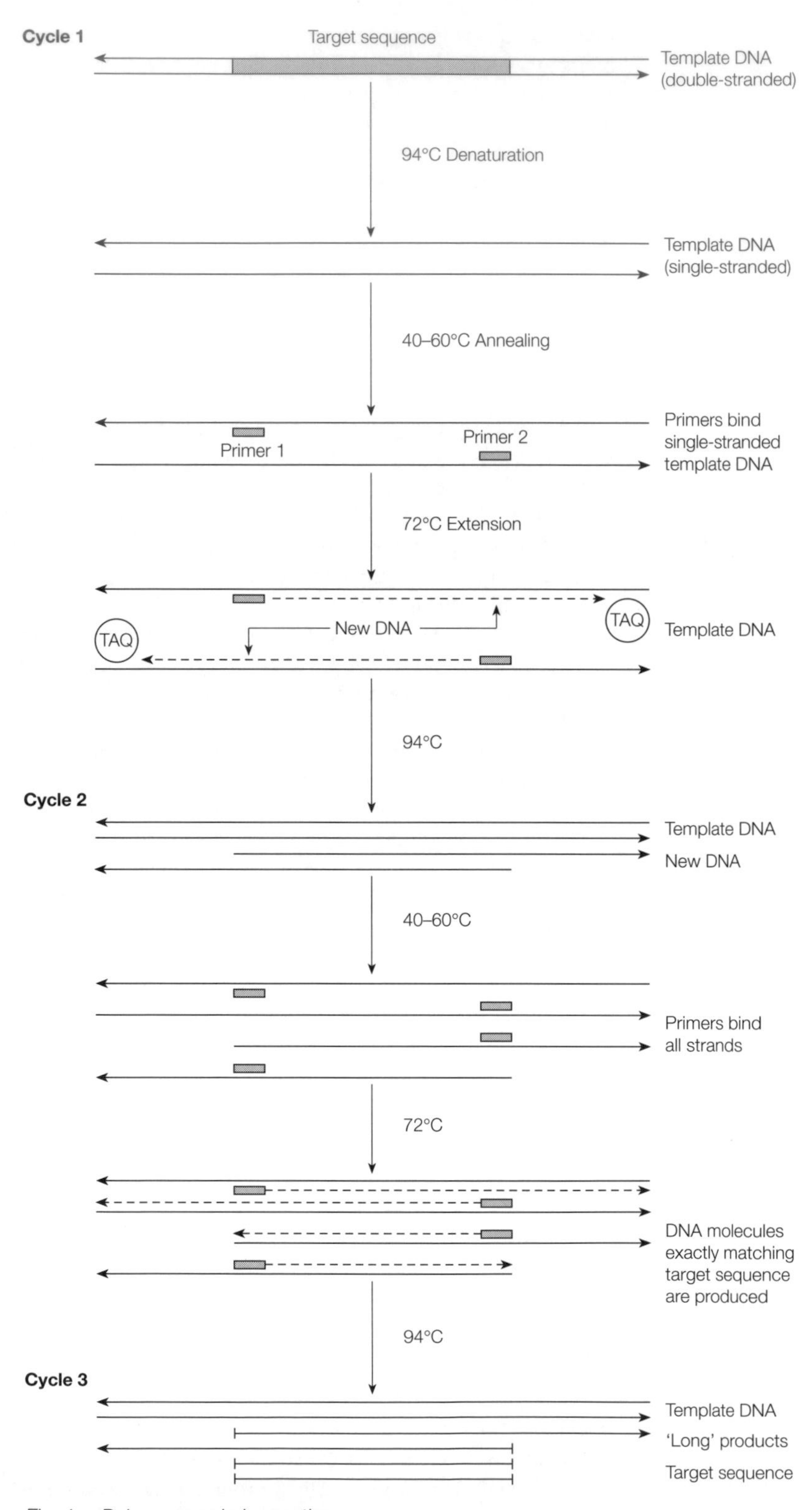

Fig. 1. Polymerase chain reaction.

Primer annealing

The reaction is cooled to a temperature that allows binding of the primers to the single-stranded DNA without permitting the double helix to reform between the template strands. This process is called **annealing**. The temperature used varies (typically 40–60°C) and is determined by the sequence and the number of bases in the primers.

Extension

This stage is carried out at the temperature at which the DNA polymerase is most active. For *Taq*, this is 72°C. The DNA polymerase, directed by the position of the primers, copies the intervening target sequence using the single-stranded DNA as a template. A total of 20–40 PCR cycles is carried out depending on the abundance of the target sequence in the template DNA. Sequences up to several thousand base pairs can be amplified. To deal with the large number of separate incubations needed, the PCR is carried out using a microprocessor-controlled heating block known as a **thermal cycler**. In the first cycle, DNA molecules are synthesized which extend beyond the target sequence. This is because there is nothing to prevent the DNA polymerase continuing to copy the template beyond the end of the target sequence. However in subsequent cycles, newly synthesised DNA molecules which end with the primer sequence act as templates and limit synthesis to the target sequence so that the amplified DNA contains only the target sequence.

Applications

PCR is used extensively as a research tool which has greatly improved our ability to study genes. Most studies in molecular genetics involve the use of PCR at some stage, normally as part of an overall strategy and in association with other techniques. For example, DNA amplified by PCR can be used for DNA sequencing, as a probe in Northern and Southern blotting, and to generate clones. PCR has applications in most areas of biology and medicine as well as in unexpected subjects such as anthropology and archaeology. It is also an important tool in the biotechnology industries.

PCR has made important contributions in many areas which include:

- **Inherited diseases.** These disorders are caused by gene mutations passed on from parents to their children. Examples include hemophilia and cystic fibrosis. PCR is used to amplify gene sequences which can then be screened for disease-causing mutations. The information obtained has dramatically improved our understanding of these disorders and has produced the important additional benefit of allowing carriers of the disorders to be identified.
- **Cancer research.** PCR has been widely used in studies of the role of genes in cancer. For example, mutations in oncogenes and tumor-supressor genes have been identified in DNA from tumors using PCR-based strategies. This has improved our understanding of how cancer develops.
- **Forensic science.** By amplifying repetitive sequences, PCR can be used to identify individuals from samples of their DNA. This is used to link individuals with forensic DNA samples from the scene of a crime. Analysis of variable sequences is also used in tissue typing to match organ donors with recipients and in anthropology to study the origins of races of people.
- **Biotechnology.** PCR has played an important role in the production of recombinant proteins such as insulin and growth hormone which are widely used as drugs and in the development of recombinant vaccines, such as that for hepatitis B virus.

E4 DNA SEQUENCING

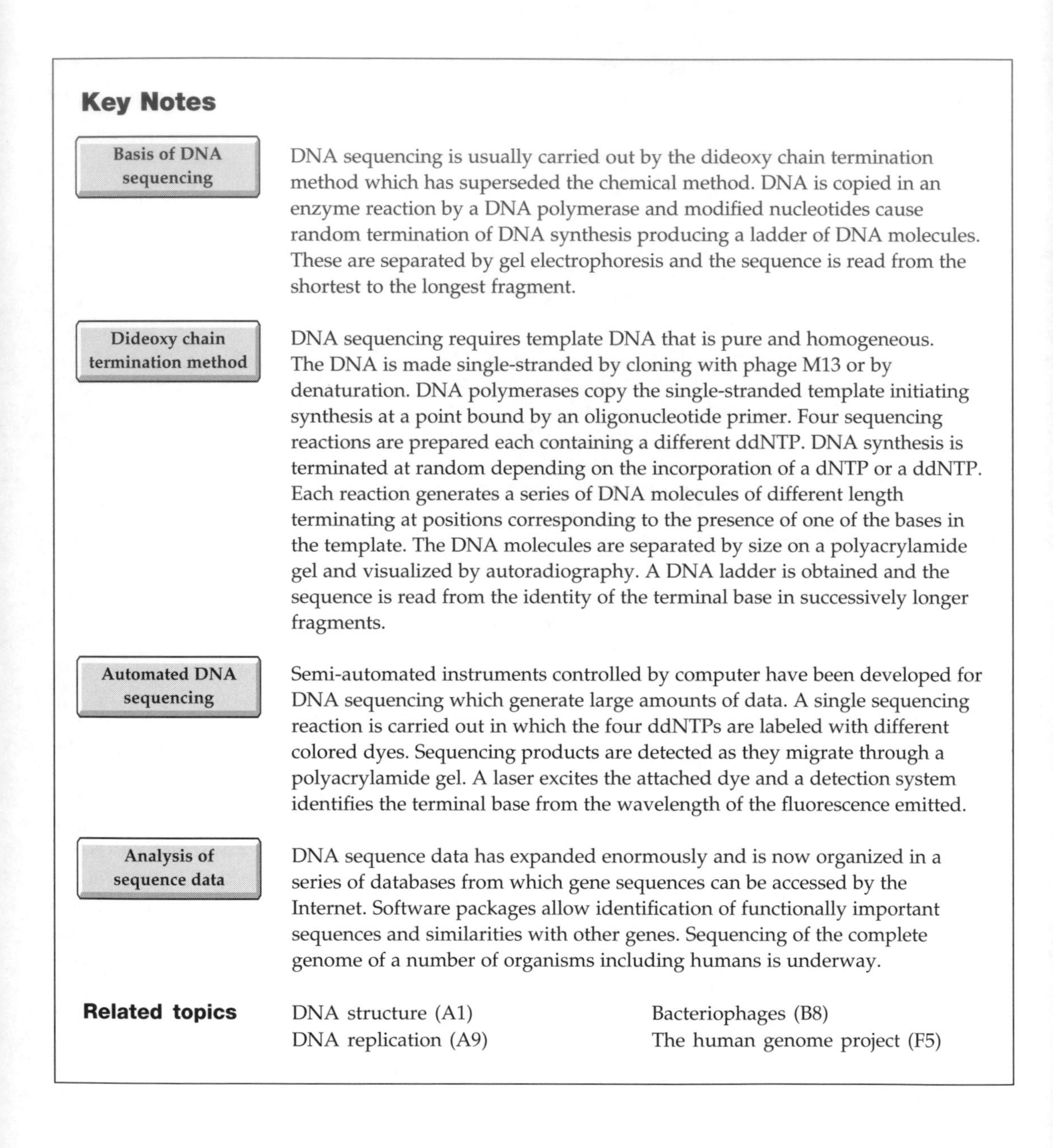

Key Notes

Basis of DNA sequencing

DNA sequencing is usually carried out by the dideoxy chain termination method which has superseded the chemical method. DNA is copied in an enzyme reaction by a DNA polymerase and modified nucleotides cause random termination of DNA synthesis producing a ladder of DNA molecules. These are separated by gel electrophoresis and the sequence is read from the shortest to the longest fragment.

Dideoxy chain termination method

DNA sequencing requires template DNA that is pure and homogeneous. The DNA is made single-stranded by cloning with phage M13 or by denaturation. DNA polymerases copy the single-stranded template initiating synthesis at a point bound by an oligonucleotide primer. Four sequencing reactions are prepared each containing a different ddNTP. DNA synthesis is terminated at random depending on the incorporation of a dNTP or a ddNTP. Each reaction generates a series of DNA molecules of different length terminating at positions corresponding to the presence of one of the bases in the template. The DNA molecules are separated by size on a polyacrylamide gel and visualized by autoradiography. A DNA ladder is obtained and the sequence is read from the identity of the terminal base in successively longer fragments.

Automated DNA sequencing

Semi-automated instruments controlled by computer have been developed for DNA sequencing which generate large amounts of data. A single sequencing reaction is carried out in which the four ddNTPs are labeled with different colored dyes. Sequencing products are detected as they migrate through a polyacrylamide gel. A laser excites the attached dye and a detection system identifies the terminal base from the wavelength of the fluorescence emitted.

Analysis of sequence data

DNA sequence data has expanded enormously and is now organized in a series of databases from which gene sequences can be accessed by the Internet. Software packages allow identification of functionally important sequences and similarities with other genes. Sequencing of the complete genome of a number of organisms including humans is underway.

Related topics

DNA structure (A1)
DNA replication (A9)
Bacteriophages (B8)
The human genome project (F5)

Basis of DNA sequencing

Two methods have been developed to determine the nucleotide sequence of DNA molecules: the **dideoxy chain termination method** of **Sanger** and the **chemical degradation** method developed by **Maxam and Gilbert**. The chain termination method has now superseded the chemical method because it is more efficient and is simpler to perform.

The basis of the chain termination method is that the DNA molecule whose sequence is to be determined is copied in an enzyme reaction by a DNA polymerase. Modified nucleotide triphosphates are included in the reaction that cause termination of DNA synthesis randomly at each of the four bases where they occur in the template DNA. The overall effect is that a series of DNA molecules each one nucleotide longer than the next is synthesized which can be separated according to size by electrophoresis. The base sequence of the template DNA can then be determined by identifying the terminal base of each synthesized DNA molecule from the shortest to the longest.

Dideoxy chain termination method

The DNA to be sequenced is called the **template**. It must be obtained in a purified form and must be homogeneous; i.e. it must contain only DNA molecules with the same sequence. Purified plasmids containing cloned DNA and DNA produced by PCR are commonly used as templates. A requirement for sequencing by the dideoxy chain termination method is that the DNA must be present in single-stranded form capable of being copied by a DNA polymerase. Previously this was achieved by cloning the DNA into the phage vector, M13 (see Topic B8). When recombinant phage were used to infect *E. coli*, single-stranded copies of the cloned sequence were produced which could be used as templates for sequencing. More recently, single-stranded DNA for sequencing has been produced in a much simpler way without cloning by denaturation of the DNA template using heat or alkali. Several different DNA polymerases have been used for the sequencing reaction. These include part of the *E. coli* DNA polymerase I known as the **Klenow fragment**, a genetically modified DNA polymerase from the phage T7, called **Sequenase** and ***Taq* DNA polymerase** which is also used in the polymerase chain reaction.

All DNA polymerases require short regions of double-stranded DNA to initiate DNA synthesis on a single-stranded template. This is provided in sequencing reactions by the addition of short, single-stranded DNA molecules called **oligonucleotide primers** which are produced by chemical synthesis. The sequence of the primer is chosen to be complementary to the template DNA such that it binds to it forming a short double-stranded region, thus determining the point at which the sequencing reaction is initiated. To sequence a DNA molecule four separate enzyme reactions are prepared. Each contains: template DNA in single-stranded form, DNA polymerase, primer, each of the four deoxynucleotide triphosphates (dNTPs) which are the building blocks for DNA synthesis and a modified nucleotide called a **dideoxynucleotide triphosphate (ddNTP)**. There are four ddNTPs that correspond to the four bases in DNA. They differ from dNTPs in that they lack a hydroxyl group on the 3′ carbon of the ribose sugar. They are incorporated into the growing DNA polynucleotide by the polymerase but their lack of a 3′ hydroxyl group prevents formation of a phosphodiester bond with the next nucleotide to be added. This prevents further elongation of the DNA polynucleotide being synthesized (*Fig. 1*). The ddNTPs thus act as specific inhibitors of DNA synthesis. Each of the four sequencing reactions contains a different ddNTP. Depending on which ddNTP is present, synthesis will terminate where that nucleotide is incorporated. At any point during DNA synthesis, the polymerase may incorporate a dNTP into the growing polynucleotide chain in which case synthesis continues or it may incorporate a ddNTP in which case chain elongation is blocked and synthesis ends at that position. The overall effect is that in each of the four reactions a series of DNA molecules of different lengths is produced each

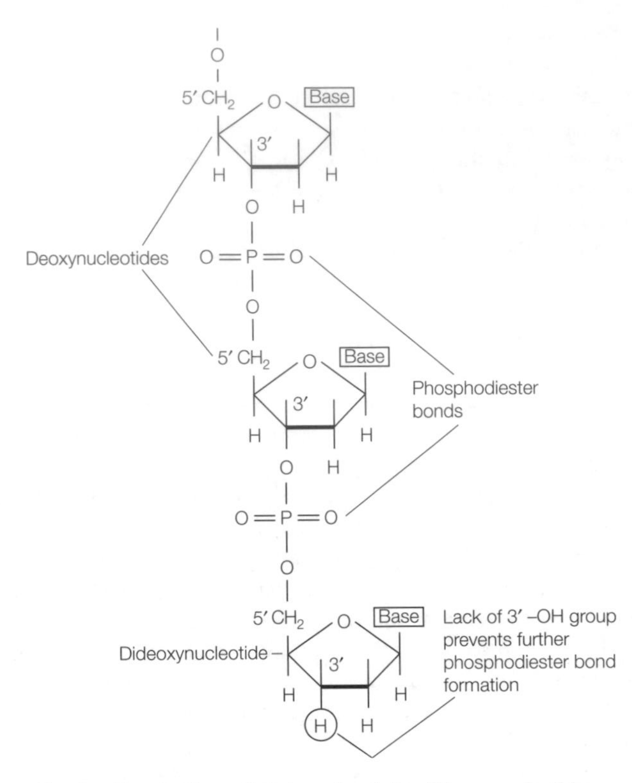

Fig. 1. Termination of DNA synthesis by dideoxynucleotides.

terminating at a position corresponding to the presence of that base in the template DNA (*Fig. 2*). For example, in the sequencing reaction containing ddATP a series of DNA molecules will be synthesized each of which ends in an A corresponding to the position of a T in the template.

When the sequencing reaction is complete the synthesized DNA molecules are separated according to size by electrophoresis on polyacrylamide gels with the four reactions run in adjacent lanes. By adding dNTP which contains radioactive phosphorus or sulfur to the sequencing reactions, the synthesized DNA becomes radioactive allowing it to be detected by placing the polyacrylamide gel against a sheet of X-ray film. This is called **autoradiography**. When the X-ray film is developed a series of bands is visible in the four lanes which form a ladder. The sequence of the template DNA can be determined by identifying the smallest band followed by successively larger bands and assigning the terminal base in each from its lane on the gel (*Fig. 3*).

Automated DNA sequencing

Although the basic principle of the sequencing reaction is unchanged, the technique has been developed to include semi-automated systems in which electrophoresis of DNA and detection and analysis of sequencing reactions is carried out by instruments controlled by computers. In this format, the four ddNTPs are labeled by the covalent attachment of a different colored dye. A single sequencing reaction is carried out and the products are a series of DNA molecules, each one nucleotide longer than the next and labeled with a different dye depending on the identity of the terminal base. The products are resolved

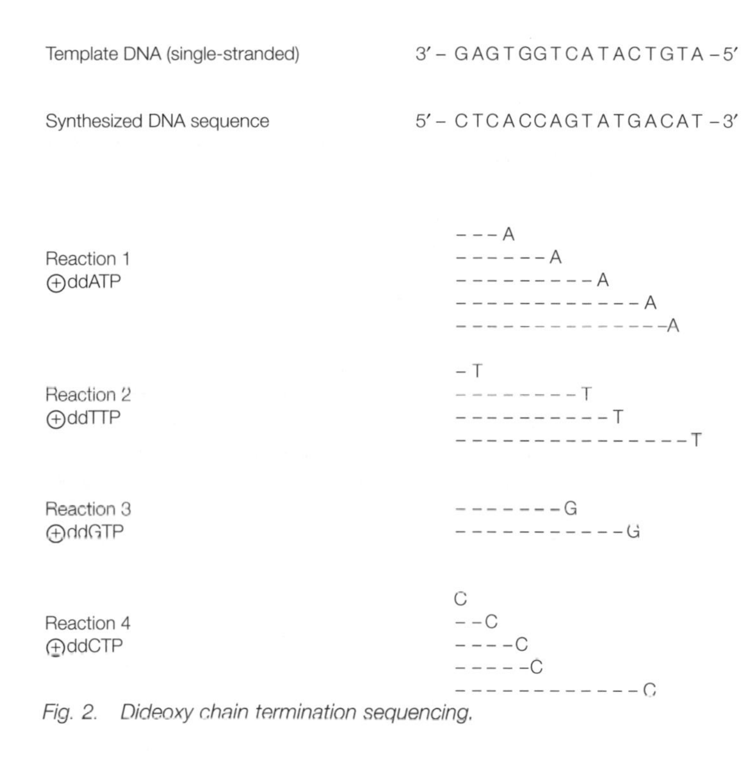

Fig. 2. Dideoxy chain termination sequencing.

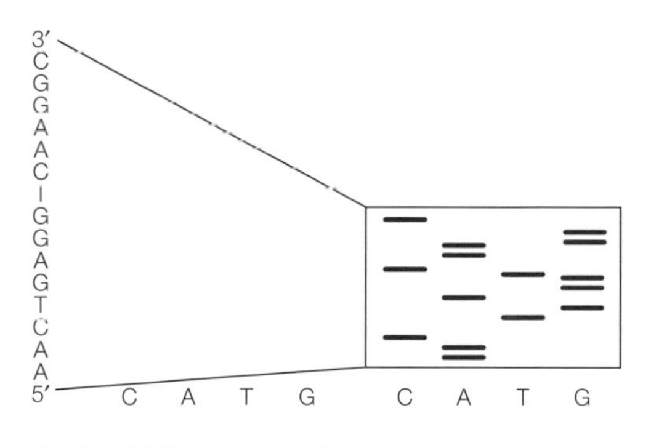

Fig. 3. DNA sequencing ladder.

by electrophoresis on polyacrylamide gels and are detected as they migrate past the end of the gel by a laser that excites the dye label to emit fluorescence which is then detected by the instrument. From the wavelength of the emitted fluorescence the instrument identifies the dye and so the terminal base on each DNA species and can build up a picture of the sequence of bases passing the detector.

Automated sequencing systems represent a significant advance over conventional or manual sequencing approaches. Each sequencing reaction requires just one tube and one lane on the gel compared with four per sample required for manual sequencing. This feature coupled with the automated detection of samples and analysis of data greatly increases the amount of sequence information that can be generated.

Analysis of sequence data

DNA sequencing is one of the core techniques of molecular genetics and is widely used in research. The importance of the technique lies in its ability to determine the information content of DNA molecules. In recent years, the amount of DNA sequence information has expanded enormously to the point where it has become necessary to organize it in a series of databases. The main databases are called **EMBL**, which is based in Europe, and **Genbank**, based in the USA. These databases contain sequences of human genes and genes from other species generated by research projects worldwide. The information is freely available over the Internet and can be down loaded for analysis.

DNA sequencing generates a large amount of data. To aid the analysis of sequences, a series of software packages have been developed. These include the **University of Wisconsin GCG package** which allows important features in DNA sequences such as restriction enzyme sites, start codons, stop codons, open reading frames, intron–exon junctions and promoter sequences to be identified. The programs also allow sequences to be compared with other sequences in the databases to look for similarities that suggest that genes are related.

The development of automated DNA sequencing systems has made it feasible to determine the entire DNA sequence of several important organisms. Already, the entire sequence of the genomes of *E. coli* (4.6×10^6 bp) and the yeast, *S. cerevisiae* (2.3×10^7 bp) has been determined. In addition, the **human genome mapping project** aims to identify and locate all human genes by determining the entire 3×10^9 bp sequence of the human genome by the year 2005 (see Topic F5).

F1 Genetic diseases

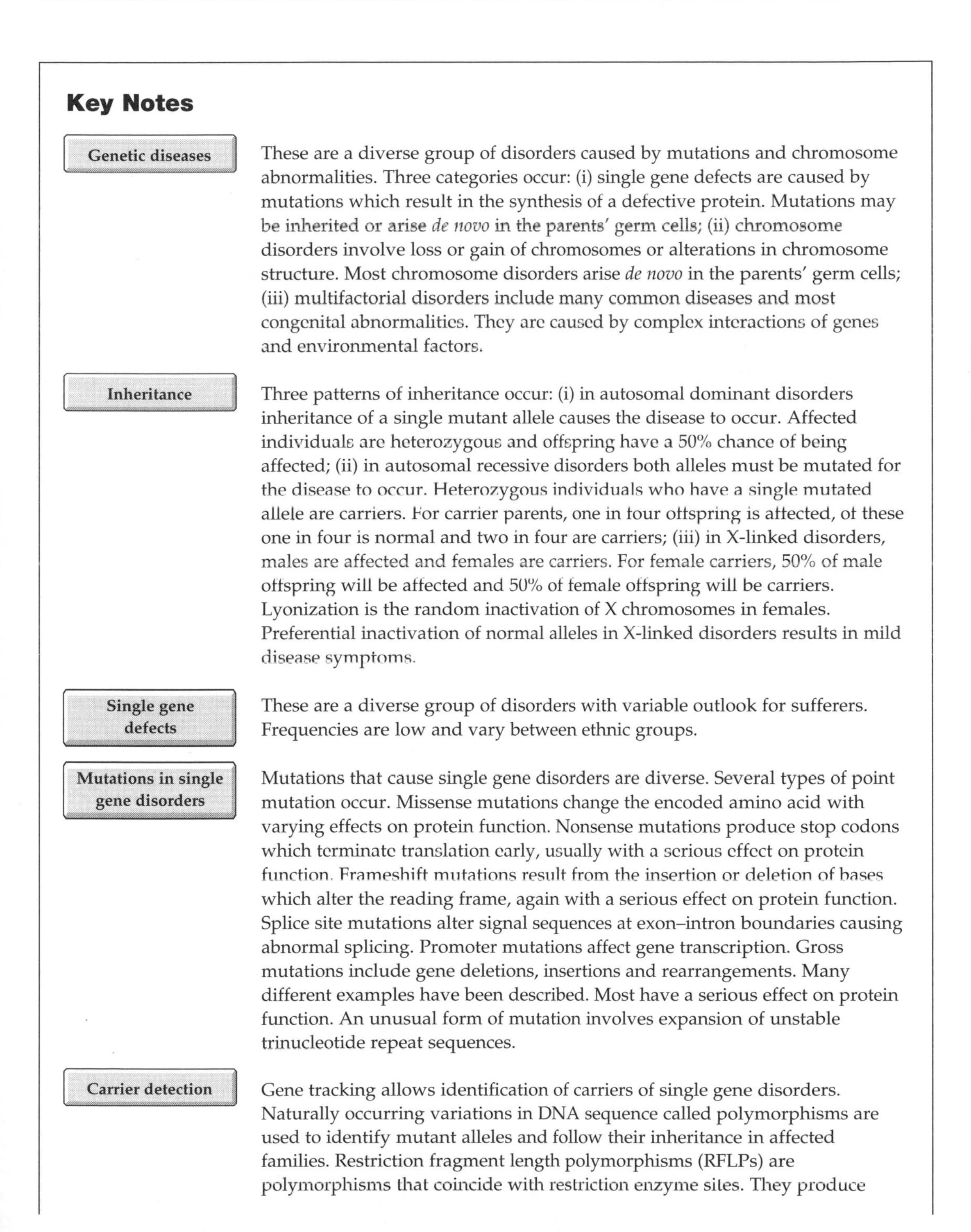

Key Notes

Genetic diseases

These are a diverse group of disorders caused by mutations and chromosome abnormalities. Three categories occur: (i) single gene defects are caused by mutations which result in the synthesis of a defective protein. Mutations may be inherited or arise *de novo* in the parents' germ cells; (ii) chromosome disorders involve loss or gain of chromosomes or alterations in chromosome structure. Most chromosome disorders arise *de novo* in the parents' germ cells; (iii) multifactorial disorders include many common diseases and most congenital abnormalities. They are caused by complex interactions of genes and environmental factors.

Inheritance

Three patterns of inheritance occur: (i) in autosomal dominant disorders inheritance of a single mutant allele causes the disease to occur. Affected individuals are heterozygous and offspring have a 50% chance of being affected; (ii) in autosomal recessive disorders both alleles must be mutated for the disease to occur. Heterozygous individuals who have a single mutated allele are carriers. For carrier parents, one in four offspring is affected, of these one in four is normal and two in four are carriers; (iii) in X-linked disorders, males are affected and females are carriers. For female carriers, 50% of male offspring will be affected and 50% of female offspring will be carriers. Lyonization is the random inactivation of X chromosomes in females. Preferential inactivation of normal alleles in X-linked disorders results in mild disease symptoms.

Single gene defects

These are a diverse group of disorders with variable outlook for sufferers. Frequencies are low and vary between ethnic groups.

Mutations in single gene disorders

Mutations that cause single gene disorders are diverse. Several types of point mutation occur. Missense mutations change the encoded amino acid with varying effects on protein function. Nonsense mutations produce stop codons which terminate translation early, usually with a serious effect on protein function. Frameshift mutations result from the insertion or deletion of bases which alter the reading frame, again with a serious effect on protein function. Splice site mutations alter signal sequences at exon–intron boundaries causing abnormal splicing. Promoter mutations affect gene transcription. Gross mutations include gene deletions, insertions and rearrangements. Many different examples have been described. Most have a serious effect on protein function. An unusual form of mutation involves expansion of unstable trinucleotide repeat sequences.

Carrier detection

Gene tracking allows identification of carriers of single gene disorders. Naturally occurring variations in DNA sequence called polymorphisms are used to identify mutant alleles and follow their inheritance in affected families. Restriction fragment length polymorphisms (RFLPs) are polymorphisms that coincide with restriction enzyme sites. They produce alternative restriction fragment patterns which can be detected by Southern blotting. RFLPs can distinguish between alleles and can be used as markers of mutant alleles. Carriers are identified by comparing their restriction fragment pattern with that of an affected relative. The use of RFLPs for carrier detection is limited because the degree of sequence variation is small. Two types of highly variable repetitive sequences called variable number tandem repeats (VNTRs) and CA repeats are now replacing the use of RFLPs. These consist of blocks of DNA in which a short sequence is repeated multiple times. The number of repeats is highly variable and can be used to follow the inheritance of mutant alleles. Carriers can also be identified by direct detection of mutations.

Related topics

The genetic code (A3)
Chromosomes (B1)
The human genome (B4)
DNA mutation (B5)
Nucleic acid hybridization (E1)
Genes and cancer (F2)
Gene therapy (F4)

Genetic diseases

These are a diverse group of diseases and conditions which result from gene mutations and chromosome abnormalities. Disorders with a genetic basis fall into three categories.

Single gene defects

Single gene defects are also known as **Mendelian disorders, monogenic disorders** or **single locus disorders**. These are a group of diseases caused by the presence, in affected individuals, of a single mutated gene. The mutation changes the coding information of the gene such that it either produces protein which is defective or fails to produce any protein at all. The resulting protein deficiency is responsible for the disease symptoms. The gene mutation may be passed between generations from parents to children or may arise spontaneously (*de novo*) in a germ cell (sperm or ovum) of a parent which, after fertilization, gives rise to a child who carries the mutation in every cell.

Chromosome disorders

These are conditions caused by the loss or gain of one or more chromosomes or by alterations in chromosome structure. Most chromosome disorders arise *de novo* in the parents' germ cells but examples of inherited chromosome disorders also exist. Abnormalities relating to the number of chromosomes may involve the presence of multiple copies of each chromosome (**polyploidy**) or the gain or loss of individual chromosomes (**aneuploidy**) (see Topic B1). Structural chromosome abnormalities result from chromosome breakage and may involve the deletion, duplication or rearrangement of chromosome segments.

Multifactorial disorders

These include many common diseases such as diabetes and coronary artery disease as well as most congenital malformations. They are influenced by genes

Table 1. Single gene disorders

Disorder	Frequency per 1000 births	Pattern of inheritance	Mutated gene	Characteristics
Hemophilia A	0.1	X-linked	Factor VIII	Abnormal bleeding
Hemophilia B	0.03	X-linked	Factor IX	Abnormal bleeding
Duchenne muscular dystrophy	0.3	X-Linked	Dystrophin	Muscle wasting
Becker muscular dystrophy	0.05	X-linked	Dystrophin	Muscle wasting
Fragile X syndrome	0.5	X-linked	FMR1	Mental retardation
Huntington's disease	0.5	Autosomal dominant	Huntingtin	Dementia
Neurofibromatosis	0.4	Autosomal dominant	NF-1,2	Cancer
Thalassemia	0.05	Autosomal recessive	Globin genes	Anemia
Sickle cell anemia	0.1	Autosomal recessive	β-globin	Anemia, ischemia
Phenylketonuria	0.1	Autosomal recessive	Phenylalanine-hydroxylase	Inability to metabolize phenylalanine
Cystic fibrosis	0.4	Autosomal recessive	CFTR	Progressive lung damage and other symptoms

in complex ways which are poorly understood but involve the interaction of multiple genes and interactions between genes and environmental factors.

During the last 15–20 years, as a result of the development of recombinant DNA technology, there have been enormous advances in our understanding of single gene disorders. The information in the following sections refers to these disorders.

Inheritance

Single gene disorders are passed on between generations from parents to their children. Three patterns of inheritance occur: **autosomal dominant**, **autosomal recessive** and **X-linked** (*Table 1*). Every human cell contains 22 pairs of homologous chromosomes known as **autosomes** and a pair of X or Y **sex chromosomes**. Females have two X chromosomes and males have a single X chromosome and a Y chromosome. Duplicate copies of genes occur on chromosome pairs and are known as **alleles**.

In autosomal dominant disorders, the inheritance of a single mutated allele is sufficient for an individual to be affected by the disease. Affected individuals have one normal allele and one mutated allele and are said to be **heterozygous**. The child of an affected individual will have a 50% chance of inheriting the mutated allele and being affected by the disorder themselves (*Fig. 1a*).

In autosomal recessive disorders, two mutant alleles (one from each parent) must be inherited for an individual to be affected by the disease. Affected individuals are said to be **homozygous** for the mutant allele. Individuals who inherit a single mutant allele are heterozygous and are not affected by the disease, but they are carriers and may pass their mutant allele on to their children. For parents who are heterozygous carriers of an autosomal recessive disorder, one in four of their children will be affected by the disorder, one in four will be normal and one in two will be carriers (*Fig. 1b*).

In X-linked disorders, the mutated gene is present on the X chromosome. Since males have a single X chromosome, inheritance of a mutated allele is sufficient for the disease to occur. Affected males are said to be **hemizygous**. Females have two X chromosomes and usually remain unaffected because most X-linked disorders are recessive. Females may be carriers of X-linked disorders

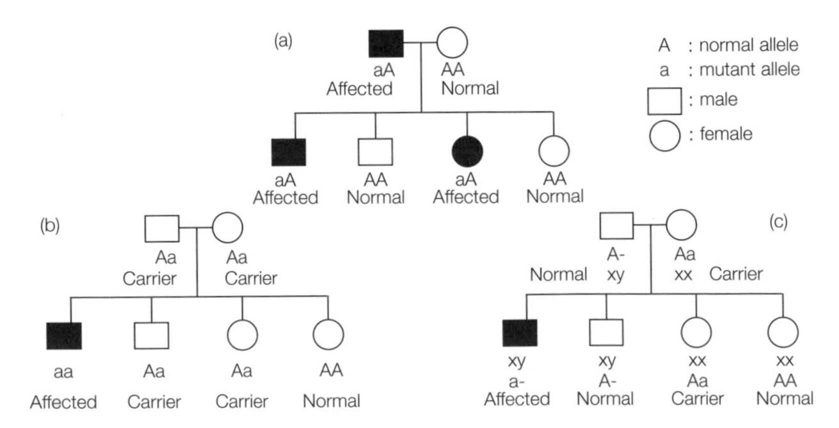

Fig. 1. (a) Autosomal dominant. Inheritance of a single mutated allele (a) results in disease. (b) Autosomal recessive. Affected individuals have two mutant alleles (aa). Heterozygotes (Aa) are carriers. (c) X-linked. For female carriers, 50% of male offspring are affected and 50% of female offspring are carriers.

with 50% of their male offspring affected and 50% of their female offspring being carriers (*Fig. 1c*).

To ensure that females produce the same amount of product from X chromosome genes as males, one of the two X chromosomes in female cells is inactivated. This process is called **Lyonization** and takes place during embryonic development. The inactivated chromosome is selected at random in each cell. However, some female carriers of X-linked diseases may show mild symptoms of the disorder due to increased inactivation of the normal X chromosome.

Single gene defects

Many different single gene disorders occur in humans (*Table 1*). These have widely different characteristics and varying consequences for the sufferers which depend on the importance of the mutated gene and the nature of the mutation present. Some disorders, such as hemophilia, produce symptoms that are treatable but others, such as Huntington's disease, have no effective treatment and lead to the premature death of sufferers.

Single gene disorders are relatively rare with frequencies between approximately 0.01 and five cases per 1000 live births (*Table 1*). The frequencies of disorders vary between ethnic groupings. For example, the incidence of cystic fibrosis is highest among northern Europeans, sickle cell anemia occurs at high frequency in Africans and β-thalassemia is most common in Asian populations. For all of the more common disorders, the defective gene has been identified and cloned and the disease-causing mutations have been identified.

Mutations in single gene disorders

Two types of mutation occur in single gene disorders: **point mutations** which involve single base changes and **gross mutations** which involve alteration of longer DNA sequences (see Topic B5). For each disorder, the type of mutation present varies. In addition, individuals affected by the same disorder may carry different mutations. For example, about 20% of cases of hemophilia A result from gross mutations. The remainder result from point mutations of which over 250 different types have been described.

Point mutations

Point mutations that cause genetic diseases fall into a number of categories.

(1) Missense mutations. These are base changes that alter the codon for an amino acid resulting in its substitution with a different amino acid. Missense mutations have a variety of consequences for the encoded protein. Due to the degeneracy of the genetic code, changes that affect the third base of a codon often do not alter the encoded amino acid. These are called **silent mutations** and have no affect on the encoded protein. In addition, many base changes result in the replacement of an amino acid with another of similar character leaving the protein unaffected. For example, mutation of the codon CTT to ATT would result in the replacement of the hydrophobic amino acid leucine with isoleucine, another hydrophobic amino acid. Many other missense mutations have been described which do affect the encoded protein and result in genetic diseases. These include an A to T mutation in the gene for β-globin, one of the polypeptides of hemoglobin. This mutation changes codon six of the gene from GAG which encodes glutamic acid to GTG which encodes valine. The mutation results in a condition called sickle cell anemia in which the red blood cells adopt an abnormal sickle shape due to aggregation of the hemoglobin molecules. The abnormal cells are short-lived, which causes anemia and become lodged in capillaries, which reduces the blood supply to organs (ischemia).

(2) Nonsense mutations. These are base changes that convert an amino acid codon to a stop codon resulting in premature termination of translation and the production of a shortened protein. Nonsense mutations usually have a serious effect on the encoded protein, especially if they occur near the 5′ end of the gene. Nonsense mutations have been identified in many different genetic diseases. Examples include a C to T mutation in codon 39 of the β-globin gene which changes the normal CAG which codes for glutamine to TAG which is a stop codon. This mutation causes premature termination of translation of β-globin mRNA leading to a complete absence of β-globin polypeptide chains and resulting in a condition called β-thalassemia which is characterized by anemia due to a failure of normal hemoglobin molecules to form.

(3) Frameshift mutations. These mutations result from the insertion or deletion of one or more bases causing the reading frame of the gene to be altered and a different set of codons to be read downstream (3′) of the mutation. Frameshift mutations usually have serious consequences for the encoded protein, especially when they occur near the 5′ end of the gene. Many different frameshift mutations have been described. Several examples have been identified in individuals affected by hemophilia A. These include a deletion of four bases which results in a change of reading frame after codon 50 and an insertion of 10 bases which alters the reading frame beyond codon 38. Both mutations were associated with severe disease symptoms.

(4) Splice site mutations. These mutations alter the signal sequences for splicing that occur at the 5′ and 3′ ends of introns leading to a failure of RNA transcribed from the mutated gene to be spliced properly. Mutations also occur in introns which result in the creation of splice sites, again causing splicing to occur abnormally. A number of splice site mutations have been described in the β-globin gene which result in a complete absence of β-globin polypeptides in homozygotes and causes β-thalassemia.

(5) Promoter mutations. These mutations occur infrequently and affect the way gene transcription is regulated, often reducing or eliminating expression of the gene. A mutation has been identified in the promoter of the gene for the blood clotting protein, Factor IX, which causes hemophilia B. Carriers of the mutation fail to produce any Factor IX protein and bleed abnormally. Unusually, the condition disappears after puberty when steroid hormones stimulate transcription of the gene.

Gross mutations

Many different gross mutations have been described. Most have a serious effect on gene function and are associated with severe disease symptoms.

(1) Deletion mutations. Gene deletions vary greatly in size from a few bases to entire genes. Complete deletion of the α-globin genes causes α-thalassemia, a blood disorder characterized by a failure to produce normal hemoglobin. Partial deletions in the dystrophin gene cause the muscle wasting disease muscular dystrophy and deletion of a single codon in the CFTR (cystic fibrosis transmembrane conductance regulator) gene accounts for about 70% of cases of cystic fibrosis.

(2) Insertion mutations. Many different insertion mutations have been reported. Examples include an unusual case of hemophilia A caused by insertion of a repetitive sequence called a LINE element (see Topic B4) into the Factor VIII gene.

(3) Rearrangements. Many different gene rearrangements have been described. Examples include a recurrent mutation that causes hemophilia A in which recombination occurs between sequences present in intron 22 of the Factor VIII gene and duplicate sequences further along the X chromosome. Due to an error of recombination the Factor VIII gene is split into two segments separated by several million base pairs, completely disrupting gene function.

(4) Trinucleotide repeat mutations. An unusual form of gene mutation has been described which involves unstable trinucleotide repeat sequences. During meiosis, a dramatic increase in the number of copies of the trinucleotide repeat occurs in germ cells leading to development of the disease in subsequent offspring. The mechanism by which the repeat expansion occurs and the way this leads to manifestation of the disease is poorly understood. Expansion of trinucleotide repeats has been found in association with a number of genetic diseases including fragile X syndrome and Huntington's disease.

Carrier detection

In recessive genetic disorders, individuals who are heterozygous for the mutated allele are not affected by the disorder but are carriers and may pass the mutated allele on to their children. If both parents are carriers, it is possible that a child will inherit two mutated alleles and will be affected by the disease. It is therefore desirable to identify carriers for the purposes of providing **genetic counseling** to allow an informed choice about parenting.

Carriers of genetic diseases can be identified by a process called **gene tracking** which allows mutant alleles to be identified and their inheritance followed among the members of a family. Gene tracking makes use of naturally occurring variations in DNA sequence that exist in the human genome. Comparison of a gene sequence in a group of individuals will often reveals variations in the DNA sequence in the form of alternative bases. For example, at a given point in the DNA sequence some individuals might have a C whereas others might have a T. These variations are called **polymorphisms** and they result from mutation of the DNA. Polymorphisms are very common occurring on average every 100–150 bases throughout the genome. When they occur within the coding sequences of genes they may cause genetic diseases and are subject to selective pressures. When they occur in noncoding sequences, they usually have no consequences for the organism and so tend to accumulate. Some polymorphisms coincide with the target sequences recognized by restriction enzymes (see Topic E2). Depending on which of the alternative sequences occurs, a restriction enzyme site may be present or absent at that point. Polymorphisms that occur at restriction enzyme sites are known as **restriction fragment length polymorphisms (RFLPs)**. RFLPs are extremely useful because they can be used as markers of the genes they occur in and can be detected by Southern blotting (see Topic E1). If a sample of genomic DNA is digested with a restriction enzyme the size of the fragments produced at the RFLP varies depending on whether a restriction site is present or absent. The pattern of fragments detected by Southern blotting will vary between individuals depending on which of the alternative sequences is present (*Fig. 2*). The restriction fragment pattern can also distinguish between two alleles carried by an individual. By analyzing DNA from an individual affected by a genetic disease it is possible to determine the polymorphism associated with a mutant allele. It is then possible to trace its inheritance by analyzing DNA from other family members. By determining from their restriction fragment pattern whether unaffected siblings have inherited normal or mutant alleles, it is possible to determine if they are carriers of the disorder.

Gene tracking can also be used for **prenatal diagnosis** to determine the status of an unborn fetus. A sample of fetal DNA is obtained and Southern blotting is used to analyze the relevant RFLP and determine if the fetus is normal, a heterozygous carrier or is homozygous and affected.

The use of RFLPs to detect carriers of genetic diseases is limited because the extent of the variation in the DNA is small. Only two possible sequences can occur at an RFLP so that in many cases the mutant allele and the normal allele will have the same sequence and cannot be distinguished. In these cases it is not possible to determine the carrier status of individuals at risk. This limitation of RFLPs has been overcome by the use of repetitive gene sequences which are much more variable.

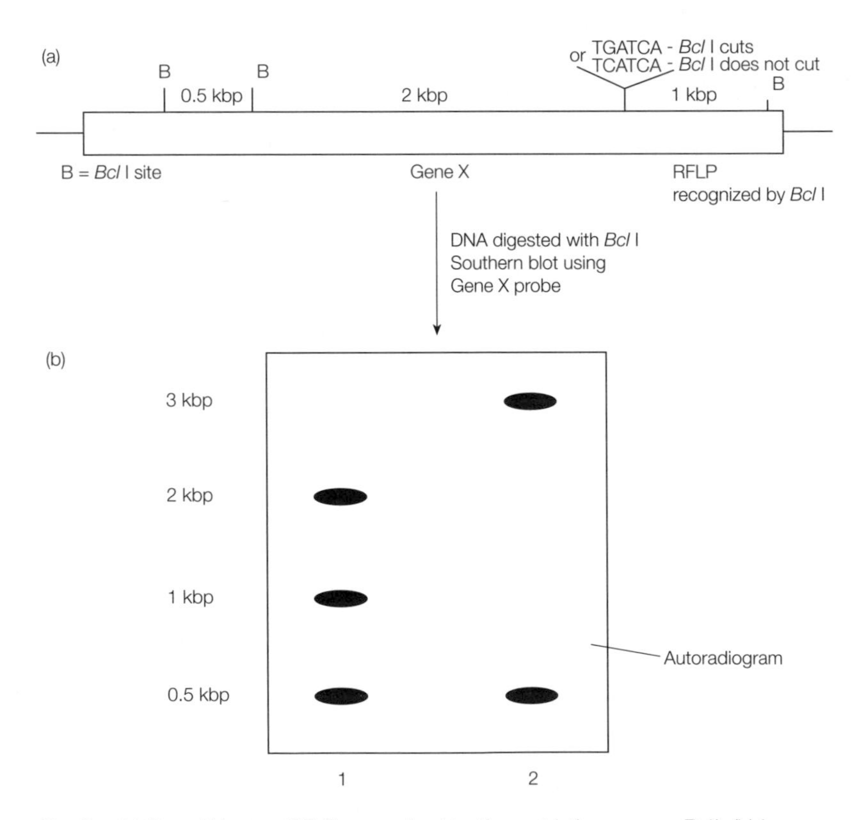

Fig. 2. (a) Gene X has an RFLP recognized by the restriction enzyme BclI. *(b) In homozygotes Southern blotting produces three fragments (lane 1: 2, 1, 0.5 kbp) when the* BclI *site is present and two fragments (lane 2: 3, 0.5 kbp) when the* BclI *site is absent.*

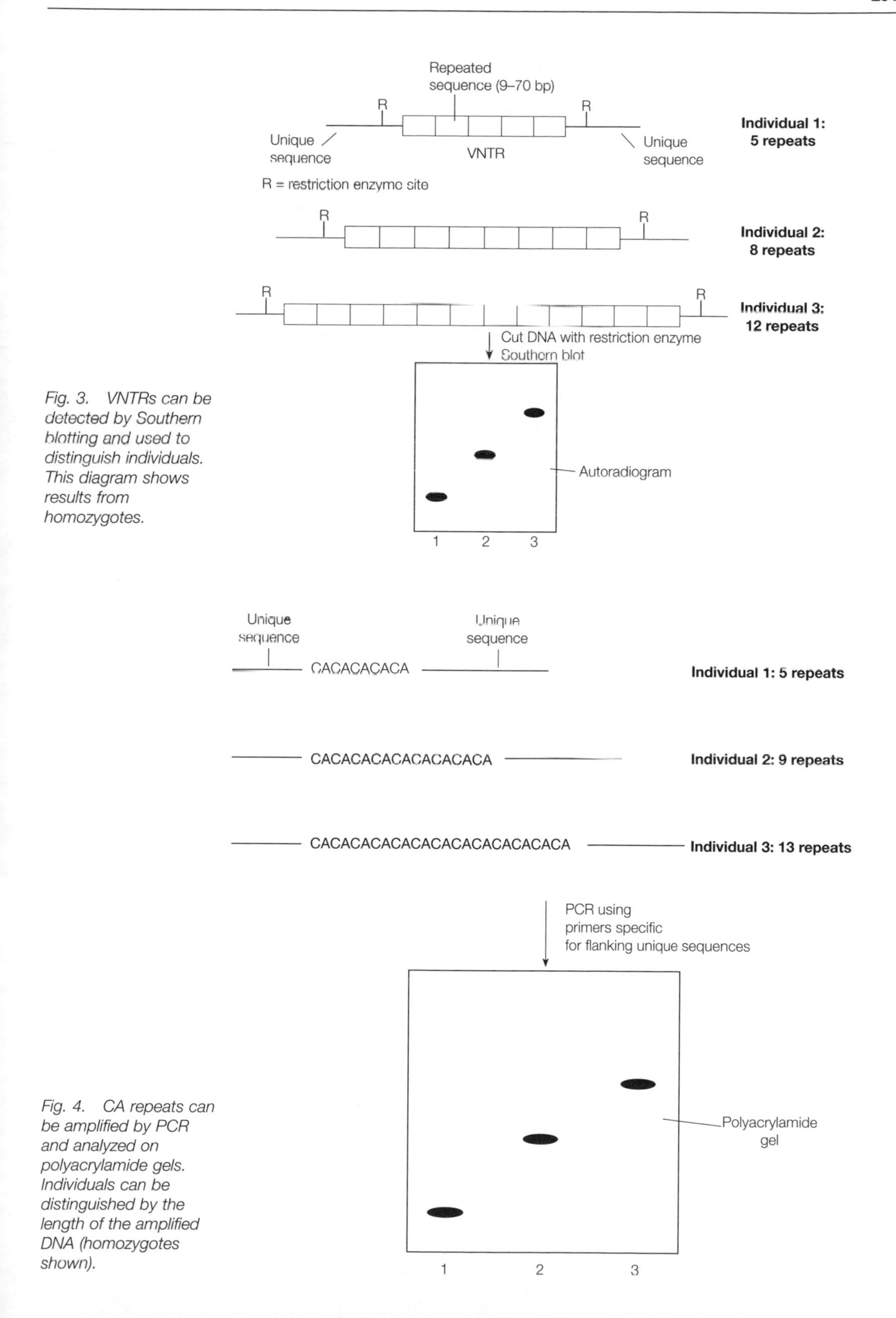

Fig. 3. VNTRs can be detected by Southern blotting and used to distinguish individuals. This diagram shows results from homozygotes.

Fig. 4. CA repeats can be amplified by PCR and analyzed on polyacrylamide gels. Individuals can be distinguished by the length of the amplified DNA (homozygotes shown).

In addition to unique sequences, the human genome contains a high proportion of sequences that are repeated and exist as blocks in which a short sequence is repeated multiple times (see Topic B4). The number of times the sequence is repeated varies between individuals. The first repetitive sequences to be identified were short sequences between 9 and 70 bp present as multiple copies arranged as head-to-tail tandem arrays. These were referred to as **variable number tandem repeats (VNTRs)** or **minisatellite DNA**. The number of repeats present at a VNTR varies between individuals from just a few copies to hundreds of copies. By digesting DNA with restriction enzymes that cut on either side of the VNTR, DNA fragments of widely varying sizes are generated that can be detected by Southern blotting (*Fig. 3*). Another useful type of repetitive sequence used are **CA repeats** (also called **microsatellite DNA**). In this case the repeated sequence is the dinucleotide CA which occurs in blocks of 10–60 copies at many locations throughout the genome. The PCR technique is used to amplify CA repeat sequences and generates DNA molecules of different lengths that can be analyzed by gel electrophoresis (*Fig. 4*). Both CA repeats and VNTRs act as gene markers and can be used to identify carriers of genetic diseases using the same strategy as that used for RFLPs.

A disadvantage of gene tracking is that it is necessary to analyze other family members to identify a carrier. This creates potential difficulties since some family members may be unavailable or unwilling to participate in the study. An alternative to gene tracking is to identify carriers by directly detecting the mutation that causes the disorder. This removes the need to analyze other family members, greatly simplifying the procedure. However, it is first necessary to identify the mutation present. This is potentially a difficult task especially if the gene involved is large.

F2 GENES AND CANCER

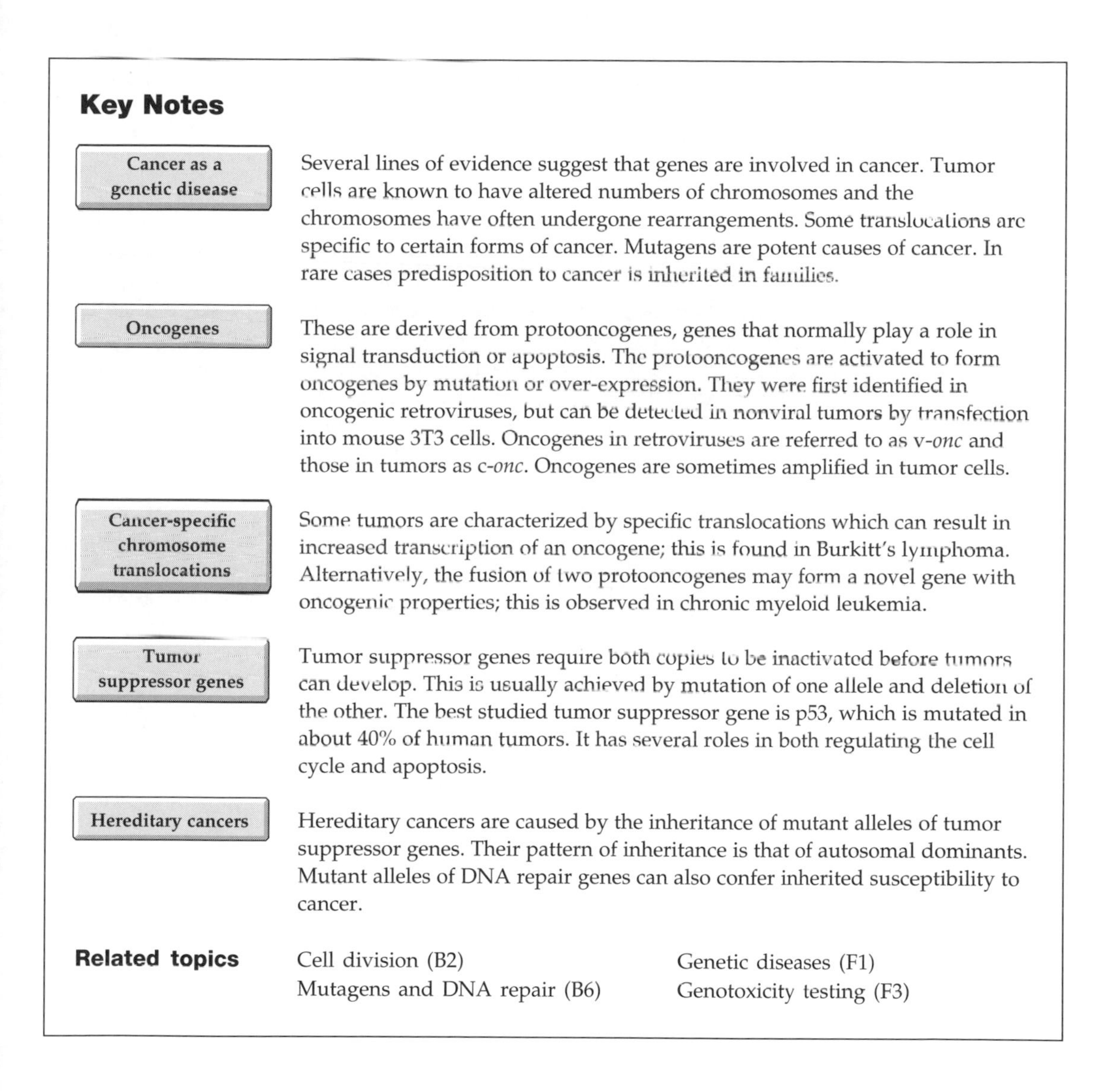

Key Notes

Cancer as a genetic disease

Several lines of evidence suggest that genes are involved in cancer. Tumor cells are known to have altered numbers of chromosomes and the chromosomes have often undergone rearrangements. Some translocations are specific to certain forms of cancer. Mutagens are potent causes of cancer. In rare cases predisposition to cancer is inherited in families.

Oncogenes

These are derived from protooncogenes, genes that normally play a role in signal transduction or apoptosis. The protooncogenes are activated to form oncogenes by mutation or over-expression. They were first identified in oncogenic retroviruses, but can be detected in nonviral tumors by transfection into mouse 3T3 cells. Oncogenes in retroviruses are referred to as v-*onc* and those in tumors as c-*onc*. Oncogenes are sometimes amplified in tumor cells.

Cancer-specific chromosome translocations

Some tumors are characterized by specific translocations which can result in increased transcription of an oncogene; this is found in Burkitt's lymphoma. Alternatively, the fusion of two protooncogenes may form a novel gene with oncogenic properties; this is observed in chronic myeloid leukemia.

Tumor suppressor genes

Tumor suppressor genes require both copies to be inactivated before tumors can develop. This is usually achieved by mutation of one allele and deletion of the other. The best studied tumor suppressor gene is p53, which is mutated in about 40% of human tumors. It has several roles in both regulating the cell cycle and apoptosis.

Hereditary cancers

Hereditary cancers are caused by the inheritance of mutant alleles of tumor suppressor genes. Their pattern of inheritance is that of autosomal dominants. Mutant alleles of DNA repair genes can also confer inherited susceptibility to cancer.

Related topics

Cell division (B2)
Mutagens and DNA repair (B6)
Genetic diseases (F1)
Genotoxicity testing (F3)

Cancer as a genetic disease

There are several distinct lines of evidence which indicate that cancer is a disease caused by alterations to genes:

- it has long been known that mitosis is less precise in tumor cells leading to variation in chromosome number between cells of the same tumor. This is referred to as **heteroploidy**;
- chromosomes in tumors frequently show structural rearrangements. Although most of the rearrangements appear to be random in origin, several cancers have specific rearrangements;

- it is clear that the majority of mutagens are also **carcinogens** (agents that cause cancer);
- predisposition to either a single or multiple forms of cancer is found to be inherited in some families.

Although this consititutes strong evidence for a major genetic involvement in cancer, it is necessary to identify specifically those genes that are involved before the hypothesis is proven. Three classes of genes have been identified, oncogenes, tumor suppressor genes and DNA repair genes.

Oncogenes

These were first identified in a group of oncogenic (cancer causing) animal viruses, the retroviruses. These viruses have an RNA genome in the virus particle, but after infecting a cell the genome is converted to DNA by the enzyme **reverse transcriptase**, and integrated into the host cell's DNA. Oncogenic strains of retroviruses differed from nononcogenic strains by the presence of an extra gene. This gene is referred to as an **oncogene**. DNA sequences closely related to viral oncogenes are also found in DNA from normal healthy animals. Oncogenic retroviruses are the result of genetic recombination between nononcogenic viral strains and cellular genes (*Fig. 1*). To differentiate between the viral and the normal cellular copies of the genes the cellular copy is called a **protooncogene**. The viral oncogene differs from the protooncogene in that it is mutated. When present in a retrovirus it is under the control of a powerful enhancer, the viral long terminal repeat (LTR), which increases its rate of transcription. Mutation and/or over-expression is referred to as **oncogene activation**.

Oncogenes are also identified in DNA from tumors where there is no viral involvement. To do this DNA is extracted from tumor cells and transfected into a mouse cell line, 3T3. These cells grow in tissue culture attached the surface of the culture dish. After they have divided sufficiently to cover the surface of the dish cells stop dividing; this is known as **contact inhibition**. The presence

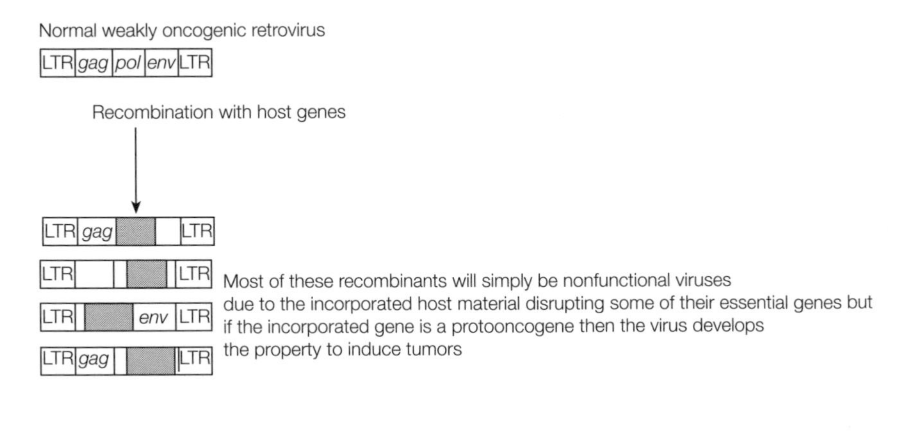

Fig. 1. Normal retroviruses contain the genes gag *(group specific antigen),* pol *(reverse transcriptase), and* env *(envelope) under the regulation of LTR, (long terminal repeat) enhancers. They can recombine with host cell genes. Recombination can take place at many sites within the virus genome.* ■, *Random genomic material:* ■, *protoncogene.*

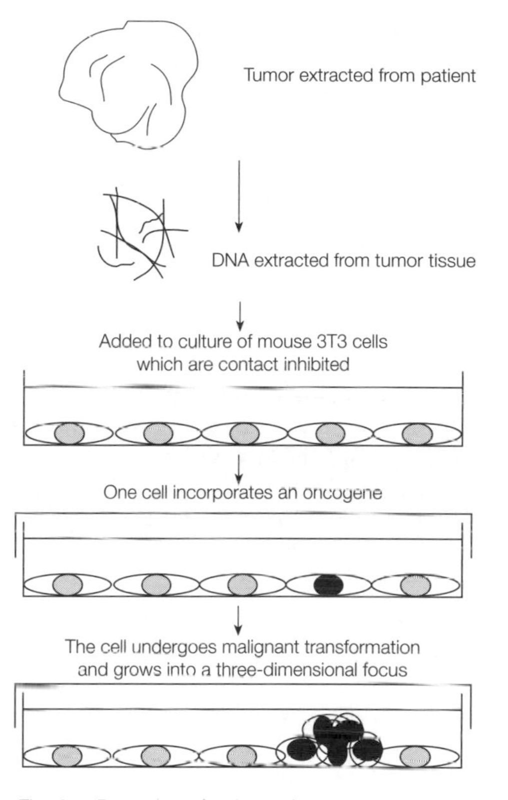

Fig. 2. *Detection of activated oncogenes in tumor DNA by transfection assay.*

of an activated oncogene overrides contact inhibition and cells that have been transfected with DNA containing an oncogene continue dividing to form multi-layered foci (*Fig. 2*). These foci can be picked, grown on, and the oncogene isolated (human DNA sequences can be identified in a mouse background by the presence of the primate-specific *alu* repeat). Oncogenes which have been activated without viral involvement are referred to as c-*onc* to distinguish them from the viral oncogenes v-*onc*.

Oncogenes are usually named after the retrovirus in which they were first identified or after the tumor from which they were first isolated by the transfection assay. Thus, v-*myc* was first isolated in avian myleocytomatosis virus, and c-*neu* was identified in DNA extracted from a rat neural tumor. The proteins encoded by oncogenes are involved mainly in **signal transduction pathways**. These are multi-step systems that receive growth signals at the cell surface and transmit the signal to the nucleus, where transcription of genes involved for cell division are initiated. Some oncogenes are involved in the regulation of **apoptosis**. This is a process in which cells become programed, by response to various triggers including viral infection or DNA damage, to commit suicide.

Activation of an oncogene results in either constitutive growth stimulation or prevention of cell death, both of which lead to an increase in cell numbers. In some cases oncogene activation arises through gene amplification. Up to several thousand copies of an oncogene can be found in each tumor cell. An example of this is the amplification of c-*myc* in neuroblastomas.

Cancer-specific chromosome translocations

In many cancers specific translocations are detected in almost every tumor. In some instances these are used by pathologists to differentiate between related types of tumor. Hemopoietic tumors such as leukemias and lymphomas are well characterized for translocations. Burkitt's lymphoma has specific translocations of chromosome 8 with chromosomes 14, 2 or 22. The translocation between chromosomes 8 and 14, t(8;14)(q24:q32) is the most commonly observed. This translocation brings the *myc* protooncogene, found on chromosome 8 into very close proximity with the immunoglobulin heavy chain gene cluster on chromosome 14. *myc* is activated by being transferred close to the immunoglobulin heavy gene enhancer (see Topic A11). It is transcribed at a greatly increased rate and the resulting increased level of its protein product stimulates cell division. The translocations to chromosomes 2 and 22 bring *myc* close to the immunoglobulin light chain gene clusters.

Translocations can also cause the 5′ and 3′ ends of two different genes to fuse to form a novel gene. This novel gene may be oncogenic. Examples of this include the so-called **Philadelphia chromosome** in chronic myeloid leukemia, where the translocation t(9;22)(q34;q11) fuses two protooncogenes *bcr* and *abl* together to form a novel oncogene.

Specific translocations are less well identified for solid tumors. This probably reflects the increased difficulties of obtaining high quality chromosome preparations from these tumors.

Tumor suppressor genes

In genes of this class, both alleles need to be inactivated in order for tumor formation to progress. The mutations are recessive. Inactivation is most frequently achieved by point mutation of one allele, and loss of the other through deletion, although mutation of both alleles is occasionally found. Inactivation of tumor suppressor genes occurs in somatic cells during the development of a tumor. The best studied tumor suppressor gene is p53. This is so named because it codes for a 53 kDa nuclear phosphoprotein, which functions as a transcription factor. The gene is mutated in over 40% of human tumors. p53 is involved in different cellular processes. It has a major role in the transition from the G_1 phase to the S phase of the cell cycle. This is a **checkpoint** where cells can arrest during the cell cycle (see Topic B2). In its native form p53 prevents the cell moving into S phase, but when phosphorylated this function is lost. This is part of the normal regulation of the cell cycle. Mutant p53 fails to arrest the cell at this checkpoint.

p53 has also been shown to regulate the response of the cell to DNA damage. This is mediated by two separate pathways: (i) the level of p53 protein in the nucleus is increased after DNA damage. This can prevent the cell entering S phase until the damage is repaired; (ii) alternatively p53 may be involved in the induction of cell death via **apoptosis**.

Both processes reduce the risk of damage to DNA resulting in mutations that could cause cells to develop into tumors. Because of these functions p53 has been described as 'the guardian of the genome'.

Transgenic mice that lack any functional p53 develop normally but produce spontaneous tumors at a high rate. Gene amplification is found at elevated levels in their cells.

Hereditary cancers

A number of rare cancers and a small proportion of the common cancers show clustering within families. These families are said to have a hereditary predisposition to cancer. This may be to a specific form of cancer e.g. breast cancer, or to a variety of different cancers.

Hereditary predisposition to cancers may arise through germline mutations of tumor suppressor genes. Here all the somatic cells of an individual carry one mutant allele, and there is a greatly increased risk of tumor formation due to inactivation of the single normal allele. Examples in humans include the retinoblastoma gene, *Rb*, on chromosome 13 and a gene for inherited predisposition to breast and ovarian cancer, *BRCA1*, on chromosome 17. The normal allele is inactivated in the cells of the tumor. Familial cancer genes are often found to undergo somatic mutation in sporadic (noninherited) cases of the same cancers.

Hereditary cancers show the inheritance pattern of autosomal dominants, because sufferers pass the predisposition on to half of their offspring (see Topic C1), however mutations in tumor suppressor genes are recessive. This appears to be a paradox. It is explained by the fact that if an individual has only one functional copy of a tumor suppressor gene in every cell then it is highly likely that, in some cells, a second mutation will cause inactivation of the normal functional copy. These doubly mutant cells can then give rise to a tumor.

DNA repair genes are also associated with hereditary cancers. Examples of these include xeroderma pigmentosum, a recessive autosomal condition, in which the affected individuals are highly sensitive to sunlight because they lack an endonuclease involved in DNA repair. Mutant alleles of several different genes can result in predisposition to colon cancer. Two of these are also involved in repair of DNA mismatches (see Topic B6). The cells of the tumor repair damaged DNA less efficiently than normal cells and produce mutations at a high frequency, and some of the mutations will be in tumor suppressor genes or protooncogenes. This is known as a **mutator phenotype**.

F3 GENOTOXICITY TESTING

Key Notes

Genotoxicity

Genotoxicity refers to the detection of agents that will damage DNA and hence cause mutations. Commercial tests can use animal experiments, but these are largely replaced by *in vitro* systems utilizing bacteria or animal cells in tissue culture.

Ames test

The Ames test assays the capacity of test agents to revert mutations in the histidine operon of *Salmonella typhimurium*. Histidine auxotrophic bacteria are plated onto agar containing very little histidine and treated with the substance under test. Only those bacteria that revert to be able to synthesize histidine will be able to form colonies. Mutagens induce increased numbers of revertants. The test can be modified to detect procarcinogens by adding rat liver extract to the medium.

Cell line mutation tests

Potential mutagens can be tested directly on animal cells. The best known test system involves the use of a mouse cell line that is grown in tissue culture. It is heterozygous for the thymidine kinase gene. Mutagens increase the frequency of thymidine kinase deficient cells.

Cytogenetic tests

Agents which can cause chromosome damage are best detected by assessing their ability to induce sister chromatid exchanges (SCEs). These occur spontaneously but the rate of induction is very sensitive to mutagens. SCEs are only detectable when two chromatids of the same chromosome stain differently. This is achieved by labeling cells for two divisions with bromodeoxyuridine prior to staining.

Related topics

Mutagens and DNA repair (B6) Genes and cancer (F2)

Genotoxicity

This is a recently developed branch of toxicology which deals with toxic agents in the environment that act through damaging DNA and hence causing mutations. Such compounds are important because they cause cancer, and are termed **carcinogens**. They also increase the mutational load in the general population. Identification of mutagenic compounds is recognized as important in many areas including, pharmaceuticals, food additives, agriculture, pollution analysis and in many industrial processes. For this reason it has been necessary to develop tests to identify dangerous compounds. Early testing systems relied on small mammals such as rats or mice, but these tests were time consuming, very expensive and attracted considerable ethical criticism. For these reasons a number of *in vitro* tests have been developed which use bacteria, or animal cells in grown in tissue culture. In many countries there are legal requirements for new chemicals to be tested by a series of different tests before being licensed by government agencies.

A great number of test systems are available but only three will be described.

Ames test

This is a very rapid test for mutagens. It utilizes bacteria of the species *Salmonella typhimurium* that have a mutation in the histidine operon, and hence cannot synthesize histidine – histidine auxotrophs (*his*$^-$). Compounds are tested to determine whether they can induce reversion of this mutation. The test can be carried out, very simply, by plating out a lawn of *his*$^-$ mutants on an agar plate that contains only a trace of histidine. A crystal, or a filter disc containing a solution of the compound to be tested, is placed on the surface of the agar. The bacteria grow for a short time until the histidine is depleted. After that time the only bacteria capable of continuing growth to form colonies are those which have undergone reversion and are capable of synthesizing their own histidine– histidine prototrophs, (*his*$^+$). If the test compound is not mutagenic a few revertant colonies will be found randomly scattered across the agar plate. If it is mutagenic then the number of colonies will be increased and they will be clustered around the point where the compound was placed on the plate. Obviously the test can be constructed in a more quantative manner to give a dose–response curve for any compound under test.

In order to identify the maximum number of mutagens two types of *his*$^-$ mutants are used, one of which is a single base substitution and the other a frameshift mutation (see Topic B5). This allows detection of mutagens which have different effects on DNA. It is also possible to alter genetically both the permeability of the bacteria to test compounds, and to decrease their ability to repair damaged DNA. This again increases the likelihood of detecting mutagenic activity.

Some compounds which are know to cause cancer are only capable of doing so after they have been altered by the action of enzymes within the body. These are known as **procarcinogens**. Enzyme action converts them to **ultimate carcinogens**. If procarcinogens are used in the Ames test they will give a negative result, however the test can be adapted to take account of this. The liver is a rich source of activating enzymes. Liver extracts, containing active enzymes, can be added together with the test compound. Activation of a procarcinogen will then occur, resulting in increased numbers of revertants.

The Ames test is rapid: it can be carried out in 48 hours. It is cheap and easily quantifiable. It has identified many compounds as mutagens including certain hair dyes, flame retardants, and food colorings.

Cell line mutation tests

One deficiency of the Ames test is that the target organism is a bacterium rather than a mammal. For this reason a number of tests have been developed using cultured animal cell lines. The most important of these is the L5178Y thymidine kinase heterozygote test. L5Y178Y is a cell line derived from mouse lymphocytes. It carries two functional alleles at the thymidine kinase (*tk*$^-$) locus. Mutation to *tk*$^-$ deficiency occurs too infrequently to be of use in a test system, but a heterozygote *tk*$^{+/-}$ has been isolated in which mutation to *tk*$^-$ can be measured. Mutants are identified by selecting treated cells in toxic analogs of thymidine. Only *tk*$^-$ cells will survive and form clones. This can be used in exactly the same way as the Ames test even allowing the addition of liver extracts. However, because of the relatively slow growth of animal cells in culture it takes up to 3 weeks to obtain results from this test.

Many other tests are available based on different endpoints in animal cells, both *in vitro* or *in vivo*.

Cytogenetic tests

Agents which damage chromosomes are known as **clastogens**. These are also identified by test systems. These tests can be carried out on cell lines, in laboratory animals or even in plants. The tests consist of scoring chromosome aberrations such as breaks, exchanges, ring chromosomes, dicentrics and translocations. The frequency of aberration is often low and detection requires highly trained personnel. An alternative to enumeration of such gross chromosomal aberrations is to count **sister chromatid exchanges** (SCE). SCEs involves exchange of material between two chromatids of the same chromosome, and is a process which takes place spontaneously at low frequencies in all cell types. It can occur both in mitosis and meiosis. Its usefulness here is that the frequency of SCEs increases much more rapidly, as a response to clastogen treatment, than do gross chromosomal aberrations. Thus clear-cut results can be obtained from analysis of small numbers of cells by relatively unskilled operators. SCEs can be detected by a number of procedures, all of which are dependent on the semi-conservative nature of DNA replication (see Topic A9). The fluorescence plus Giemsa method is described here. The essence of the technique is to make sister chromatids stain differently so that exchanges can easily be observed. Cells are grown for two rounds of replication in low, nontoxic, concentrations of the thymidine analog bromodeoxyuridine (BrdU). This is incorporated into the newly synthesised DNA in place of thymidine and alters the staining properties of the chromatid. After one round of replication both chromatids contain DNA double helices in which one

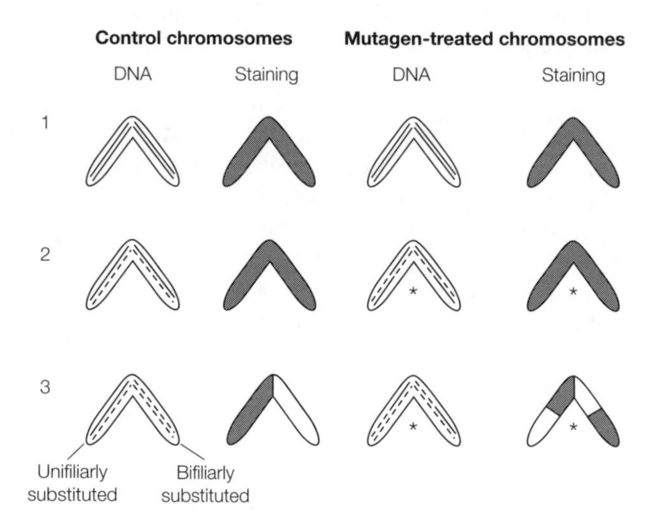

Fig. 1. Two sets of chromosomes are shown. On the right are chromosomes that have been treated with mutagens to induce SCE, and on the left untreated chromosomes. Each chromosome is represented twice, to show the status of the DNA or the staining pattern. Solid lines represent DNA single strands containing thymine, shading represents DNA single strands containing bromodeoxyuridine. 1. Prior to labeling with bromodeoxyuridine; both the chromatids contain DNA with only thymine, and stain darkly. 2. After one cell cycle in bromodeoxyuridine, in each chromosome the DNA contains one strand containing bromodeoxyuridine and one strand containing thymine; the chromatids all stain darkly. An SCE takes place in the mutagen-treated chromosome during this cell cycle at the point noted by the asterisk. Note the change in labeling of the DNA. 3. After a second cell cycle in the presence of bromodeoxyuridine the control chromosome contains differently labeled DNA in each chromatid. One chromatid is unifiliarly substituted with bromodeoxyuridine, and stains darkly; the other is bifiliarly substituted with bromodeoxyuridine and stains pale. The same is true in the mutagen-treated chromosome and thus the SCE can now be visualized.

strand contains BrdU and the other thymidine. Both are said to be **unifiliarly substituted**. The staining of the chromatids is still the same as chromosomes from untreated cells. After the second round of replication each chromosome consists of two chromatids, one of which is still unifilliarly substituted with BrdU but the other is **bifiliarly substituted** (both strands contain BrdU). When these are stained the unifiliarly stained chromatid stains darkly while the bifiliarly substituted chromatid is very lightly stained. Any exchanges that have occurred are easily detected, (*Fig. 1*). The staining process is complex, involving exposure to a fluorochrome followed by ultraviolet light irradiation and finally treatment with Giemsa stain.

F4 GENE THERAPY

Key Notes

Overview

It is now possible to treat genetic disorders by the addition of functional copies of the gene responsible for the disorder. This is, at present, limited to treating somatic cells. The major problem in developing gene therapies is finding the optimal vector to transfer the therapeutic gene into the recipients cells. Gene therapy can be carried out on the whole patient (*in vivo* therapy) or on cells which have been removed from the patient and which are then returned after the gene has been added (*ex vivo* therapy). In some disorders the therapeutic gene will have to be targeted to a particular cell type, while for other syndromes several tissues may make suitable targets.

Vectors

Viruses offer the best potential as vectors in gene therapy. The bulk of viral genes are removed to prevent virus replication in the patient. Retroviruses infect most cell types, but only if the cells are dividing. This limits their use to *ex vivo* application. Lentiviruses may prove a more useful alternative. Adenoviruses can also be used but their DNA does not integrate into the target cell's chromosomes and their effect is transient. They also may induce immune response in the patient. An alternative to the use of viral vectors involves encapsulating DNA in liposomes. This causes no side effects, but is much less efficient in transferring DNA to target cells.

Adenine deaminase deficiency

The gene for adenine deaminase can be introduced into T lymphocytes of patients using a retrovirus vector. However these cells are eventually lost from the blood and the procedure has to be repeated. The eventual aim is to be able to transfer the gene to precursor cells present in the bone marrow, thus effecting a permanent change.

Cystic fibrosis

Gene therapy for this disease has to be carried out *in vivo* because the major target tissue is the lining of the lungs. Adenovirus vectors have been used. These cause side-effects and may be of limited use. An alternative approach is to use liposomes.

Cancer

This is currently the major area of medicine in which gene therapy is being used. There are a number of different approaches. These include introducing genes to kill tumor cells, or to stimulate the response of the host. One potentially useful method involves the use of the herpes simplex thymidine kinase gene to sensitize the tumor cells to drug treatment.

Related topics

Genetic diseases (F1)
Genes and cancer (F2)

Overview

The major technical advances in molecular genetics made in the past 20 years have enabled us to isolate, clone and sequence genes from all species including humans. In addition to this the ability to map human genes (see Topic F5), in particular those for inherited disorders, has allowed us to identify unambiguously the precise genetic reason for ill health in many individuals. This has raised the question of whether it would be possible to use genetic methods to cure such individuals, or at least to increase their quality of life. This would require the necessary gene to be isolated and targeted into the cells of the affected individual – **gene therapy**. In the case of inherited syndromes there is a choice of approaches: a normal copy of a gene can be introduced into either the germ line, or into the somatic cells of the affected individual. The former would prevent transfer of a deleterious allele to the next generation whereas the latter would ameliorate the effects of the allele in the affected individual. Current studies are concerned only with treatment of somatic cells. In the majority of cases the intention is introduce a functional copy of a gene into cells which have two nonfunctional alleles, so as to correct the defect. However, in proposed gene therapies for cancer it is often the intention to introduce genes into tumor cells that will induce cell death. Assuming that problems involving cloning of the relevant gene and regulating its expression have been largely solved (see Topics E1 and A11), the major obstacle to the development of a gene therapy lies in finding suitable vectors to introduce genes into target cells efficiently while not causing serious side-effects. Paradoxically the most promising candidate vectors for gene therapy are derived from viruses that cause disease in humans, precisely because they have evolved efficient systems for transmitting their genomes into human cells.

Gene therapy can be carried out by introducing genes into the patient and hoping they will find an appropriate target cell – ***in vivo* therapy**. This is obviously an inefficient process, and in some cases it is possible to remove cells from the patient, manipulate them in cell culture, and then return them to the patient. This is known as ***ex vivo* therapy**.

It is important to realize that targets for therapeutic genes will differ. In certain cases the gene must be delivered to a specific cell type, for example replacement of globin genes in inherited hemoglobinopathies in which the gene must be delivered to erythroid cells. In other cases it is only necessary to increase the level of a specific protein, for example a hormone, in the body. Here any cell type that will allow the product to be secreted into the blood system will be a suitable target.

Vectors

Before a virus can be used as a vector to carry therapeutic genes those genes necessary for viral functions must be removed. This prevents spread of the virus within the body, and reduces the response of the patient's immune system. Viruses deleted in this way are referred to as **gutted viruses**. The desired gene is inserted into the remaining virus genome along with DNA sequences to promote and regulate its transcription, and this recombinant DNA molecule is packaged into the appropriate viral coat so that it can infect human cells. Several different virus vectors can be used.

Retroviruses (see Topic F2) are easily adapted to carry a human gene insert. These have the advantage that they will infect most, if not all, cell types and are efficient at integrating their genome into the host cell. Their main disadvantage is that they will only infect actively dividing cells. For this reason their use is largely limited to *ex vivo* use with cells grown in culture. Another problem is that the inserted gene is often only expressed for a short period, usually less than 2 weeks. A subgroup of the retroviruses, the **lentiviruses**, which includes the human immunodeficiency virus (HIV), are being developed for use in gene therapy. These can infect nondividing cells and hence are useful for *in vivo* procedures.

One DNA virus group that has been adapted for gene therapy is the **adenovirus** group. Several adenoviruses are associated with common upper respiratory tract infections in humans. Their genome can accept large insertions of human DNA, but they have several major disadvantages, including the fact that their genomes do not integrate into the host cell's genome. This means that they cannot engender a permanent change in any cell they infect. They also induce an immune response which may result in the killing of infected cells. This, however, may be useful in the treatment of tumors. Other DNA viruses which are under examination as potential vectors include the much smaller adeno-associated viruses and herpes simplex virus, which has a large genome and can thus incorporate large inserts.

All of the viral vectors have drawbacks, either in the efficiency of their delivery system, their triggering of an immune response, or because they are closely related to viruses that may have a capacity to induce tumors. For this reason considerable effort has been put into the development of nonbiological delivery systems in gene therapy. Of these the most promising appears to be **liposomes**, lipid micels, that can be used to carry naked DNA across the cell membrane. The phospholipid binds DNA and because it is of a similar composition to the cell membrane the two can fuse, and thus the DNA is carried into the cytoplasm of the target cells. Liposomes have several advantages: they elicit no immune response; they can operate *in vivo* or *ex vivo* and they can carry any size of DNA fragment. However, the DNA they transfer has a low frequency of integration into the host chromosome. Over 200 clinical trials are underway using gene therapy to treat human disease, and animal model systems are also used to determine the efficacy of novel strategies. A number of particular examples are dealt with below.

Adenine deaminase deficiency

This is a rare, recessive, autosomal inherited disease in which the patient has two mutant alleles for the adenine deaminase (ADA) gene. As for other defects in nucleoside metabolizing enzymes deficiency of this enzyme has differential effects in a variety of tissues. ADA deficiency has its most serious effect in the cells of the immune system. Lack of the enzyme results in the development of severe combined immuno-deficiency syndrome. The gene that encodes ADA was cloned and initial attempts to treat the disease by gene therapy were first carried out in 1990. The gene was inserted into a retroviral vector, T lymphocytes were removed from the patient, stimulated to grow in culture and transfected with the gene. Cells carrying the functional gene were then grown up, before being tranferred back to the patient. This procedure has shown success in some cases. However it is not a cure for the disease: because the T lymphocytes are mature, differentiated cells they are progressively replaced in the body by new cells from the bone marrow, hence the process has to be repeated on a regular basis to maintain the patient's level of ADA. If precursor cells in the bone marrow could be transfected efficiently by the gene then the treatment might only need to be given once.

Cystic fibrosis

Cystic fibrosis, like ADA, is the result of recessive mutation of a single gene, however the frequency of mutant alleles of this gene is much higher, particularly in Caucasian populations. The defect is in a gene responsible for the transport of chloride ions across the cell membrane. Lack of this function has major effects on the cells that line the lungs and gut. The gene therapy approach was initially similar to that for ADA in that the gene was cloned and engineered into a vector. Because of the cell types that are involved it is not possible to transfect cells *ex vivo*. For this reason adenovirus has been used as a vector. The engineered virus containing the functional gene can be delivered to patients as an aerosol. Some virus particles are able to infect cells of the respiratory tract. Problems in this approach include the difficulty of infecting cells through the mucus lining of the tissue, the fact that adenovirus does not integrate into host cell chromosomes, and possible immunologic reaction to virus infected cells. An alternative approach is to use liposomes to transfer the gene. This introduces the gene into fewer cells, but may have longer-term effects.

Cancer

The majority of clinical trials involving gene therapy have been directed at tumor cells. Given that tumors arise as a result of gene mutation (see Topic F2) it is theoretically possible to prevent further growth of tumor cells by replacing defective tumor suppressor genes. Another approach is to use exogenous genes either to induce death in tumor cells or to render them more sensitive to the normal defence mechanisms of the body.

One potentially useful approach to specific killing of tumor cells is transfection with the thymidine kinase gene of the herpes simplex virus. This enzyme differs considerably from its human equivalent, and can phosphorylate a wider range of substrates. One substrate that cannot be utilized by human thymidine kinase is gancyclovir. This compound is nontoxic to human cells but its phosphorylated derivatives are highly toxic. Transfer of the herpes thymidine kinase gene to tumor cells thus renders them specifically sensitive to gancyclovir treatment. There are two other advantages in this methodology: (i) thymidine kinase is active only in cells that are replicating DNA and hence nontumor cells are less likely to be affected; and (ii) the phosphorylated gancyclovir molecules, which are produced only in those tumor cells that have taken up the gene, are exported to surrounding tumor cells through gap junctions. The result is that the proportion of tumor cells killed by gancyclovir is greatly increased. This approach has been used with brain and ovarian tumors, and is set out schematically in *Fig. 1*.

As yet no gene therapy approach has been successful in eradicating all tumor cells from a patient. It may be that gene therapy will never be totally successful on its own but will become an extra treatment regime to be used in conjunction with surgery and conventional drug and radiation therapies.

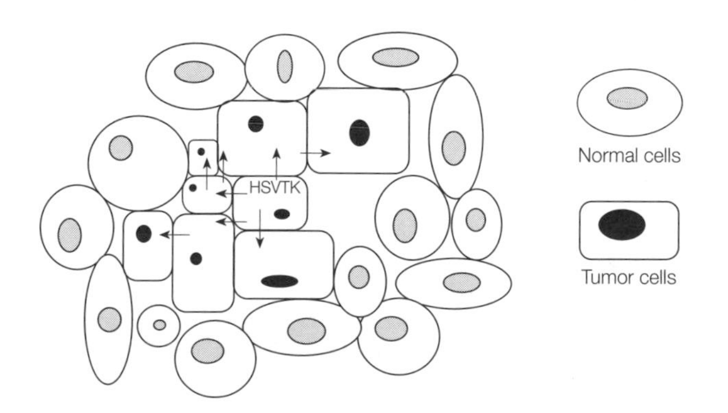

Fig. 1. Killing of tumor cells containing the herpes simplex thymidine kinase gene. A single tumor cell which has been transfected with the herpes simplex virus thymidine kinase gene (HSVTK) can export the toxic phosphorylated products of gancyclovir to other tumor cells. Normal cells are relatively unaffected because they are not undergoing DNA synthesis.

F5 THE HUMAN GENOME PROJECT

Key Notes

Background

Before beginning a sequencing project of the human genome it is first necessary to produce good framework maps. This can be done by physical or genetic mapping. It is important to have high numbers of polymorphic markers to construct genetic maps. Enzyme polymorphisms can be used but DNA polymorphisms such as restriction fragment length polymorphisms and mini- and microsatellites are more useful.

Genetic maps

Genetic maps are based on recombination frequencies between markers. They are good for ordering genes but do not give accurate measurements of distances between markers. The most refined genetic map gives data for over five thousand microsatellite markers.

Physical maps

These are produced using radiation hybrids or contigs. The former are somatic cell hybrids in which an irradiated human cell is fused to a rodent cell. The fragments of human DNA are randomly integrated into the rodent chromosomes. Each clone carries a different set of human fragments. Screening a panel of hybrid clones with a human marker allows the marker to be mapped. The linkage between markers on a chromosome is directly proportional to the number of base pairs of DNA between them. Contigs are groups of cloned DNA sequences that can be aligned in an overlapping fashion to cover a region of the human genome. Yeast artificial chromosomes (YACs) are used most for this purpose as they carry large inserts of human DNA; however, smaller vectors are becoming more popular because the YACs are somewhat unreliable.

Fluorescence *in situ* hybridization

Hybridizing DNA probes onto human chromosome preparations allows genes to be mapped directly to their chromosomal location. The probes are labeled with fluorochromes or are detected by immunofluorescence techniques.

Placing genes on the map

Framework maps are composed largely of noncoding markers. Genes are identified by positional cloning, where an inherited disorder is mapped to a region on a genetic framework map. The region is then analyzed for coding DNA. Known cDNAs can be mapped directly using radiation hybrids or contigs. Transcript maps are produced by mapping fragments of anonymous cDNA molecules to physical maps.

Summary

The construction of framework maps is completed. The major effort now is to place genes on the map and to obtain sequence data.

Related topics

The human genome (B4)
Linkage (C4)
Somatic cell fusion (C10)
Genetics in forensic science (F6)

Background

During the 1980s the complete DNA sequences of several viral genomes were determined. This raised the possibility that species with much larger genomes, such as humans, could also be sequenced. The proposal was controversial. Even amongst those in favor of sequencing the human genome there were arguments as to whether the entire genome or only coding regions should be sequenced. In 1991 an international collaborative project was initiated with the aim of sequencing the entire human genome by the year 2005. Responsibility for coordinating the work done was placed in the hands of HUGO, the Human Genome Organisation. In parallel to this several genome projects were begun for different species. Some of these, which dealt with bacterial and yeast genomes, have already been completed. Before sequencing technologies can have any significant impact the human genome must be mapped well. This is achieved by two

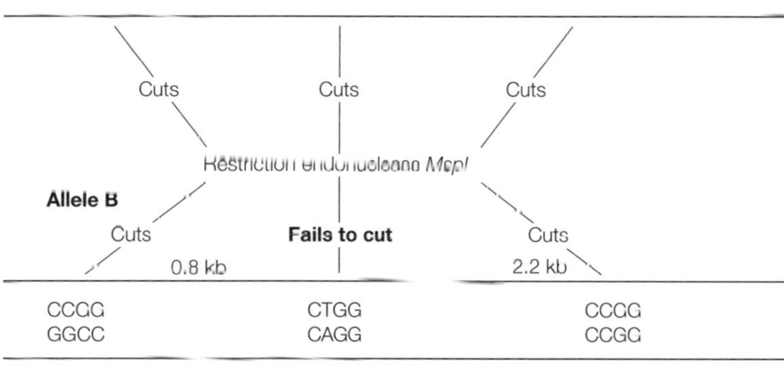

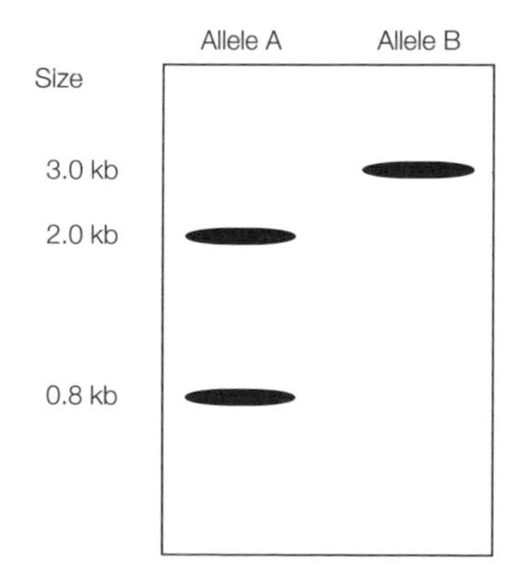

*Fig. 1. Detection of restriction length fragment polymorphism in two alleles of a hypothetical locus restricted with the restriction endonuclease (*MspI*). (a) In two variants of the same region of the human genome allele A is restricted to yield two fragments of 0.8 kb and 2.2 kb. A mutation in the restriction site in allele B results in loss of the ability of* MspI *to cut the DNA and a 3.0 kb fragment is produced. (b) These are detected on a Southern blot.*

Table 1. Sequences of some VNTR minisatellite loci

Locus	Chromosome	Basic repeat sequence[a]
YNZ2	1	GAGGCTCATGGGGCACA
THH59	17	CTGGGGAGCCTGGGGACTTTCCACACC
α-Globin	18	AACAGGGACACGGGGG

[a]These sequences represent the simple sequence that is repeated different numbers of times in the various alleles of the locus.

related, but distinct, processes, **genetic and physical mapping**. The production of **framework maps** was the initial goal of the human genome project. The rate of progress of mapping depends on the availability of genetic markers; a marker is any defined segment of DNA, protein or phenotype. If such a marker varies within the population it is said to be polymorphic (see Topic D4). Ideally, these should be distributed approximately evenly throughout the genome.

The earliest crude maps of the human genome had relied on identifiable phenotypes such as blood groups and inherited syndromes such as Nail–Patella syndrome which show linkage on chromosome 9 (see Topic C4). Prior to the introduction of molecular DNA technology these markers had been augmented by enzyme polymorphisms detected by electrophoresis, but the number and distribution of genetic markers were still totally unsatisfactory for the production of a framework map on which to build a sequencing project. The first significant improvement was the production of the Donis–Keller map in 1987. This used 180 marker loci (a locus is any position on a chromosome) defined by anonymous restriction fragment length polymorphisms (RFLPs). RFLPs are detected as alterations to Southern blot patterns produced when a specific cloned DNA fragment is used to probe genomic DNA (*Fig. 1*) (see Topic E4): different patterns are produced if a mutation occurs in the sequence of DNA which is recognized by the restriction endonuclease. Thus any cloned DNA can be used to detect RFLPs even if nothing is known about its sequence, or its

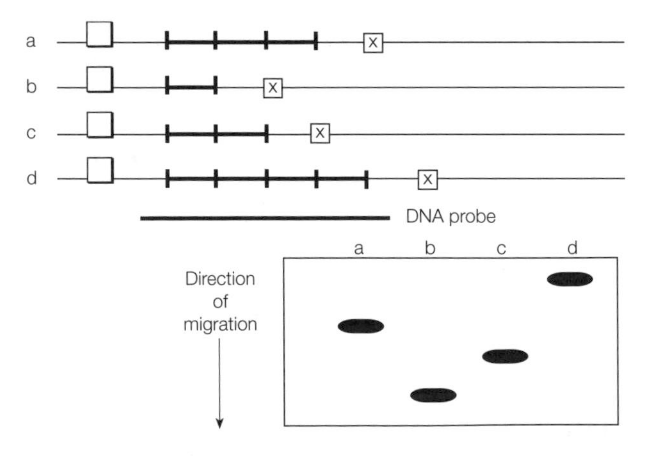

Fig. 2. Detection of variable number tandem repeats. a, b, c, and d represent four different copies of the same chromosome in which the region denoted by —— has undergone different levels of duplication. □ and ⊠ are restriction sites flanking the region of variable number tandem repeat (VNTR). When the DNA is cut at these sites and a Southern blot is probed with the DNA fragment shown, bands will be observed whose sizes are determined by the number of repeats.

map position (an anonymous fragment). Where a fragment detects a polymorphism this can be used directly in gene mapping.

RFLPs are relatively rare, and often the polymorphism is not frequent enough to be useful. A significant improvement over RFLPs was made by the discovery of **variable number tandem repeat sequences (VNTRs)** in eukaryote DNA (see Topic B4). These are repeated sequences that exist in the genome as tandemly repeated clusters. The number of repeats per cluster frequently differs between copies of the same chromosome, hence they are highly polymorphic. VNTRs are divided into two groups: (i) repeats of larger sequences, between 10–25 bases are referred to as **minisatellites** (*Table 1*). The variation in these is detected as bands of different size on Southern blots as shown in *Fig. 2*. Minisatellites greatly increased the number of polymorphic markers, but their distribution is not random; they are found more frequently in telomeric regions of human chromosomes; (ii) VNTRs with a repeat unit of 1–4 bases are known as **microsatellites**. Dinucleotide repeats of $(CA)_n$ are now the most frequently used DNA markers because of their high polymorphism and random distribution throughout the genome. Mono-, tri- and tetranucleotide microsatellites are also useful but these occur less frequently. Microsatellites are detected by polymerase chain reaction (PCR) (see Topic E2) rather than by Southern blotting: this reduces the amount of DNA required and is a more rapid process. PCR analysis requires a region of unique DNA on either side of the repeat to be sequenced to provide primers. The number of repeats present will change the size of the PCR product. The sequence of a typical microsatellite is shown in *Fig. 3*.

CCACCTCCCCTCAACATGGTTCCCACTCACATAACTCCCCCCACACTCTG
TACCCTCCATGGCCCCCAGCACACACATGCACATGCACACCTACACGAA
CACATTTGCACACATT**CACACACACACACACACACACACACACACACAC**
ACACACAGAGCCATGGCCTCTGCTGGCAG

Fig. 3. Underlined sequences represent the position of the two primers used to amplify this microsatellite by PCR. The region of CA repeats between the two primers is emboldened. The number of CA doublets varies between the different alleles, and the PCR products can be identified as differently sized fragments after electrophoresis.

Genetic maps

Genetic maps are maps based on recombination frequency and are constructed in a manner similar to the linkage analysis described in Topic C4; however, because DNA polymorphisms are co-dominant (see Topic C1) the genotype of each individual is detected directly as opposed to being inferred from phenotypes. Although they are extremely useful in ordering genes and markers along a chromosome genetic maps do not give an accurate measurement of distance between loci. This is because the frequency of crossing-over is not constant throughout the genome (see Topic C4): it is high in regions close to the telomeres and reduced near centromeres and there are occasional localized hot spots for recombination. However, genetic maps will give an accurate indication of gene order. A number of framework maps have been produced by following the inheritance of microsatellite polymorphisms through several generations of sets of families. Work by the French Genethon group on the CEPH families (a group of large pedigrees which are used as a standard in human map construction) has produced a framework map detailing 5264 microsatellite loci. The average distance between markers is 1.6 cM, and the largest gap between markers is 11 cM. This has the necessary definition to act as a framework map for accurate mapping of disease genes, and to act as the basis for genome sequencing.

Physical maps

Physical maps depend not on the likelihood of crossing-over between loci but on their physical distance apart on a DNA molecule. Essentially they are produced by dividing the human genome into smaller and smaller fragments and determining which loci remain attached to each other. A simple form of physical mapping has been dealt with in Topic C10 where somatic cell hybrids between human and rodent cells can be used to assign genes to specific human chromosomes. The chromosomes represent large fragments of the human genome.

The development of **radiation hybrids** proved essential for the production of fine structure physical framework maps. In these, the donor human cells are heavily X-irradiated prior to fusion, causing breakage of the chromosomes. The dying cells are then rescued by fusion to a rodent cell line. In the hybrid cell fragments of human chromosomes will integrate at many sites in the rodent chromosomes, and any fragments that do not integrate will be lost from the culture. Because integration is essentially a random process each hybrid clone will contain a different set of human chromosomal fragments. The radiation hybrid clones can then be screened for the presence or absence of human genes or anonymous DNA segments. Human loci that are strongly linked will rarely be split by irradiation and will tend to be retained together or lost together in the hybrid clones, whereas genes which lie well apart will be retained or lost independently. Thus the order of genes in the physical map can be determined. Higher doses of radiation cause more breaks and produce higher resolution maps.

The system can be used to produce maps of the whole genome or of individual chromosomes. The latter are obtained by using rodent × human somatic cell hybrids containing only a single human chromosome as the source of human material for the radiation hybrids. DNA from individual radiation hybrid clones is screened either by probing on Southern blots or by PCR with primers for a specific cloned DNA marker. Depending on the pattern of positive and negative clones the new marker can be assigned to a region on a specific chromosome. Physical distances to other markers can be calculated because the frequency of radiation-induced breaks in DNA is directly proportional to the physical distance between markers.

Another form of physical mapping is developed by organizing cloned fragments of the human genome into contiguous overlapping arrangements, **contigs** (*Fig. 4*). This can be done for large elements of the genome for long-range mapping or for smaller areas for intense analysis. In the former the usual cloning vectors are yeast artificial chromosomes (YACs) which can carry up to two megabases of DNA. However they have several drawbacks in that they are relatively unstable, and can lose all or part of the human insert. They are also notoriously chimeric; sequences from different parts of the genome can be incorporated into the same YAC. YAC contigs have been constructed for large regions of all the human chromosomes. They are anchored to the framework map by identification of sequence tagged sites – elements of defined sequence which have been mapped precisely onto framework maps and can be shown to be present on specific YACs. A good example of this approach is a YAC contig that spans the entire length of chromosome 21.

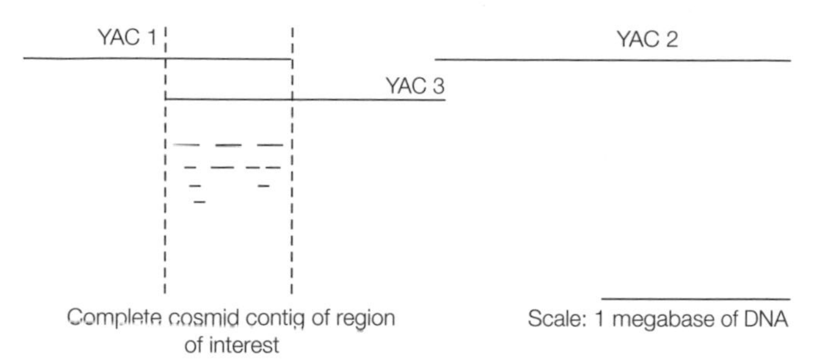

Fig. 4. The YAC contig for the region of the chromosome provides a crude outline for further analysis. If it is subsequently shown that the gene under study is present in YAC 1 and YAC 3, it must lie in the overlap between these two. This region of interest is then subcloned into cosmids and these are tested to show which overlap with each other. A fine-scale contig of overlapping cosmids is then produced for this region, and this will facilitate screening for the gene.

Cosmids, vectors that carry much less DNA (30–45 kb), are much more stable, but the small size of insert means that contigs take much more time and effort to construct. Very often the two are combined in searching a specific region for new genes. The YACs provide a low level outline map and the detail is made up by a cosmid contig, as shown in *Fig. 4*.

Recently two alternative cloning vectors, bacterial artificial chromosomes (BACs) and P1 artificial chromosomes (PACs), have been developed; these have stabilities similar to those of cosmids, and low incidences of chimerism, but can accommodate up to 300 kb of DNA.

Fluorescence *in situ* hybridization

Fluorescence *in situ* hybridization now has a major role in the mapping of human genes. *In situ* hybridization is the process of hybridizing a labeled probe to a DNA target on a cytological preparation rather than on a filter. Fragments of genomic DNA or cDNA molecules can be hybridized to chromosome preparations to identify their specific position on a chromosome. Originally radiolabeled probes were used, but they gave low resolution. Probes are now either labeled directly with fluorochromes or with modified bases that can be detected by fluorescent antibodies. Fine scale mapping is carried out on stretched chromosomes or on purified DNA.

Placing genes on the map

The framework maps consists almost entirely of polymorphic markers from noncoding DNA. Eventually when the genome is fully sequenced it will be straightforward to detect the position of all genes but at present one of the major tasks for the human gene program is to indicate the locations of genes. This is done in a number of ways. **Positional cloning** is a method that depends on locating the gene responsible for a particular inherited syndrome by first mapping to loci on the framework map. If the gene can be placed in a sufficiently small region (usually under two megabases), the region can be examined for the presence of coding sequences, and any that are found are evaluated to see if they could be responsible for the syndrome. Genes that have been already isolated from mRNA as cDNA can be mapped by screening YACs that have been previously placed in mapped contigs. Alternatively the locus of the gene can be determined using radiation hybrid panels.

Only a small proportion of the genes in the human genome are known, however it is now possible to map genes that have not yet been identified. This process depends entirely on the isolation of **expressed sequence tags** (ESTs), and is known as **transcript mapping**. ESTs are unique sequences belonging to anonymous mRNA molecules which can be used to design primers for PCR to search for the relevant coding sequences in genomic DNA. The procedure commences with establishing cDNA libraries made from polyA-containing mRNA. This can be done for several tissues to increase the number of messages in the cDNA pool. The cDNA molecules are then checked to exclude multiple copies of the same mRNA. A portion of each cDNA is sequenced, to allow the identification of PCR primer sequences unique to that message. These are then positioned by screening radiation hybrids and YAC contigs. This is a most powerful approach as over half the genes expressed in humans are now codified as cDNA molecules. The aim is to define ESTs for all regions of the genome so that whenever the gene for an inherited disorder is mapped to a chromosomal location a series of ESTs will be already known for the region. These can then be assessed to find the gene responsible.

Summary

The human genome mapping project has now reached the end of the stage concerned with the production of framework maps. The main activities now concentrate on placing genes on the map and achieving the ultimate goal of completing the entire nucleotide sequence. Already several regions of the genome have been sequenced. The procedure for continuing this work will be as follows. Regions of the genome that have been covered by contigs will be subcloned into smaller vectors and these mapped to span the region. Then each of these will be further subcloned into sequencing vectors. The base sequence will be built up from these and overlaps will be used to orient the sequences. This will be a major task generating huge amounts of data and necessitating the development of novel databases to manage the information. Major problems will arise in areas of repeated sequence where ordering the sequences will prove difficult.

F6 GENETICS IN FORENSIC SCIENCE

Key Notes

Unique correlations

Sexually reproducing organisms are genetically unique. All the individual's cells carry the same genetic information, and any tissue samples of unknown origin can be compared to those of candidate individuals. The power of the technique is in exclusion. A single difference between a scene-of-crime sample and a suspect can prove an absence of connection, and exclude the innocent. If a perfect match is found, the probability of finding that same genotype by chance in the population can be calculated, provided that the frequencies of the alleles involved are known for that population. Genetic relatedness can also be used to trace family connections and associate children with their natural parents. The same techniques are also used in nonhumans, for example to trace lost or stolen horses and hawks. Using the polymerase chain reaction (PCR), a genetic profile can be obtained from a few cells, and will often establish genetic identity beyond reasonable doubt.

Protein comparisons

Proteins may be compared by immunological techniques (e.g. using antibodies to identify blood group or tissue type) or by electrophoresis to screen for charge : mass variation caused by amino acid substitutions. These techniques may exclude most of the population.

DNA comparisons

All cells in an individual carry copies of the same DNA sequence, and in sexually reproducing organisms, each individual has a unique sequence of DNA in their genome. If DNA sequences from a tissue sample found at the scene of a crime are compared (in sufficient detail) with DNA sequences from a suspect, differences can be found, unless the tissue sample came from the suspect (or his or her clone)! The variable sequences used for forensic purposes are micro- and minisatellites. These are arrays of tandem repeats of units, each unit being 2–30 base pairs long. The arrays frequently change their length (repeat number) by mutation, hence their name: variable number tandem repeats (VNTRs). Variation was first detected in restriction fragment length polymorphisms (RFLPs). If two samples are identical then all the fragment lengths between adjacent restriction sites will be the same length. This has been superseded by PCR-based techniques.

RFLPs: genetic fingerprints

RFLPs occur when the distance between successive restriction enzyme sites varies by mutation. The variation is caused by loss or gain of restriction enzyme sites, or by changes in the copy number of VNTRs (micro- and minisatellites). Genetic fingerprinting involves cutting genomic DNA with a restriction enzyme, separating fragments by electrophoresis, then Southern blotting with a multilocus probe. This detects all occurrences of a repeated sequence which is found at many places in the genome, and so provides many bands in a unique combination. The disadvantages are that it requires a relatively large sample of DNA, and the wealth of data that is obtained is difficult to quantify and communicate numerically in court.

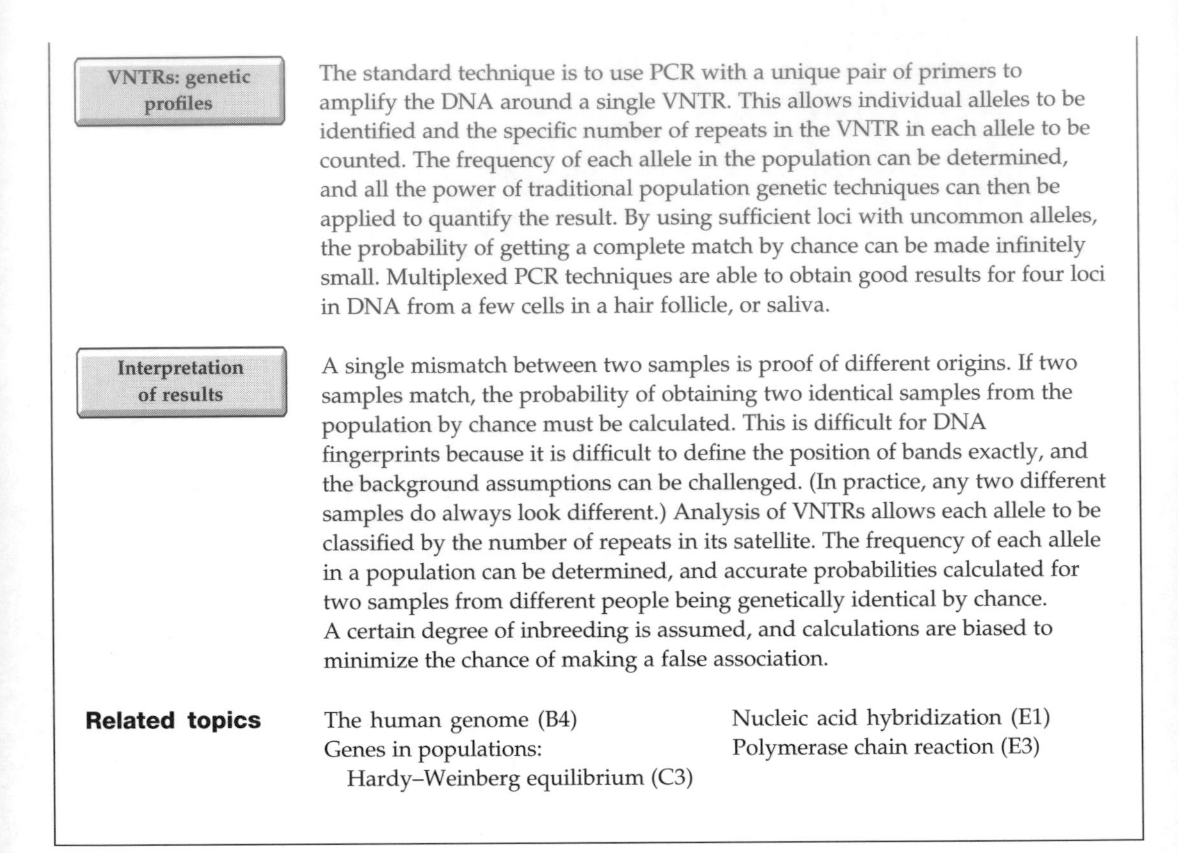

VNTRs: genetic profiles	The standard technique is to use PCR with a unique pair of primers to amplify the DNA around a single VNTR. This allows individual alleles to be identified and the specific number of repeats in the VNTR in each allele to be counted. The frequency of each allele in the population can be determined, and all the power of traditional population genetic techniques can then be applied to quantify the result. By using sufficient loci with uncommon alleles, the probability of getting a complete match by chance can be made infinitely small. Multiplexed PCR techniques are able to obtain good results for four loci in DNA from a few cells in a hair follicle, or saliva.
Interpretation of results	A single mismatch between two samples is proof of different origins. If two samples match, the probability of obtaining two identical samples from the population by chance must be calculated. This is difficult for DNA fingerprints because it is difficult to define the position of bands exactly, and the background assumptions can be challenged. (In practice, any two different samples do always look different.) Analysis of VNTRs allows each allele to be classified by the number of repeats in its satellite. The frequency of each allele in a population can be determined, and accurate probabilities calculated for two samples from different people being genetically identical by chance. A certain degree of inbreeding is assumed, and calculations are biased to minimize the chance of making a false association.

Related topics

The human genome (B4)
Genes in populations: Hardy–Weinberg equilibrium (C3)
Nucleic acid hybridization (E1)
Polymerase chain reaction (E3)

Unique correlations

The uniqueness of each individual (in a sexually reproducing species) can be used to identify that individual, or any cellular trace part of the individual. Using the polymerase chain reaction (PCR), a genetic profile can be obtained from a few cells, and will often establish genetic identity beyond reasonable doubt. Any tissue containing intact DNA provides useful samples: semen (in rape cases), blood, hair, skin fragments, mouth epithelial cells (in saliva), or bones in skeletons.

Forensic (law-related) science can compare the genetic characteristics of biological fragments found at the scene of a crime (presumed to come from the criminal) to a suspect to exclude or associate them. Any biological sample from a person will match that person in all genetically determined characteristics. The great power of the technique is to exclude the innocent. A single genetic character which does not match can prove that the suspect did not deposit the sample. If enough suitable genetic characteristics do match, that is evidence 'beyond reasonable doubt' that the suspect did deposit the sample. The court can then test the suspect's explanation of this 'contact' with the crime scene.

The genetic similarity between parents and offspring can also be used to confirm or reject their relationship. A child must get one allele of each gene from each parent, so half the child's genotype must match each parent. Genetic tests can be used in disputes over paternity to identify or exclude potential fathers. Cases of child-stealing and mix-ups between babies in hospitals can also be resolved. Genetic tests in nonhumans can also be used forensically, for example to detect kangaroo meat in minced beef, and to identify stolen horses.

Murderers have been convicted by genetic fingerprinting of nonhuman material. In one case a seed found in a suspect's vehicle came from the specific individual tree beside the murder victim's body, and in another case cat hairs at a crime scene were shown to be genetically identical to the suspect's cat. Both cases were used to show, with a probability of error of less than one in a million, an association between the suspect and the scene-of-crime which was denied by the suspects. In both cases the suspects were first suspected because of routine police investigations, the DNA evidence was confirmation.

Protein comparisons

The amino acid sequence of proteins is determined genetically, and can be examined directly by examining the molecules, or indirectly by the effects of different variants of a protein on the phenotype. Describing someone as blue-eyed describes a genetically determined characteristic. It is not very useful in Scandinavia where most people are blue-eyed, but could be very useful in a country where most people are brown-eyed. The usefulness of a characteristic for identification depends upon it being uncommon. The first scientific tests were for particular cellular molecules identified immunologically. **Antibodies** are produced by the immune system in response to foreign molecules. It was found that mammals, usually rabbits, sheep, goats or horses, which were injected with human blood (inoculated with an inoculum) would produce antiserum containing antibodies which would bind to red blood cells of the type found in the inoculum. These antisera could distinguish the various classes in the A-B-O blood groups, and also several other characters, including several types of rhesus blood group and the M-MN-N groups, according to the type of blood used for the initial inoculum. These antibodies can be used to identify the type of a drop of blood. If blood similar to the victim is found on a suspect, or vice versa, that suggests a physical contact between them. The less common the alleles, the more useful they are in exclusion. Blood group N is restricted to 1.1% of Navaho Indians, 21% of Caucasians, but is found in 67% of Australian Aborigines. It would be useful for excluding 98.9% of Navaho, 80% of Caucasians, but only 33% of Aborigines.

Tissue-type alleles show a great deal of variation, and were used extensively in Argentina to identify the 'children of the disappeared'. Children were taken

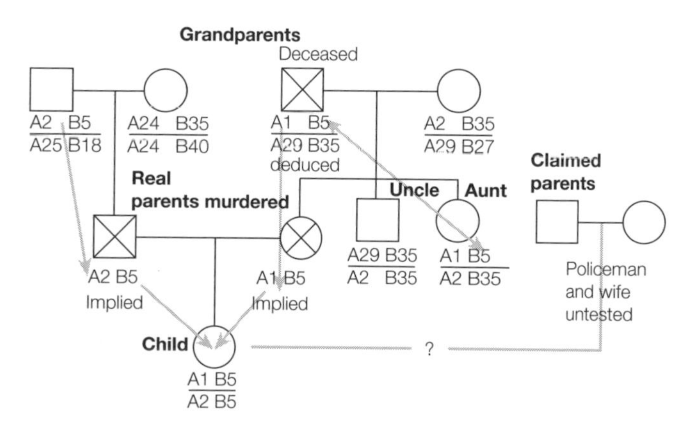

Fig. 1. Family tree of HLA (tissue type) alleles of presumptive grandparents and abducted grandchild. Gray arrows trace alleles from grandchild to relatives, confirming the presence of the child's alleles in his dead parents family, and showing 99.9% probability of a genuine relationship.

as babies from political dissidents who were murdered. When suspected stolen children were located, their tissue type alleles were compared with their supposed grandparents (their supposed parents being dead) and where possible with the people claiming to be their parents. This often showed that there was a very high probability that the child was the grandchild of the deceased dissident's parents (*Fig. 1*).

Another technique is **protein electrophoresis**, a process analogous to electrophoresis of nucleic acids (see Topic D6). A mutation can cause a change in a single charged amino acid in a protein which in turn can change the protein's charge to mass ratio, and change the rate at which it migrates in a gel. Variation for such mutations is common. Again the principle is to compare the scene-of-crime sample with one from the suspect.

DNA comparisons

Every sexually produced individual has a unique combination of sequences in their DNA, each individual has a unique genotype. When the DNA from tissue recovered at a crime scene matches the sequence of DNA from a suspect, it is as certain as it can be that the sample came from the suspect (providing sufficient sequences are compared). The development of techniques to manipulate DNA, particularly the use of restriction enzymes, made it possible to look directly at DNA for comparisons (see Topics E4 and E3).

The techniques used for forensic work are all based on separating DNA fragments by length using electrophoresis. The fragments are initially produced by digesting genomic DNA with restriction enzymes (see Topics B4 and F5). **Restriction endonucleases** cut DNA at specific sequences (called **restriction sites**) which occur fairly randomly within the genomic DNA (e.g. the enzyme *Eco*RI cuts within the sequence GAATTC which occurs every 2000–5000 bp on average in species with high and low AT : GC ratios respectively). The fragments produced are called restriction fragments.

If the sequence is always the same, corresponding restriction fragment lengths will always be the same from all people, but fortunately they differ. There are two ways that mutations can change the length of a restriction fragment: (i) by changes in restriction sites. A mutation can create a new restriction site within the fragment so that it is cut shorter (e.g. GAAATC could mutate to GAATTC) or mutation can destroy an existing site so that it does not cut, and two adjacent fragments now make one longer fragment; (ii) base pairs may be inserted or deleted, making the fragment longer or shorter. This commonly happens by changes in the number of copies present in a block of **tandemly repeated sequences** (see Topic A14) which may be highly unstable and so very polymorphic (variable) within the population. These are called **variable number tandem repeats (VNTRs)**. If the individual units which are repeated are less than about 6 bp long, they are called microsatellites, larger units 6–30 bp long are minisatellites. Longer ones are just called VNTRs.

Most of the variation that is used for forensic purposes is in VNTR loci which are not subject to selection to maintain their sequence. Their high level of variation makes it less likely that two people will be identical. There are two basic techniques used by forensic scientists to study this length variation. In the original technique all occurrences of a particular satellite sequence were studied, and because this happened at many places in the genome it was a **multilocus** system. This gave a DNA fingerprint with a unique combination of many bands. More recently examination of the microsatellites at a few particular sites (**single loci**) detected by PCR primers has become the favored technique. This is because each individual only has two alleles of each **single locus**, and the defined data is more easily quantifiable.

RFLPs: genetic fingerprints

The first forensic technique used on DNA was to examine **restriction fragment length polymorphisms (RFLPs)** (see Topics B4, E1 and F5). The restriction fragments can be separated according to length by **gel electrophoresis** (see Topic E1). Digests of genomic DNA (total nuclear DNA) from cellular organisms contain a complete range of lengths, and individual fragments are not distinguishable from slightly longer or shorter ones. It is therefore necessary to highlight a selection of the bands. This is done by **Southern blotting** the gel using a **multilocus probe** (see Topic E1). These are sequences that occur at multiple sites in the genome, ideally producing as many distinct bands as can be clearly measured on the autoradiograph (*Fig. 2*). If digests of two good samples of DNA from the same person are run in adjacent lanes on the same gel (but separated by an empty lane to avoid the possibility of cross-contamination) it is clear that they are identical and unique, with bands of similar intensity clearly appearing side-by side. When they are run on separate gels, however, the small unavoidable variations may make this comparison more difficult, and if the band's position on the gel has to be measured relative to markers of known length, then errors of measurement must also be considered, and information on band intensity is lost. The data is much more difficult to present in court, and its reliability can be questioned.

Disadvantages of RFLPs

RFLPs are no longer favored for forensic purposes because: (i) they require relatively large samples of DNA, sufficient to act as a target for the probes in Southern blots; (ii) there are difficulties in interpretation of the data in terms of calculating the probability of falsely obtaining a match between two samples from different people. There are limitations on the accuracy with which the lengths of restriction fragments in individual bands can be measured, and this is compounded by small differences between different gels. It is impossible to say which bands represent different alleles of the same genetic locus, so classical population genetics cannot be used to calculate the frequency of specific genotypes. These difficulties are largely overcome by using single locus probes and techniques based on the PCR.

VNTRs: genetic profiles

Forensic investigation usually uses length variation at a set of single loci, each containing one VNTR. About 0.5% of the human genome is composed of dCA (=dTG) repeats, (dCT.dAG repeats compose 0.2%). These are dinucleotide repeats, but tetranucleotide repeats (e.g. CAGA) are preferred for forensic work because of the greater difference in length caused by a gain or loss of one repeat, and the greater stability of the sequence when amplified by PCR. VNTRs provide much of the variation detected by multilocus probes, but for the purposes of forensic science they are usually examined at an individual locus by a **single locus probe**, and the exact number of copies of the repeat in the specific alleles present in the sample is found. Because the sequence is defined, it can be studied in the population and allele frequencies can be determined. A particular VNTR may, for example, have between 12 and 23 repeats. The frequency of each length allele in different racial groups can be determined by sampling volunteers from the general population. This allows population genetics to be applied to calculating genotype frequencies (see Topic D3). This

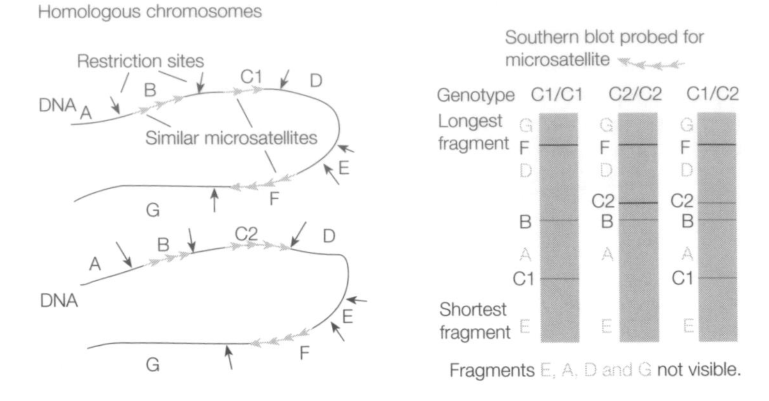

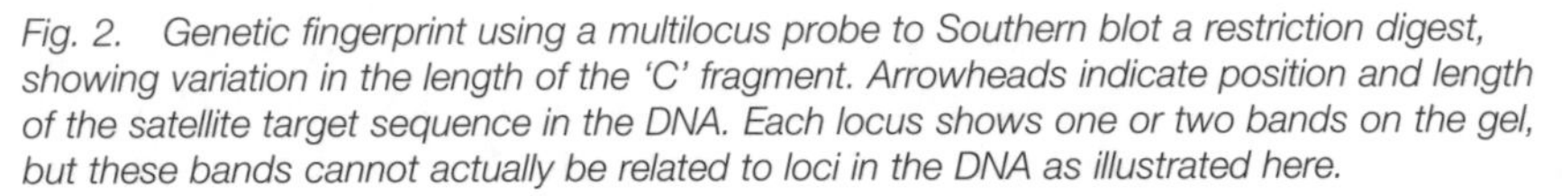
Fig. 2. Genetic fingerprint using a multilocus probe to Southern blot a restriction digest, showing variation in the length of the 'C' fragment. Arrowheads indicate position and length of the satellite target sequence in the DNA. Each locus shows one or two bands on the gel, but these bands cannot actually be related to loci in the DNA as illustrated here.

is an important advance and RFLP data (multilocus, band lengths) is presented to the court quite differently from VNTR data (single allele repeat numbers), although scientifically they are two ways of examining a similar genetic phenomenon.

The alleles of a single locus could be detected on Southern blots by using a probe to the unique sequences adjacent to the variable site, but it is simpler to use the PCR (see Topic D3). This can amplify DNA from a very low copy number, and a single hair follicle is sufficient to determine genotype. Unique pairs of primers are used to amplify the locus containing one block of tandem repeats. Using the example in *Fig. 2*, primers would be used to amplify only fragment C. One of the primers has a fluorescent label attached, and the length of the bands produced is measured by an automated gel device using the same technique as automated sequencing (see Topic E4). A heterozygote will produce two bands for one locus. In practice, four pairs of primers are used together (**multiplexed**) to detect four loci simultaneously, producing eight bands from a diploid heterozygous at all four loci. The loci used are chosen from different chromosomes to avoid linkage and ensure independent assortment. They also have a high degree of variability in the population, and have different overall lengths between loci to avoid confusion, although different color labels on different primers allows each locus to be identified.

Interpretation of results

In some countries, trained judges try to establish the facts from the evidence, but in the UK and the USA the adversarial system (derived from trial by combat) relies on convincing the jury of the rightness of one side of the case and the wrongness of the other. The jury is composed of 12 laypeople, (that is people with no qualifications in interpreting the evidence) and the judge is there to uphold the law, not to investigate the case. This leads to a failure to use evidence sensibly, and some of the first cases using genetic fingerprinting evidence were dismissed because the defense raised scientifically unwarranted doubts in the minds of the jury, while prosecutors made extravagant claims which they could not support. For these reasons forensic evidence is presented very carefully in a formalized way which is not open to criticism. There have always been cases where scientific proof was ignored. Charlie Chaplin was accused by a young actress of fathering her child. He was convicted despite the blood group evidence that he could not possibly be the child's father. Conversely, a suspect was convicted because a technician spilled the suspect's DNA into the adjacent well used for the scene-of-crime sample, producing identical bands in both lanes after many negative results. Such an effect of technical incompetence should have been obvious to anyone familiar with the technique. It should be clear from this that presentation of evidence in court is more important than scientific facts.

Genetic fingerprinting works on the principle of exclusion, a single difference between sample and suspect rules out the suspect. The problem is to work out the probability of not excluding someone who is really innocent, but whose sample matches the scene-of-crime sample by chance. For example, suppose that a suspected rapist has an identical genotype to semen from a crime scene. If the frequency of that genotype in the population is 0.01 (1/100) then the probability of the suspect and criminal having the same genotype by chance is 1 in 100. Convicting the suspect on that evidence alone would convict one person in every hundred innocent members of the public. However, if the frequency of the genotype is 10^{-8} then the probability of getting the match by chance is one in 10^8. It is unlikely that there is another person on the same continent who could be mistaken for the rapist.

Traditional fingerprinters did not use statistics, they simply determined that if enough points matched, then two fingerprints came from the same finger. Genetics has relied on statistics since Mendel, and attempts are made to calculate the numerical probability that a sample from the scene-of-crime matches the suspect by chance. By merely questioning these calculations, some defense councils managed to win cases by causing doubt. Multilocus data requires the measuring of band position to calculate fragment lengths. If 200 mm of a gel can be measured with an accuracy of one millimeter, then only 200 band positions (bins) can be measured. Bands differing in position by 0.5 mm will be classed together (put in the same bin, hence **binned**), reducing the useful information. The probability that all the bands in two different samples fall into the same bins by chance must calculated, and realistically it is likely to be one in many millions. By raising doubts about the validity of the procedures, arguing that a probability might be less than the value calculated by the prosecution and that the figure could be wrong, the method was initially discredited. This was largely overcome by forensic witnesses being cautious, erring on the side of supporting innocence, and learning what questions they would be asked, and how to answer them.

Using single locus probes, individual alleles can be identified. Their frequency in the population can be measured, and the Hardy–Weinberg equation (see Topic D3) can be used to calculate quite accurately the probability of two unrelated samples being the same genotype by chance. The problem with this is the existence of subpopulations, a form of inbreeding. These have proved to be the biggest objection to genetic profile (VNTR) evidence. If a suspect and the scene-of-crime sample both have a rare allele with a frequency of 0.001, then only about 1 in 500 people will have it, but one of the suspect's parents must have it, and also half of his brothers, sisters and children. If the family has lived in the community for a long time there will be cousins and many distant relatives who will also have a much greater than 1/500 chance of having that allele, so the chance of mistaken identity is greater. This can be overcome by giving rare alleles a minimum frequency of 0.05. The court is then given the lower probability of not excluding the innocent suspect if someone quite closely related, (e.g. a first or second cousin) was the actual culprit. Probabilities of 1/10 000 to 1/1 000 000 are typically produced. Other loci can be used to improve this, and if there is other evidence implicating the accused, these values are strong supporting evidence.

Races are also subpopulations. Alleles of human locus *WVA* contain between 12 and 21 repeats. Alleles with 12 repeats have only been identified in one racial group. The 15-repeat allele has a frequency around 0.09 in Asians and Caucasians, and 0.2 in Afro-Caribbeans. The highest frequency appropriate to the suspect's racial group or the population must be used to give the best chance to the accused. Again, if there is other evidence of racial group, the population genetics calculations can be more accurate.

F7 GENETIC ENGINEERING AND BIOTECHNOLOGY

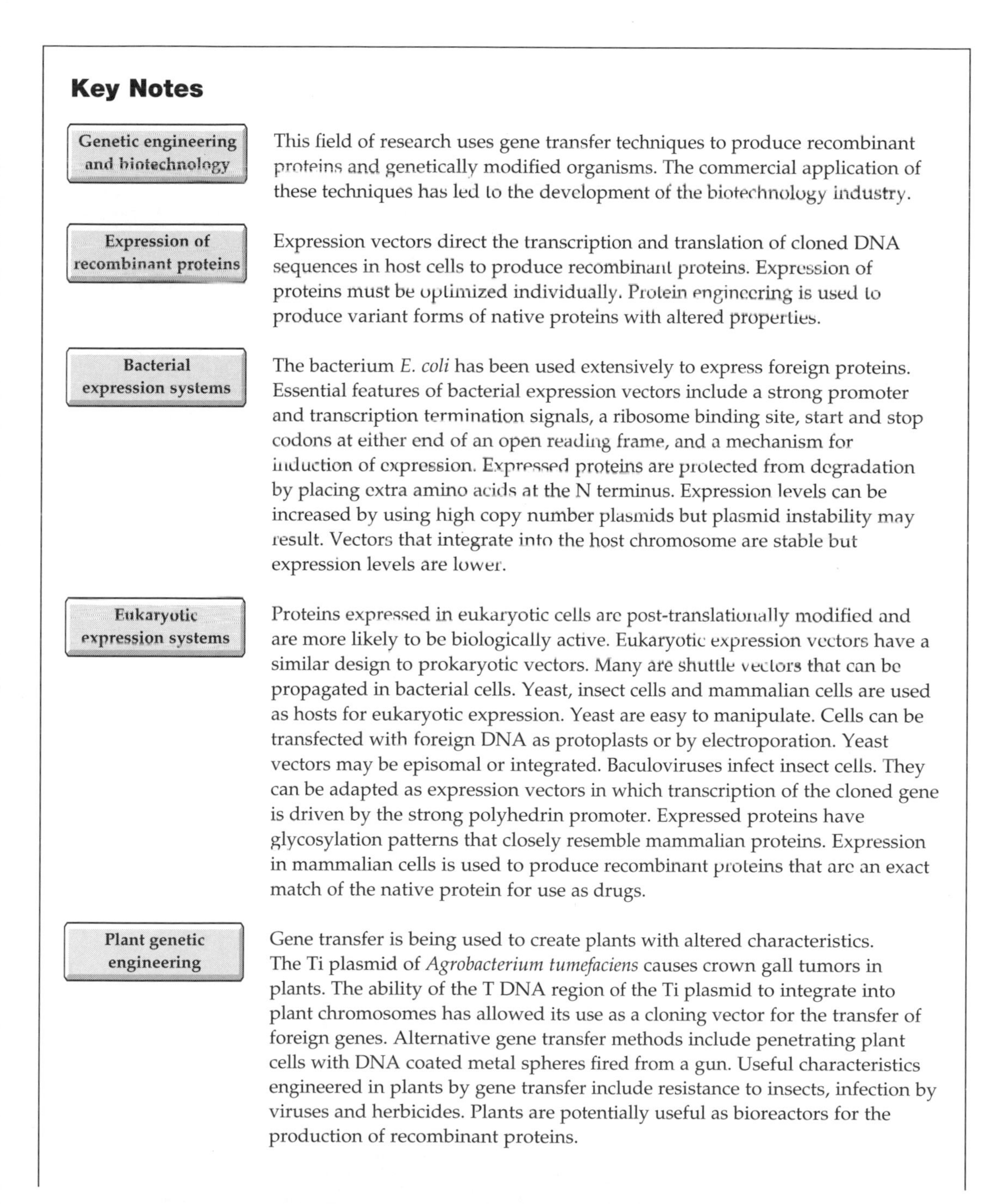

Key Notes

Genetic engineering and biotechnology

This field of research uses gene transfer techniques to produce recombinant proteins and genetically modified organisms. The commercial application of these techniques has led to the development of the biotechnology industry.

Expression of recombinant proteins

Expression vectors direct the transcription and translation of cloned DNA sequences in host cells to produce recombinant proteins. Expression of proteins must be optimized individually. Protein engineering is used to produce variant forms of native proteins with altered properties.

Bacterial expression systems

The bacterium *E. coli* has been used extensively to express foreign proteins. Essential features of bacterial expression vectors include a strong promoter and transcription termination signals, a ribosome binding site, start and stop codons at either end of an open reading frame, and a mechanism for induction of expression. Expressed proteins are protected from degradation by placing extra amino acids at the N terminus. Expression levels can be increased by using high copy number plasmids but plasmid instability may result. Vectors that integrate into the host chromosome are stable but expression levels are lower.

Eukaryotic expression systems

Proteins expressed in eukaryotic cells are post-translationally modified and are more likely to be biologically active. Eukaryotic expression vectors have a similar design to prokaryotic vectors. Many are shuttle vectors that can be propagated in bacterial cells. Yeast, insect cells and mammalian cells are used as hosts for eukaryotic expression. Yeast are easy to manipulate. Cells can be transfected with foreign DNA as protoplasts or by electroporation. Yeast vectors may be episomal or integrated. Baculoviruses infect insect cells. They can be adapted as expression vectors in which transcription of the cloned gene is driven by the strong polyhedrin promoter. Expressed proteins have glycosylation patterns that closely resemble mammalian proteins. Expression in mammalian cells is used to produce recombinant proteins that are an exact match of the native protein for use as drugs.

Plant genetic engineering

Gene transfer is being used to create plants with altered characteristics. The Ti plasmid of *Agrobacterium tumefaciens* causes crown gall tumors in plants. The ability of the T DNA region of the Ti plasmid to integrate into plant chromosomes has allowed its use as a cloning vector for the transfer of foreign genes. Alternative gene transfer methods include penetrating plant cells with DNA coated metal spheres fired from a gun. Useful characteristics engineered in plants by gene transfer include resistance to insects, infection by viruses and herbicides. Plants are potentially useful as bioreactors for the production of recombinant proteins.

Transgenic animals

Gene transfer can be used to engineer useful traits in animals. The transgene is introduced into a fertilized ovum or cells of an early stage embryo by microinjection, manipulation of embryonic stem cells or by using retroviral vectors. Transgenic offspring result following implantation of the modified embryo in the uterus of a host female. Transgenic animals are used as research tools to study gene function or to produce animal models of human disease. Transgenic sheep and goats are being used to produce recombinant proteins secreted in milk.

Related topics

Gene transcription (A4)
Translation (A8)
Prokaryotic genomes (B3)
DNA cloning (E2)
Gene therapy (F4)

Genetic engineering and biotechnology

These terms refer to a field of research that involves using recombinant DNA techniques to transfer genes from one organism to another. Gene transfer between organisms has two important applications: firstly, to produce large amounts of biologically useful proteins. These include proteins that are used as drugs to treat diseases (*Table 1*) and proteins or enzymes used in industrial processes (*Table 2*). Secondly, gene transfer is used to create organisms (plants, animals and microorganisms) with altered characteristics, for example plants that are resistant to disease. These important applications of genetic engineering and biotechnology have enormous commercial potential and have led to the establishment of the biotechnology industry.

Table 1. Recombinant human proteins used as drugs

Protein	Used to treat
α1-antitrypsin	Emphysema
Calcitonin	Rickets
Chorionic gonadotropin	Infertility
Erythropoietin	Anemia
Factor VIII	Hemophilia
Insulin	Diabetes
Interferons (α, β, γ)	Viral infections, cancer
Interleukins	Cancer
Tissue plasminogen activator	Blood clots
Growth hormone	Growth retardation

Table 2. Recombinant enzymes with industrial uses

Protein	Industrial use
Rennin	Cheese making
α-amylase	Beer making
Bromelain	Meat tenderizer, juice clarification
Catalase	Antioxidant in food
Cellulase	Alcohol and glucose production
Lipase	Cheese making
Protease	Detergents

Expression of recombinant proteins

One of the primary aims of genetic engineering and biotechnology is to produce biologically useful proteins. This is achieved by expressing cloned genes in a variety of host cells. Bacteria and yeast as well as animal and plant cells are used. The expression of recombinant proteins requires the use of specialized vectors called **expression vectors** which contain DNA signal sequences that direct the transcription and translation of the cloned sequence by the host cell. Other vector sequences are included which influence the stability of the expressed proteins or direct the host cell to secrete the expressed protein. To ensure high levels of expression, plasmid vectors are used that exist in host cells as multiple copies. The conditions required for efficient expression vary from protein to protein and must be optimized individually. This has led to the development of many diverse expression systems.

The use of recombinant DNA technology to produce proteins allows variant forms of natural proteins to be produced. This is known as **protein engineering**. Cloned gene sequences can be altered by a procedure called ***in vitro* mutagenesis** and expressed to produce proteins with modified amino acid sequences that have altered characteristics and properties. Protein engineering can be used to produce variants of normal proteins that have properties such as improved heat stability or, for enzymes, altered substrate specificity.

Bacterial expression systems

The first protein expression systems to be developed were based on the use of bacteria as host cells. The bacterium *E. coli* allows many proteins to be produced rapidly and inexpensively and has been used extensively. Bacterial expression systems remain important and are used for the production of most commercially important proteins (*Tables 1 and 2*).

Vectors for expression of proteins in bacteria have a number of essential features (*Fig. 1*):

- A promoter placed upstream of the cloned sequence to initiate transcription. Selection of a promoter that binds RNA polymerase strongly is necessary to ensure high levels of expression.
- A transcription termination signal placed downstream of the cloned sequence.
- A ribosome binding site (Shine–Dalgarno sequence) placed downstream of the transcription start site. The level of protein synthesis is influenced by how strongly the ribosome binds this sequence.
- The cloned DNA sequence should have an AUG translation initiation codon at the start and a termination codon at the end. The intervening sequence should be an open reading frame. It is very important to establish the correct reading frame to ensure that the correct amino acid sequence is translated. Vectors usually allow sequences to be cloned in different ways so that each of the three reading frames can be used.
- Vectors should allow expression of the cloned sequence to be induced. Continuous expression creates a drain on the energy resources of the cell which is detrimental to other cell functions. Expression is more efficient if it can be switched on for a limited period then switched off again. Most systems are induced by the addition of a small molecule, often a metabolite, to the bacterial culture. Some systems are induced by altering the culture temperature.

Proteins expressed in bacteria are often unstable due to degradation by protease enzymes in the host cell. Expressed proteins may be protected from degradation by altering the sequence of the vector to place one or more amino acids at the N terminus of the expressed protein. The added amino acids may also be used as a tag which allows purification of the protein using antibodies to the tag immobilized on a gel. Vectors are usually designed to allow enzymatic or chemical removal of the tag from the expressed protein following purification.

The level of protein production can often be increased by using plasmid vectors that exist as multiple copies in the host cell. However, the presence of plasmids at high copy number tends to create instability and some cells lose the plasmid: these cells tend to grow faster than the cells that have retained their plasmid and can take over the culture leading to an overall decrease in expression. To overcome this problem of plasmid instability, some vectors are designed to integrate into the host cell chromosome where they are stable, although expression levels are reduced because of the lower copy number.

Eukaryotic protein expression systems

Although many eukaryotic proteins can be successfully expressed in bacterial systems, some expressed proteins are unstable or are biologically inactive due to the inability of bacteria to carry out post-translational modifications, such as glycosylation, that occur in eukaryotic cells. In addition, proteins expressed in bacteria are sometimes contaminated with toxins making them unsuitable for use with humans or animals. Expression systems based on eukaryotic host cells have been developed to produce versions of expressed proteins that more closely resemble the native form.

Eukaryotic expression vectors have many features equivalent to those of their prokaryotic counterparts (*Fig. 1*). These include a promoter and transcription and translation signal sequences. Additional features specific to eukaryotic systems, such as signal sequences for polyadenylation, are also included.

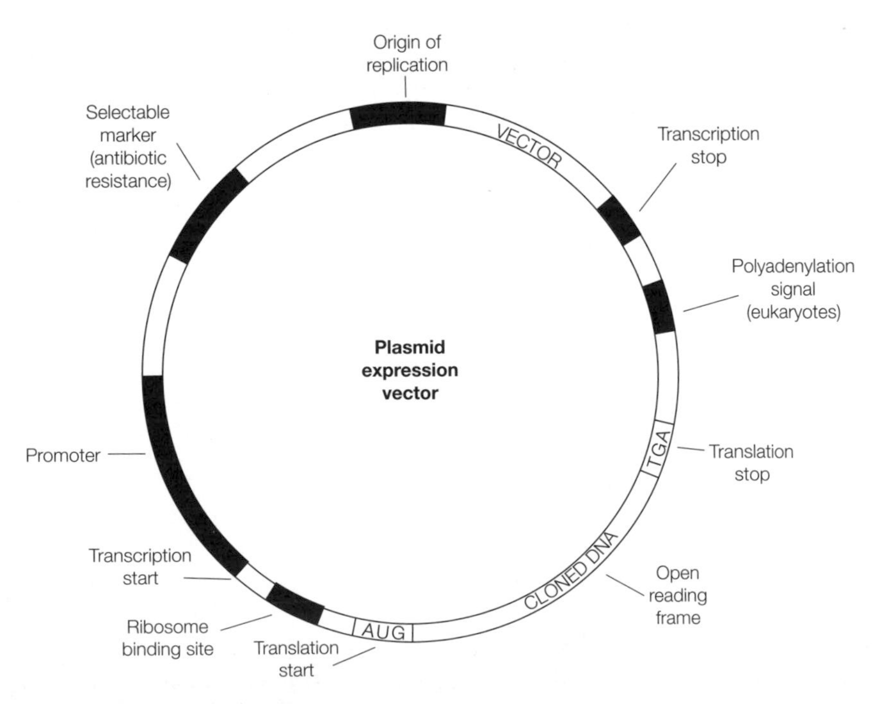

Fig. 1. Features of plasmid expression vectors.

Vectors can exist independently of the host cell genome as plasmids or can be integrated into a host cell chromosome. Some vectors called **shuttle vectors** have prokaryotic sequences that allow them to be manipulated and propagated in bacteria and then transferred to eukaryotic cells for protein expression.

Three types of eukaryotic host cells are used: yeast, insect cells and mammalian cells.

Yeast

The yeast *S. cerevisiae* is widely used for expression of eukaryotic proteins. A number of features make it especially suitable as a host. Yeast are easy to culture and their genetics and physiology have been well characterized. They carry out many post-translational modifications and expressed proteins are easily purified following secretion. In addition, yeast have an endogenous plasmid called the **2μ plasmid** which can be used as a cloning vector. DNA can be introduced into yeast by a variety of methods. Some involve removal of the cell wall to form **protoplasts**. Another method called **electroporation** involves treating cells with pulses of electric current. Yeast expression vectors may exist in cells as plasmids (**episomal vectors**) or integrated into the host cell genome. Episomal vectors give the highest levels of expression but tend to be unstable in large cultures. Examples of recombinant proteins produced by expression in yeast include insulin which is used to treat diabetes, and the hepatitis B virus surface antigen which is used as a vaccine.

Insect cells

Baculoviruses are viruses that infect invertebrates including many insects. They have been adapted for use as eukaryotic expression vectors. Insect cells are used as hosts and are particularly suitable because they produce proteins with glycosylation patterns that are very similar to those of mammalian proteins. The baculovirus contains a gene for a protein called **polyhedrin** which is transcribed from an exceptionally strong promoter. The gene is not essential for viral replication and can be deleted allowing a foreign gene to be inserted. The baculovirus vector containing the cloned gene is transferred into the host cells where it is expressed at high levels late in infection. Recombinant proteins produced using the baculovirus system include the hormone erythropoietin which is used to treat anemia and the anti-virus agent β-interferon.

Mammalian cells

Recombinant proteins used to treat disease in humans must be an exact match of the native form of the protein. In many cases the recombinant protein is only synthesized and properly modified post-translationally when expressed in mammalian cells. As a consequence, expression vectors that use mammalian cells as hosts have been developed. Mammalian expression vectors are shuttle vectors and can be conveniently propagated in bacteria. Transfer into host cells is achieved by adding the DNA as a precipitate with calcium phosphate or by electroporation. A range of human recombinant protein drugs are produced by expression in mammalian cells. These include growth hormone, the blood clotting protein, Factor VIII, β-interferon and erythropoietin.

Plant genetic engineering

Conventional plant breeding has been used for many years with great success to produce crops with increased yield and nutritional value. Genetic engineering, however, provides a more direct method for producing plants with altered characteristics. Plants are especially suitable for genetic modification because most plant cells are **totipotent**. This means that an entire plant can be generated from a single genetically modified cell. If the plant is fertile, the modification will be present in the plant seeds allowing further modified plants to be propagated.

A number of systems have been developed for transferring genes into plants. A very successful example is based on the **Ti plasmid** which occurs in the soil bacterium *A. tumefaciens*. This bacterium infects dicotyledenous plants (all agricultural crops except cereals) and induces formation of a cancerous growth called a **crown gall tumor**. Transformation of plant cells is due to the effect of the Ti plasmid carried by the bacterium. Part of the Ti plasmid called the **T DNA** integrates into a plant chromosome and expression of T DNA genes causes cell transformation. Vectors for the transfer of genes into plants have been developed based on the use of the Ti plasmid and the *A. tumefaciens*. The system is very effective but is restricted by the limited number of plants that are infected by the bacterium. Alternative methods are now being developed for the transfer of genes into plant cells. One unusual but effective method involves coating gold and tungsten spheres with DNA which are then fired into the plant tissue from a special gun. This is known as **biolistics**. The spheres penetrate the cell wall and enter the cells. The DNA is then released from the sphere and integrates into the host cell genome.

A number of useful characteristics that are being engineered into plants are described below.

Resistance to attack by insects

Attempts are being made to engineer plants with insecticidal activity. Two strategies have been used. One involves transfer of a gene from the bacterium *Bacillus thuringiensis* which encodes a protein called **protoxin** which is toxic to some insects. The gene was transferred into tomato plants but only weak protection against insects was observed due to low levels of expression. Another approach involved engineering expression of **protease inhibitor proteins** which interfere with the insects' ability to digest plant tissue.

Resistance to infection by viruses

Genetic engineering is being used to develop novel mechanisms of virus resistance. Genes encoding antisense copies of viral genes have been transferred to plants but have had only limited success in preventing infection. A more successful strategy involves transfer of a gene encoding a viral coat protein. It is not certain how expression of the viral coat protein inhibits infection.

Resistance to herbicides

Despite the use of herbicides, about 10% of global crop production is lost due to infestation by weeds. In addition, herbicides are expensive, potentially toxic to the environment, and can kill crops as well as weeds. Gene transfer has been used to engineer resistance to herbicides in plants to allow selective killing of weeds. **Glyphosate** is a widely used herbicide that acts by inhibiting an enzyme involved in the synthesis of aromatic amino acids in plants. Resistance to glyphosate has been engineered in plants by transfer of a bacterial form of the enzyme that is unaffected by the herbicide.

Other useful characteristics that have been engineered into plants include: delayed ripening of fruits to increase shelf life; tolerance of environmental stresses such as drought, altered pigmentation in flowers; and improved nutritional quality of seeds. Because plants are easy to grow and generate considerable biomass, they are potentially useful for the production of recombinant proteins. Research is being carried out into the possibility of using plants as **bioreactors**. Some success has been obtained and transgenic plants that produce monoclonal antibodies have been created.

Transgenic animals

Selective breeding has been used extensively to produce domesticated animals with desirable characteristics such as high milk yield or high growth rate. Although selective breeding has been very successful, it becomes difficult to introduce new characteristics without altering existing traits. It is now possible to engineer traits directly in animals by gene transfer. The genetically modified animal is called a **transgenic animal** and the transferred gene is called a **transgene**.

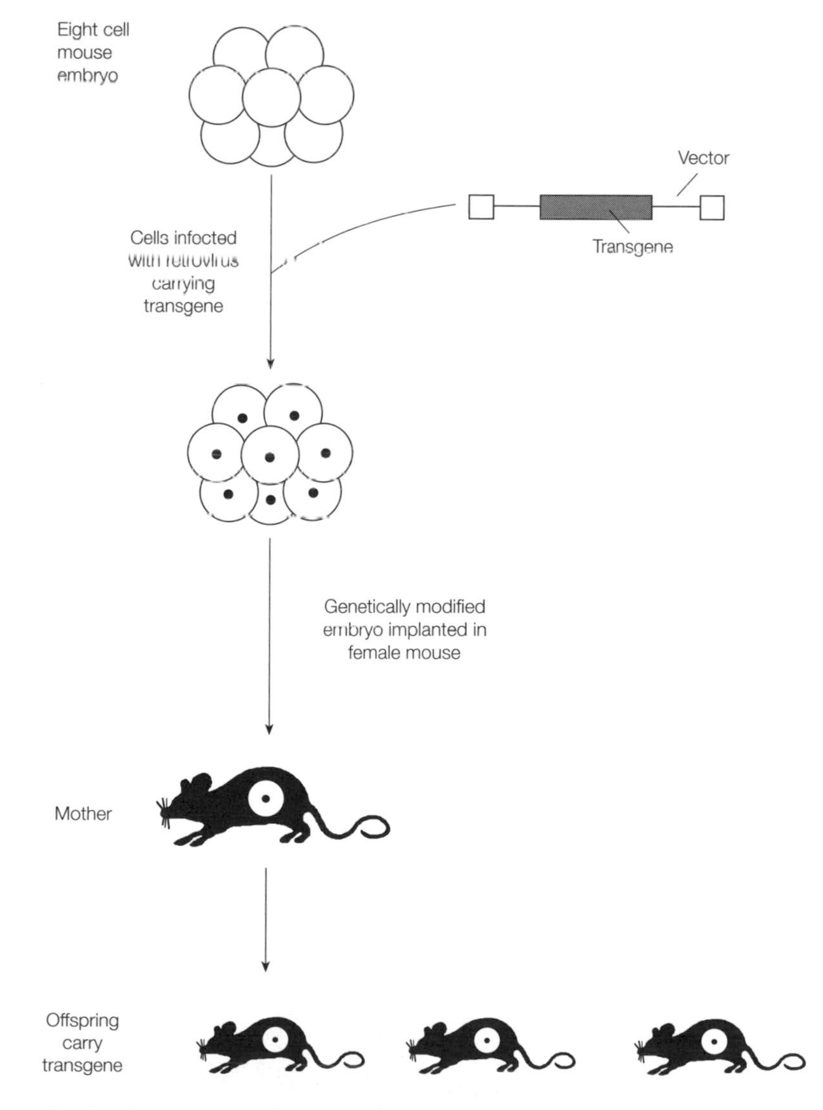

Fig. 2. Gene transfer using a retroviral vector.

Transgenes can be introduced into animals by three methods. Each involves gene transfer into a fertilized ovum or into cells of an early stage embryo. Modified embryos are then implanted into the uterus of a host animal where they develop into genetically modified offspring.

Retroviral vectors
Retroviruses can be used to infect cells of an early stage embryo prior to implantation (*Fig. 2*). The transgene carried by the retroviral vector is efficiently integrated into the host genome but the size of the gene that can be transferred is limited and the use of retroviruses has implications for safety.

Microinjection
This method involves injecting DNA directly into the nucleus of a fertilized ovum viewed under the microscope (*Fig. 3*). Although technically challenging, this is now the most widely used method of producing transgenic animals.

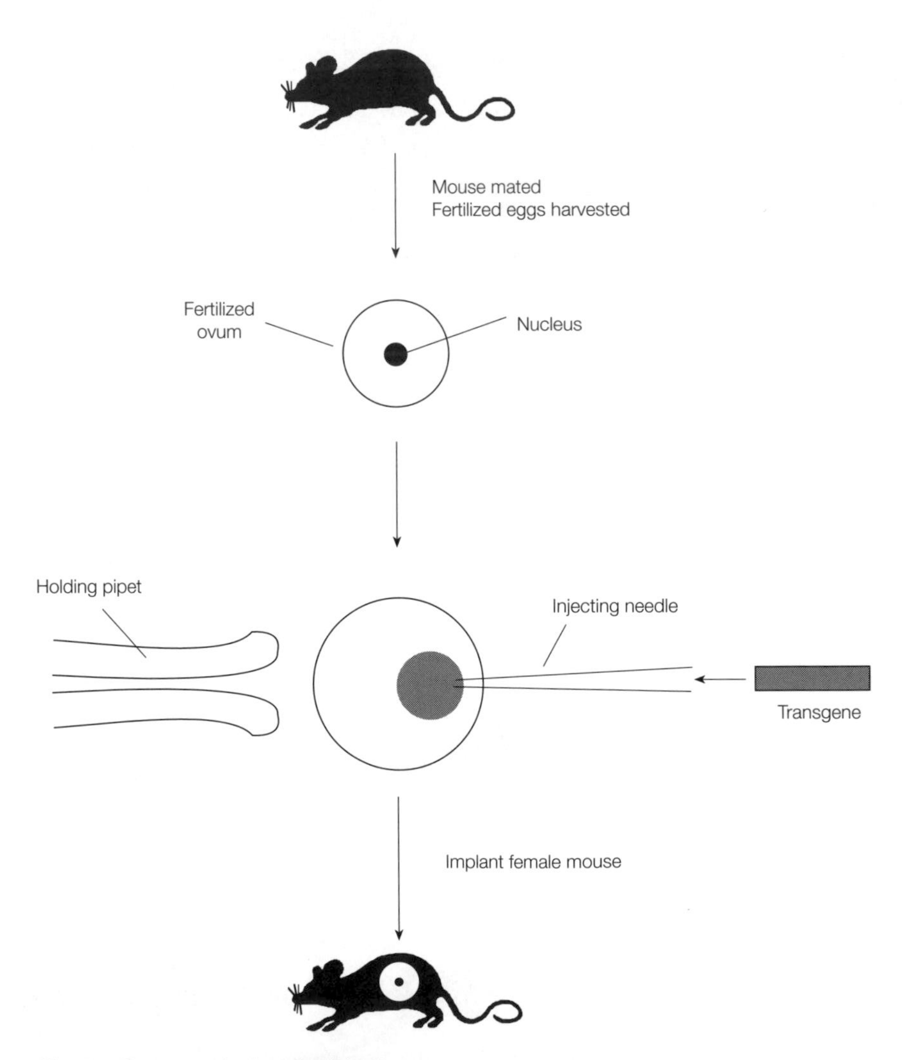

Fig. 3. Gene transfer by microinjection.

Embryonic stem cells

Cells from the **blastocyst** stage of early mouse embryos can be removed and grown in culture. These are called embryonic stem (ES) cells and have the ability to differentiate into all other cell types. ES cells can be genetically modified in the laboratory and returned to the blastocysts for implantation (*Fig. 4*).

Transgenic animals are used as tools in research and for the production of recombinant proteins. They have three main applications.

Studying gene function

Transgenic technology was perfected using mice in the early 1980s. Since then, hundreds of genes have been introduced into mice. By examining the characteristics of the genetically modified animal, it is possible to obtain information about the function of the transferred gene. This has contributed greatly to our understanding of gene regulation and function. A related technique involves using gene transfer to disrupt genes. **Knockout mice** which lack functional

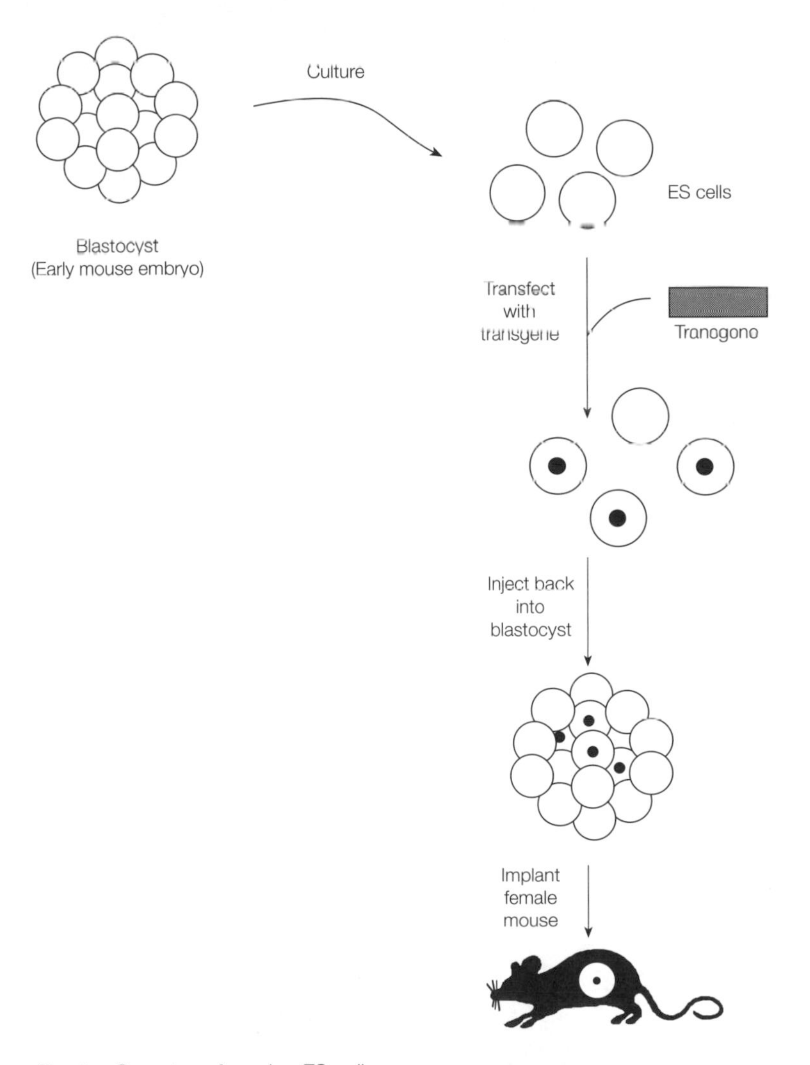

Fig. 4. Gene transfer using ES cells.

forms of specific genes are produced which provide information about the function of the knocked out gene.

Model systems for human disease

Transgenic animals can be created which simulate human diseases in which defective genes play a role. Whole animal models can be used to follow the onset and progression of diseases and provide a system for testing new drugs that may be useful for treatment. This approach has been used to develop models for diseases such as Alzheimer's disease and arthritis.

Recombinant protein production

Transgenic sheep and goats have been produced which secret recombinant human proteins in milk. This method of producing recombinant proteins has a number of advantages. Milk is renewable, it is produced in large quantities and can be collected without harm to the animal. The expressed proteins closely resemble their human versions and are easily purified because milk contains only a small number of different proteins. Proteins produced by this method include the blood clotting protein Factor IX and the plasma protein α1-antitrypsin.

FURTHER READING

There are many comprehensive textbooks of molecular biology and biochemistry and no one book that can satisfy all needs. Different readers subjectively prefer different textbooks and hence we do not feel that it would be particularly helpful to recommend one book over another. Rather we have kisted some of the leading books which we know from experience have served their student readers well.

General reading

Brown, T.A. (1992) *Genetics a Molecular Approach*, 2nd edn., Chapman and Hall, London.

Lewin, B. (1997) *Genes VI*. Oxford University Press, Oxford.

Maxon, L.R. and Daugherty, C.H. (1992) *Genetics a Human Perspective*, 3rd Edn., William Brown.

Primrose, S.B. (1995) *Principles of Genome Analysis*. Blackwell, Oxford.

Ridley, M. (1996) *Evolution*, 2nd edn., Blackwell, Oxford.

Singer, M. and Berg, P. (1991) *Genes and Genomes*. University Science Books, Mill Valley, CA, USA.

Strachan, T. and Read, A.P. (1996) *Human Molecular Genetics*. BIOS Scientific Publishers, Oxford.

Weaver, R.F and Hendrick, P.W. (1992) *Genetics*, 2nd edn., WC Brown, Dubuque, IA.

More advanced reading

The following selected articles are recommended to readers who wish to know more about specific subjects. In many cases they are too advanced for first year students but are very useful sources of information for subjects that may be studied in later years.

Section A

Cedergren, R. and Miramontes, P. (1996) The puzzling origin of the genetic code. *Trends Biochem. Sci.* **21** (6), 199–200.

Foiani, M., Lucchini, G. and Plevani, P. (1997) The DNA polymerase α-primase complex couples DNA replication, cell cycle progression and DNA damage response. *Trends Biochem. Sci.* **22** (11), 424–427.

Grieder, C.W. and Blackburn, G.H. (1996) Telomeres, telomerase and cancer. *Sci. Am.* **274** (2), 80–85.

Johnson, H.M., Bazer, F.W., Szente, B.G. and Jarpe, M.A. (1994) How interferons fight disease. *Sci. Am.* **270** (6), 40–49.

Kable, M.L., Heidmann, S. and Stuart, K.D. (1997) RNA editing: getting U into RNA. *Trends Biochem. Sci.* **22** (5), 162–166.

Kamakaka, R.T. (1997) Silencers and locus control regions: opposite sides of the same coin. *Trends Biochem. Sci.* **22** (4), 124–127.

Kodadek, T. (1998) Mechanistic parallels between DNA replication, recombination and transcription. *Trends Biochem. Sci.* **23** (2), 79–83.

Mackay, J. B. and Crossley, M. (1998) Zinc fingers are sticking together. *Trends Biochem. Sci.* **23** (1), 1–4.

McGinnis, W. and Kuziova, M. (1994) The molecular architects of body design. *Sci. Am.* **270** (2), 36–43.
Ramakrishnan, V. and White, S.W. (1998) Ribosomal protein structures: insights into the architecture, machinery and evolution of the ribosome. *Trends Biochem. Sci.* **23** (6), 208–212.
Reeder, R.T. and Lang, W.H. (1997) Terminating transcription in eukaryotes: lessons learned from RNA polymerase I. *Trends Biochem. Sci.* **22** (12), 473–477.
Rennie, J. (1993) DNA's new twists. *Sci. Am.* **266** (3), 88–96.
Rhodes, D. and Klug, A. (1993) Zinc fingers. *Sci. Am.* **268** (2), 32–39.
Roca, J. (1995) The mechanisms of DNA topoisomerases. *Trends Biochem. Sci.* **20** (4), 155–160.
Scott, W.G. and Klug, A. (1996) Ribozymes: structure and mechanism in RNA catalysis. *Trends Biochem. Sci.* 21 (6), 220–224.
Strachan, T. and Read, A.P. (1996) *Human Molecular Genetics*. BIOS Scientific Publishers, Oxford.
Tarn, W.Y. and Steitz, J.A. (1997) Pre-mRNA splicing: the discovery of a new spliceosome doubles the challenge. *Trends Biochem. Sci.* **22** (4), 132–137.
Tjian, R. (1995) Molecular machines that control genes. *Sci. Am.* **272** (2), 38–45.
White, R.J. (1997) Regulation of RNA polymerases I and III by the retinoblastoma protein: a mechanism for growth control. *Trends Biochem. Sci.* **22** (3), 77–80.
Wahle, E. and Keller, W. (1996) The biochemistry of polyadenylation. *Trends Biochem. Sci.* **21** (7), 247–250.
Weijland, A. and Parmeggiani, A. (1994) Why do two EF-Tu molecules act in the elongation cycle of protein biosynthesis. *Trends Biochem. Sci.* **19** (5), 188–193.
Trends in Biochemical Sciences. Vol. 21, No. 9, 1996. Whole issue devoted to articles on RNA Polymerase II and Control of Transcription.

Section B

Brown, W. and Tyler-Smith, C. (1995) Centromere activation. *Trends Genet.* **11:** 337–339.
Cann, A.J. (1997) *Principles of Molecular Virology*, 2nd edn., Academic Press, London.
Dale, J.W. (1994) *Molecular Genetics of Bacteria*, 2nd edn., John Wiley, Lewes.
Gorman, M. and Baker, B.S. (1994) How flies make one equal two: dosage compensation on *Drosophila*. *Trends Genet.* **10:** 376–380.
Grieder, C.W. and Blackburn, E.H. (1996) Telomeres, telomerase and cancer. *Sci. Am.* **274:** 80–85.
Haseltine, W.A. (1997) Discovering genes for new medicines. *Sci. Am.* **276** (3), 78–83.
Kippling, D. and Warburton, P.E. (1997) Centromeres, CEN-P and Tigger too. *Trends Genet.* **13:** 141–144.
Lefell, D.J. and Brash, D.E. (1996) Sunlight and skin cancer. *Sci. Am.* **275** (1), 38–49.
Liljas, I. (1996) Viruses. *Curr. Opin. Struct. Biol.* **6** (2) 151–156.
Madigan, M.T. and Marrs, B.L. (1997) Extremophiles. *Sci. Am.* **276** (4), 66–71.
Naysmyth, K. (1996) At the heart of the budding yeast cell cycle. *Trends Genet.* **12:** 405–411.
Strachan, T. and Read, A. P. (1996) *Human Molecular Genetics*. BIOS Scientific Publishers, Oxford.

Scientific American. Vol. 279, No. 1, 61–86. 1998. Several Articles on HIV.

Trends in Biochemical Sciences. Vol. 20, No. 10, 1995. Whole issue devoted to articles on DNA Repair.

Wolffe, A.P. and Pruss, D. (1996) Deviant nucleosomes: the functional specialization of chromatin. *Trends Genet.* **12:** 58–62.

Zinn, A.R., Page D.C. and Fisher, E.M.C. (1993) Turner syndrome: the case of the missing sex chromosome. *Trends Genet.* **9:** 90–93.

Section C

Capel, B. (1995) New bedfellows in the mammalian sex-determination affair. *Trends Genet.* **11:** 161–163.

Chaffe, M. (1997) Is the traditional way of teaching three-factor mapping sufficient? *Trends Genet.* **13:** 94–95.

Charlesworth, D. and Gilmartin, P.M. (1998) Lilly or Billy – Y the difference. *Trends Genet.* **14:** 261–262.

Cline, T.W. (1993) The *Drosophila* sex determination signal: how do flies count to two? *Trends Genet.* **9:** 385–389.

Egel, R. (1995) The synaptonemal complex and the distribution of meiotic recombination events. *Trends Genet.* **11:** 206–208.

Lightowlers, R.N., Chinnery, D.F., Turnbull, D.M. and Howell, N. (1997) Mammalian mitochondrial genetics: hetedity, heteroplasmy and disease. *Trends Genet.* **13:** 450–454.

Mitochondrial DNA in aging and disease. *Sci. Am.* **277:** 22–29.

Plonim, R. and Defries, J.C. (1998) The genetics of cognitive abilities and disabilities. *Sci. Am.* **278** (May): 40–47.

Siracusa, L.D. (1994) The agouti gene: turned on to yellow. *Trends Genet.* **10:** 423–427.

Section D

Cavalli-Sforza, L.L. (1998) The DNA revolution in population genetics. *Trends Genet.* **14:** 60–65.

Leaky, M. and Walker, A. (1997) Early hominid fossils from Africa. *Sci. Am.* **276** (6): 74–79.

Mallet, J. (1995) A species definition for the modern synthesis. *Trends Ecol. Evol.* **10:** 294–299.

Tattersall, I. (1997) Out of Africa again . . . and again? *Sci. Am.* **276** (4): 60–67.

Wilson, A.E. and Cann, R.L. (1992) The recent African genesis of humans. *Sci. Am.* **266** (4): 68–73.

Section E

Brown, T.A. (1995) *Gene Cloning: An Introduction*, 3rd edn., Chapman and Hall, London.

Erlich, H.A., Gelfand, D. and Sninsky, J.J. (1991) Recent advances in the polymerase chain reaction. *Science.* **252**, 1643–1651.

Mullis, K.B. (1990) The unusual origin of the polymerase chain reaction. *Sci. Am.* 262 (4), 36–43.

Old, R.W. and Primrose, S.B. (1994) *Principles of Gene Manipulation: An Introduction to Genetic Engineering*, 5th Edn., Blackwell Scientific Publications, Oxford.

Sambrook, J., Fritsch, E.F. and Maniatis, T. (1989) *Molecular Cloning: a Laboratory Manual*, 2nd edn., Cold Spring Harbor Laboratory Press, Cold Spring Harbor, NY.

Strachan, T. and Read, A. P. (1996) *Human Molecular Genetics*. BIOS Scientific Publishers, Oxford.

Section F

Boucher, R.C. (1996) Current status of CF gene therapy. *Trends Genet.* **12:** 81–85.

Capecchi, M.R. (1994) Targeted gene replacement. *Sci. Am.* **270** (3), 34–41.

Freidman, T. (1994) Gene therapy for neurological disorders. *Trends Genet.* **10:** 210–214.

Gasser, C.S. and Fraley, R.T. (1992) Transgenic crops. *Sci. Am.* **266** (6), 34–39.

Gilboa, E. and Smyth, C. (1994) Gene therapy for infectious diseases: the AIDS model. *Trends Genet.* **10:** 139–144.

Glick, B.R. and Pasternak, J.J. (1998) *Molecular Biotechnology*, 2nd edn., ASM Press, Washington.

Helin, K. and Peters, G. (1997) Tumor suppressors: from genes to function and possible therapies.

Hoheisel, J.D. (1994) Application of hybridization techniques to genome mapping and sequencing. *Trends Genet.* **10:** 79–83.

Horgan, J. (1994) High profile – The Simpson case raises the issue of DNA reliability. *Sci. Am.* **241** (4): 33–36.

Jansen-Durr, P. (1998) How viral oncogenes make the cell cycle. *Trends Genet.* **14:** 8–10.

Jiricny, J. (1994) Colon cancer and DNA repair: have mismatches met their match? *Trends Genet.* **10:** 164–168.

Lehman, A.R. and Carr, A.M. (1995) The ataxia-telangiesstasia gene: a link between checkpoint controls, neurodegeneration and cancer. *Trends Genet.* **11:** 375–377.

Lichter, P. (1997) Multicolor FISHing: what's the catch. *Trends Genet.* **13:** 475–479.

Rennie, J. (1994) Grading the gene tests. *Sci. Am.* **270** (6), 66–74.

Ronald, P.C. (1997) Making rice disease resistant. *Sci. Am.* **277** (5), 68–73.

Sudbery, P. (1998) *Human Molecular Genetics*. Longman, UK

Strachan, T. and Read, A. P. (1996) *Human Molecular Genetics*. BIOS Scientific Publishers, Oxford.

Velander, W.H., Lubon, H. and Drohan W.N. (1997) Transgenic livestock as drug factories. *Sci. Am.* **276** (1), 54–59.

Walter, M.A. and Goodfellow, P.N. (1993) Radiation hybrids: irradiation and fusion gene transfer. *Trends Genet.* **9:** 352–355.

Welsh, M.J. and Smith, A.E. (1995) Cystic fibrosis. *Sci. Am.* **273** (6), 36–43.

INDEX